Synthese Library

Volume 498

Studies in Epistemology, Logic, Methodology, and Philosophy of Science

Editor-in-Chief

Otávio Bueno, Department of Philosophy, University of Miami, Coral Gables, USA

Editorial Board Members

Berit Brogaard, University of Miami, Coral Gables, USA

Steven French, University of Leeds, Leeds, UK

Catarina Dutilh Novaes, VU Amsterdam, Amsterdam, The Netherlands

Darrell P. Rowbottom, Department of Philosophy, Lingnan University, Tuen Mun, Hong Kong

Emma Ruttkamp, Department of Philosophy, University of South Africa, Pretoria, South Africa

Kristie Miller, Department of Philosophy, Centre for Time, University of Sydney, Sydney, Australia

The aim of *Synthese Library* is to provide a forum for the best current work in the methodology and philosophy of science and in epistemology, all broadly understood. A wide variety of different approaches have traditionally been represented in the Library, and every effort is made to maintain this variety, not for its own sake, but because we believe that there are many fruitful and illuminating approaches to the philosophy of science and related disciplines.

Special attention is paid to methodological studies which illustrate the interplay of empirical and philosophical viewpoints and to contributions to the formal (logical, set-theoretical, mathematical, information-theoretical, decision-theoretical, etc.) methodology of empirical sciences. Likewise, the applications of logical methods to epistemology as well as philosophically and methodologically relevant studies in logic are strongly encouraged. The emphasis on logic will be tempered by interest in the psychological, historical, and sociological aspects of science. In addition to monographs *Synthese Library* publishes thematically unified anthologies and edited volumes with a well-defined topical focus inside the aim and scope of the book series. The contributions in the volumes are expected to be focused and structurally organized in accordance with the central theme(s), and should be tied together by an extensive editorial introduction or set of introductions if the volume is divided into parts. An extensive bibliography and index are mandatory.

Roman Frigg • J. McKenzie Alexander •
Laurenz Hudetz • Miklos Rédei • Lewis Ross •
John Worrall
Editors

Proofs and Research Programmes: Lakatos at 100

 Springer

Editors
Roman Frigg
Department of Philosophy, Logic and
Scientific Method
London School of Economics and
Political Science
London, UK

Laurenz Hudetz
Bundesgymnasium Zaunergasse
Salzburg, Austria

Lewis Ross
Department of Philosophy, Logic and
Scientific Method
London School of Economics and
Political Science
London, UK

J. McKenzie Alexander
Department of Philosophy, Logic and
Scientific Method
London School of Economics and
Political Science
London, UK

Miklos Rédei
Department of Philosophy, Logic and
Scientific Method
London School of Economics and
Political Science
London, UK

John Worrall
Department of Philosophy, Logic and
Scientific Method
London School of Economics and
Political Science
London, UK

ISSN 0166-6991   ISSN 2542-8292   (electronic)
Synthese Library
ISBN 978-3-031-88212-8   ISBN 978-3-031-88213-5   (eBook)
https://doi.org/10.1007/978-3-031-88213-5

This work was supported by London School of Economics and Political Science.

© The Editor(s) (if applicable) and The Author(s) 2025. This book is an open access publication.
Open Access This book is licensed under the terms of the Creative Commons Attribution-NonCommercial-NoDerivatives 4.0 International License (http://creativecommons.org/licenses/by-nc-nd/4.0/), which permits any noncommercial use, sharing, distribution and reproduction in any medium or format, as long as you give appropriate credit to the original author(s) and the source, provide a link to the Creative Commons license and indicate if you modified the licensed material. You do not have permission under this license to share adapted material derived from this book or parts of it.
The images or other third party material in this book are included in the book's Creative Commons license, unless indicated otherwise in a credit line to the material. If material is not included in the book's Creative Commons license and your intended use is not permitted by statutory regulation or exceeds the permitted use, you will need to obtain permission directly from the copyright holder.
This work is subject to copyright. All commercial rights are reserved by the author(s), whether the whole or part of the material is concerned, specifically the rights of translation, reprinting, reuse of illustrations, recitation, broadcasting, reproduction on microfilms or in any other physical way, and transmission or information storage and retrieval, electronic adaptation, computer software, or by similar or dissimilar methodology now known or hereafter developed. Regarding these commercial rights a non-exclusive license has been granted to the publisher.
The use of general descriptive names, registered names, trademarks, service marks, etc. in this publication does not imply, even in the absence of a specific statement, that such names are exempt from the relevant protective laws and regulations and therefore free for general use.
The publisher, the authors and the editors are safe to assume that the advice and information in this book are believed to be true and accurate at the date of publication. Neither the publisher nor the authors or the editors give a warranty, expressed or implied, with respect to the material contained herein or for any errors or omissions that may have been made. The publisher remains neutral with regard to jurisdictional claims in published maps and institutional affiliations.

This Springer imprint is published by the registered company Springer Nature Switzerland AG
The registered company address is: Gewerbestrasse 11, 6330 Cham, Switzerland

If disposing of this product, please recycle the paper.

Preface

Imre Lakatos was one of the most significant philosophers of science and mathematics of the twentieth century, and his ideas remain important and relevant today. As the entry on Lakatos in the *Stanford Encyclopedia of Philosophy* attests "Lakatos's influence, particularly in the philosophy of science, has been immense". November 2022 saw the centenary of Lakatos's birth, and the event was marked by an international conference held at the LSE—where Lakatos made his career after he had emigrated from Hungary to England—the conference focussing on the continuing influence and relevance of his work. With the exception of two papers, this volume consists of a selection of papers that were presented at the conference.

We are immensely grateful to Dr Spiro Latsis, without whose generous financial support the conference would not have been possible and this book would not have been published open access. We would also like to thank our colleagues in the Department of Philosophy, Logic and Scientific Method and the Centre for Philosophy of Natural and Social Science for logistical support with the organisation.

London, UK	Roman Frigg
London, UK	J. McKenzie Alexander
Salzburg, Austria	Laurenz Hudetz
London, UK	Miklos Rédei
London, UK	Lewis Ross
London, UK	John Worrall

Contents

1. **Introduction** 1
   Roman Frigg, J. McKenzie Alexander, Laurenz Hudetz,
   Miklós Rédei, Lewis Ross, and John Worrall

2. **Mathematical Methodology** 7
   Philip Kitcher

3. **Proofs as Dialogues: The Enduring Significance of Lakatos
   for the Philosophy of Mathematical Practice** 27
   Catarina Dutilh Novaes

4. **Lakatos and the Euclidean Programme** 47
   A. C. Paseau and Wesley Wrigley

5. *Proofs and Refutations*, **Non-classically and Game Theoretically** 69
   Can Başkent

6. **Extending Heuristics: Discovery in Logic, Mathematics,
   and the Sciences** 91
   Otávio Bueno

7. **The Case of Early Copernicanism: Epistemic Luck** *vs.*
   Predictivist Vindication 105
   Vincenzo Crupi

8. **The Bayesian Research Programme in the Methodology of
   Science, or Lakatos Meets Bayes** 127
   Stephan Hartmann

9. **Lakatos's Naturalism(s): Distinguishing Between Rational
   Reconstructions and Normative Explanations** 145
   Thodoris Dimitrakos

10. **Heuristic, Physics Avoidance and the Growth of Knowledge** 165
    Jack Ritchie

11	Beyond Footnotes: Lakatos's Meta-philosophy and the History of Science .... Samuel Schindler	183
12	Cholesterol and Cardio-Vascular Disease: Degenerating Research Programmes in Current Medical Science .... John Worrall	205
13	Trade-offs and Progress in Cancer Science .... Anya Plutynski	231
14	Epilogue: Scientific Theory-Change and Rationality – Lakatos and the "Popper-Kuhn Debate" .... John Worrall	247

Index 265

Chapter 1
Introduction

Roman Frigg, J. McKenzie Alexander, Laurenz Hudetz, Miklós Rédei, Lewis Ross, and John Worrall

Abstract This chapter provides an introduction to the book. The twelve essays in the book fall into three groups. Essays in the first group address problems in the philosophy of mathematics; essays in the second group investigate foundational questions concerning Lakatos's philosophy of science; and essays in the third group apply Lakatos's concept of Methodology of Scientific Research Programmes (MSRP) to medicine. The book ends with an epilogue.

Although Lakatos is nowadays primarily known for his work in philosophy of natural science, and in particular for his Methodology of Scientific Research Programmes (MSRP), his first major contribution was his *Proofs and Refutations*—a groundbreaking work in the philosophy of mathematics. The central thesis of *Proofs and Refutations* is that the development of mathematics does not consist in the steady accumulation of eternal truths, as conventional philosophy of mathematics suggests. Mathematics develops, according to Lakatos, in a much more dramatic and exciting way, via a process of conjecture, followed by attempts to "prove" the conjecture (in his view, to reduce it to other conjectures) followed by criticism via attempts to produce counterexamples both to the conjectured theorem and to the various steps in the proof, resulting in the proof of a much modified version of the original conjecture.

Among the still open questions about Lakatos's views are: Does Lakatos's account really amount to a fully "quasi-empirical" view of the epistemology of mathematics to rival the traditional philosophies of logicism, formalism and intuitionism? Or is it instead "merely" an—albeit fascinating—account of how mathematical theorems are arrived at, an account which has no consequences for

---

R. Frigg (✉) · J. M. Alexander · M. Rédei · L. Ross · J. Worrall
Department of Philosophy, Logic and Scientific Method, London School of Economics and Political Science, London, UK
e-mail: r.p.frigg@lse.ac.uk

L. Hudetz
Bundesgymnasium Zaunergasse, Salzburg, Austria

© The Author(s) 2025
R. Frigg et al. (eds.), *Proofs and Research Programmes: Lakatos at 100*, Synthese Library 498, https://doi.org/10.1007/978-3-031-88213-5_1

the epistemological status of those eventually arrived-at theorems? Is Lakatos's central example—the Descartes-Euler conjecture about polyhedra—itself too "quasi-empirical" to be representative of mathematics in general? Finally, did Lakatos outgrow his Hegelian roots? Or is *Proofs and Refutations* best, or perhaps even *only*, understandable as a thoroughly Hegelian work? Some of these issues are touched on in the contributions to the philosophy of mathematics section of this volume.

Turning, then, to his philosophy of science, Lakatos famously presented MSRP as a synthesis of the views of Karl Popper and Thomas Kuhn—preserving from the former the claim that theory change in science is a rational process, while allowing that the latter's account of how scientists regard and deal with experimental difficulties is altogether more true-to-scientific-life than Popper's. There is no consensus as to whether or not this "synthesis" succeeds. Nor is there any consensus about how to interpret Lakatos's central notion of progress and the associated concept of "novel fact". Another open issue is whether the insights underlying Lakatos's MSRP can be captured and thereby given a more solid foundation by the Bayesian approach to scientific reasoning. The view that those insights can be given a Bayesian justification was argued by Howson and Urbach in their *Scientific Reasoning: The Bayesian Approach*.

A large part of the continuing influence of Lakatos's ideas consists in attempts to apply his MSRP to identify and evaluate research programmes in special sciences such as medicine, psychology, economics, and sociology, as well as in disciplines like educational theory, informatics, and international relations, which otherwise receive scant attention in philosophy of science. These attempts often originate in the sciences themselves and are driven by practitioners' desire to understand developments in their fields, rather than by traditional philosophical concerns. Being relevant beyond the confines of professional philosophy is probably the best marker of a lasting influence.

The book consists of 12 essays, which fall into three groups. Essays in the first group address problems in the philosophy of mathematics; essays in the second group investigate foundational questions concerning Lakatos's philosophy of science; and essays in the third group apply his MSRP to medicine. The book ends with an epilogue.

The first group of essays begins with Philip Kitcher's "Mathematical Methodology". Lakatos regarded his *Proofs and Refutations* as a study in the "methodology of mathematics" or the logic of mathematical discovery. Philip Kitcher agrees that philosophy of mathematics has—both before and after Lakatos—concentrated on issues about the status of mathematical results and ignored issues about how those results emerged in the first place; and it has done so to its cost. Accordingly, Kitcher's contribution develops a mathematical methodology. He outlines the major changes that resulted in the mathematics of the late nineteenth century, indicates how those results emerged, and appraises them in terms of a notion of pragmatic progress (progress *from*) as opposed to any notion of teleological progress (progress *to*). Kitcher's methodology transcends Lakatos in many ways but is recognisably Lakatosian in spirit.

In her "Proofs as Dialogues: The Enduring Significance of Lakatos for the Philosophy of Mathematical Practice", Catarina Dutilh Novaes focuses on what Lakatos's ideas have to offer for contemporary philosophical work on mathematical practice. In particular, she highlights the influence of Lakatos's *Proofs and Refutations* on the development of her dialogical account of deduction and mathematical proof, which relies on so-called Prover-Skeptic dialogues. Similarities and differences between Prover-Skeptic dialogues and Lakatosian Prover-Refuter dialogues are discussed with special attention to the roles of cooperation and adversariality. The article closes with a reflection on the broader philosophical differences between Lakatos's "Hegelian" approach and Dutilh Novaes's dialogical pragmatism: Lakatos aims at the dialectic development of mathematical *concepts,* disregarding individual human activities, while Dutilh Novaes's dialogical account of mathematical proof is primarily about *human agents* and their interactions.

In their "Lakatos and the Euclidean Programme", Alexander Paseau and Wesley Wrigley critically examine and revise Lakatos's account of the Euclidean Programme (EP), which is a foundationalist account of mathematical knowledge inspired by Euclid's Elements. In Lakatos's view, a system of mathematical knowledge that is organised according to the EP starts from a finite set of trivially true axioms with perfectly well-understood primitive terms, and truth then "flows" from axioms to theorems via deductive channels. The authors critically examine various aspects of Lakatos's account and suggest modifications that lead to an improved characterisation of the EP, consisting of seven principles. The proposed characterisation inherits some core ideas from Lakatos's account (e.g. the idea of flow) but differs in various other respects. The outcome is an updated reconstruction of the EP in the spirit of Lakatos.

In "Proofs and Refutations, Non-Classically and Game Theoretically", Can Başkent argues that the reasoning in Lakatos's *Proofs and Refutations* is not governed by the rules of classical logic but instead exemplifies paraconsistent logic—a type of logic where it is not the case that everything follows from a contradiction. Başkent points out that inconsistencies play a fundamental role in the Lakatosian method of proofs and refutation. Crucially, when contradictions arise (e.g. due to counterexamples to a conjecture), one is not permitted to draw arbitrary conclusions. How one can move forward in the face of a contradiction is precisely what defines the method of proofs and refutations. Furthermore, Başkent argues that the strategic way in which inconsistencies should be handled according to Lakatos can be fruitfully analysed through the lens of game theory. This is illustrated using concrete examples from *Proofs and Refutations.*

Vincenzo Crupi's "The Case of Early Copernicanism: Epistemic Luck versus Predictivist Vindication" is the first contribution of the second group, which concerns foundational questions about Lakatos's philosophy of science. In his paper, Crupi investigates the issue of whether the adoption of the Copernican theory by Kepler and Galileo (as well as by Copernicus himself) was, as many have claimed, a matter of "epistemic luck": these luminaries happened to make what was by later lights the correct choice but had no empirical justification for that choice at the time when they initially made it. The idea that Kepler and Galileo were 'lucky' has generally been based on the claim that—allegedly—any empirical phenomenon that might be taken to support Copernican theory could in fact equally well be accounted for on the rival Ptolemaic theory. In a widely read paper, Lakatos and Zahar argued that, to the contrary, once the notion of prediction is properly understood, the initial Copernican theory is seen to have enjoyed predictive successes not shared by its Ptolemaic rival and hence Kepler's and Galileo's theory-choices are vindicated. Crupi investigates whether Lakatos and Zahar's view stands up to historical and philosophical analysis.

The paper "The Bayesian Research Programme in the Methodology of Science, or Lakatos Meets Bayes" by Stephan Hartmann argues that, when understood correctly, Bayesianism is an instance of a progressive Lakatosian research programme in the methodology of science. This stands in stark contrast to Lakatos's own rather sceptical view about Bayesianism. To support its claim, the paper considers and then dismisses three challenges to Bayesianism. These arise in connection with indirect evidence, new types of evidence, and genuinely new evidence. Hartmann shows how these challenges can be met within the Bayesian Research Programme. He also shows that in order to be able to handle these challenges, one has to abandon a core tenet of traditional Bayesianism: that belief change has to be made via standard conditionalization. Instead of relying on standard conditionalization, belief change should be based on the "Principle of Conservativity": the requirement that belief change should minimize a certain distance between the probability measures representing beliefs.

Thodoris Dimitrakos' "Lakatos's Naturalism(s): Distinguishing between Rational Reconstructions and Normative Explanations" examines Lakatos's concept of "rational reconstruction" in the philosophy of science, defending its use against critics like Kuhn who claim it distorts historical records. After briefly discussing, and setting aside, some uncharitable criticisms of Lakatos's account, Dimitrakos identifies the real problem it faces: that Lakatos's attempt to provide both a historically informed philosophy of science and an account of scientific rationality led to problems of circularity. Dimitrakos argues that these problems can be resolved in three steps. First, one needs to distinguish between rational reconstruction, a philosophical tool for evaluating different theories of scientific rationality, and normative explanation, a historiographical category. Second, one has to reject Popper's "three worlds" conception, situating Lakatos's approach within a liberal naturalism. And, finally, one must replace Lakatos's inter-methodology evaluation process with a suitable intra-methodology process. In doing so, the chapter aims to show how Lakatos's work remains relevant to contemporary debates about the relationship between history and philosophy of science.

In his "Heuristic, Physics Avoidance, and the Growth of Knowledge", Jack Ritchie examines the notion of positive heuristic in Lakatosian philosophy of science, particularly in *Methodology of Scientific Research Programmes*. He begins by setting aside an alternative view of heuristic due to John Worrall (claiming that it departs too far from the source text), and then offers a different interpretation inspired by the work of Mark Wilson. On Ritchie's account, the positive heuristic fosters the growth of knowledge through a process often best understood as "model-making and improving". On this view, a central driver of progress is the construction and refinement of scientific models. The aim of these models is to convert empirical difficulties into mathematical difficulties. These difficulties include the construction of mathematically tractable models and providing plausible bridges between higher and lower-level models of the same phenomena. On the view that Ritchie provides, refutation is less essential to Lakatosian progress than sometimes supposed, with the incremental improvement of models playing a more central role.

Samuel Schindler's "Beyond Footnotes: Lakatos's Meta-Philosophy and the History of Science" revisits Lakatos's approach to historical facts. Lakatos infamously claimed that the actual history of science could be recorded in the footnotes of rational reconstructions of science. Schindler points out that Lakatos's approach to actual history was more reasonable than that, not least because he argued that a philosophical methodology of science should aim to maximise rationally explainable facts, even though there should be no expectation that all historical facts will turn out to be rational. Schindler examines this idea in the context of the contemporary discussion about meta-philosophy. The paper then compares Kuhn's and Lakatos's approaches to science and argues that Lakatos's account, contrary to what he himself thought, doesn't have a more legitimate claim to rationality than Kuhn's.

The next two contributions form the third group of papers, which are dedicated to the philosophy of medicine. In his "Cholesterol and Cardio-Vascular Disease: Degenerating Research Programmes in Current Medical Science", John Worrall argues that the mini research programmes built to defend two extremely influential claims in current medicine have both consistently degenerated. If so, as he remarks, one would have expected those two claims to have been rejected as not evidence-based. But in fact, although the consensus on the first claim now shows some signs of breaking up, it remained in place for many years after degeneration set in; while the second remains almost universally accepted in medicine and remains the basis for accepted medical advice and treatment. The second part of his paper analyses this clash between expectation and reality, leading to a re-examination of Lakatos's distinction between internal and external history.

Anya Plutynski's "Trade-offs and Progress in Cancer Science" begins with the observation that almost all examples of research programmes analysed in terms of progress and degeneration by Lakatos and those influenced by him were from physics (or occasionally chemistry). One might therefore be tempted to object that MSRP, while a useful tool for analysing developments in basic sciences like physics and chemistry, is not usefully employed in other, more "special" or applied sciences. Plutynski raises this question and concludes that appropriately analysing developments in Cancer Science may require replacing Lakatos's notion of progress in science with one that recognizes the prevalence of trade-offs intrinsic to the culture of science.

The book ends with an epilogue, John Worrall's Scientific Theory-Change and Rationality: Lakatos and the "Popper- Kuhn Debate" in which he takes a look back at the "Popper-Kuhn" debate and Lakatos's attempt to resolve it. The Popper-Kuhn debate was one of the foci of attention at the famous Bedford College Colloquium held in the summer of 1965. What exactly was at issue in this debate? Was Lakatos right that Kuhn's account of theory-change in science denies that change is a rational affair by reducing change to "a matter of mob psychology"? Was Lakatos right that his MSRP provides a satisfactory "synthesis" of the views of Popper and Kuhn— one that preserves the rationality of theory-change? How has the debate progressed since 1965 and where does it currently stand? The chapter is the written version of a lecture Worrall gave at the conference *Centenário Imre Lakatos: matemática e ciência* in Sao Paolo in November 2022. We have kept the lecture in its original form. The points come across most vividly in this talk by a PhD student supervised by Lakatos himself.

Open Access This chapter is licensed under the terms of the Creative Commons Attribution-NonCommercial-NoDerivatives 4.0 International License (http://creativecommons.org/licenses/by-nc-nd/4.0/), which permits any noncommercial use, sharing, distribution and reproduction in any medium or format, as long as you give appropriate credit to the original author(s) and the source, provide a link to the Creative Commons license and indicate if you modified the licensed material. You do not have permission under this license to share adapted material derived from this chapter or parts of it.

The images or other third party material in this chapter are included in the chapter's Creative Commons license, unless indicated otherwise in a credit line to the material. If material is not included in the chapter's Creative Commons license and your intended use is not permitted by statutory regulation or exceeds the permitted use, you will need to obtain permission directly from the copyright holder.

Chapter 2
Mathematical Methodology

Philip Kitcher

Abstract In *Proofs and Refutations*, Imre Lakatos proposed to reorient the philosophy of mathematics. He suggested abandoning the search for a foundation for mathematics in favor of providing a methodology for mathematics. This essay attempts to pursue the methodological project in a different fashion. It sketches the long history of mathematics, and reflects on the kinds of benefits attained in a number of important transitions. It argues that these advances embody quite different gains. While diverging from Lakatos's quasi-Popperian framework for mathematical progress, I conclude that his proposed reorientation of the philosophy of mathematics would be an advance that is long overdue.

2.1 Introduction

Imre Lakatos began his career in philosophy with some remarkably original essays in the philosophy of mathematics. His Cambridge Ph.D. dissertation was published almost in its entirety as a four-part article, "Proofs and Refutations", in the *British Journal for the Philosophy of Science* (Lakatos, 1963–1964). After his death, this became a free-standing book (Lakatos, 1976), edited by John Worrall and Elie Zahar, in which further, previously unpublished, material was added. Two years later, the first half of the second volume of Lakatos's *Philosophical Papers* (Lakatos, 1978), edited by John Worrall and Gregory Currie, contained articles on the philosophy of mathematics (some in revised versions).

I am most grateful to the organizers of the Lakatos Centennial Conference at LSE for inviting me to deliver a paper, and to contribute to this volume, rightly celebrating the work of a major philosopher of science. I am also indebted to a knowledgeable and perceptive reviewer for comments that have enabled me to improve the final version.

P. Kitcher (✉)
Department of Philosophy Columbia University, New York, NY, USA
e-mail: psk16@columbia.edu

Lakatos plainly hoped to reform this area of philosophy. The Introduction to the book version of *Proofs and Refutations* is forthright:

> The purpose of these essays is to approach some problems in the *methodology of mathematics*. ... The recent expropriation of the term 'methodology of mathematics' to serve as a synonym for 'metamathematics' ... indicates that in formalist philosophy of mathematics there is no proper place for methodology qua logic of discovery. (Lakatos, 1976, p. 3)

Despite his firm understanding of then-current debates about the merits of various programs in the "foundations of mathematics", and of the set-theoretic and proof-theoretic results around which those debates centered, Lakatos contended that the discussions among logicians and philosophers of mathematics neglected important questions—indeed *the* most important philosophical questions about mathematics. Strenuous efforts to *reconstruct* mathematical knowledge in some preferred fashion completely bypassed issues about how the body of knowledge to be set in order had emerged. Obsessed with tidying up the corpus of mathematical knowledge, philosophers seemed completely uninterested in how mathematicians had obtained that knowledge in the first place. Hygienic proposals for arranging the corpse took precedence over understanding the ways in which the life of mathematics had proceeded.

Were philosophers of mathematics embracing an odd variety of skepticism, Lakatos wondered, tacitly denying any genuine mathematical knowledge before the advent of formal logic? With characteristically acerbic wit, he pointed out that "Newton had to wait four centuries until Peano, Russell, and Quine helped him into heaven by formalising the calculus" (Lakatos, 1976, p. 2). Understanding the growth of mathematical knowledge, he thought, might relieve Newton of any strain on his patience. To achieve that understanding would require specifying the methodology of mathematics, that is, identifying the standards governing the rational progress of mathematics. For Lakatos, at this stage of his career, that meant adapting the correct normative theory of the natural sciences to the mathematical case. Convinced that Popper had supplied that theory, his title, and his four-part article echoed Sir Karl.

Despite the brilliance of his discussion of his major example—the historical development of ideas about the relations of the number of vertices, edges and faces of polyhedra, initiated by the Descartes-Euler conjecture—I don't think Lakatos solved the problem he had posed. Nor do I believe that the essays he wrote about mathematics identified the problem in its full generality. Nonetheless, this part of his philosophical work has always seemed to me a major achievement, one that remains underrated to this day. For he had formulated the right question, and offered a convincing treatment of some instances of it. Today's philosophy of mathematics would be far richer and far healthier if its practitioners had paid attention.

Perhaps their indifference stemmed, in part, from Lakatos's tendency to draw provocative corollaries. Viewing mathematics as growing according to a quasi-Popperian methodology led him to dispute the common emphasis on and understanding of proofs, and to wage a campaign on the search for foundations. Were card-carrying philosophers of mathematics so threatened by these implications that they dismissed his central project, and thus failed to confront the methodological question he had raised (and, apparently, thought he had answered)? Historians of science certainly bristled at the suggested role for history—consigned to footnotes, while "rational reconstruction" filled the text; and they could meet the suggestion with a compelling critique (Arabatzis, 2017).

Although I have always been far more sympathetic to Lakatos' approach to the philosophy of mathematics than most of those who have ventured into that area, I must confess to a previous distortion that has caused me to underrate his achievement (Kitcher, 1977). I subordinated the question of elaborating a methodology for mathematics to the project of disputing the standard view—then and now—that mathematical knowledge is *a priori*. Half a century on, though I continue to maintain my (mad?) heterodox views about the apriority of mathematical knowledge, I no longer take that question to lie at the center of philosophical interest. An account of mathematical methodology, drawn from historical studies of the growth of mathematics is far more important.[1] Lakatos saw that clearly, I did not.

What follows is a belated effort to correct one of the mistakes of my philosophical youth.

2.2 Large-Scale Transitions: Conceptual Innovation

Between the earliest mathematical practice about which we have a relatively clear and detailed vision, and the mathematics for which Frege, Russell, Peano, and Hilbert hoped to supply foundations, a vast number of changes occurred. Even more would have to be considered if we were to study the further evolution of the discipline in the twentieth century and in our own. I shall terminate my explorations of history with the *status quo* as of 1879. I do so for two reasons. First, I would be woefully out of my depth in trying to understand, let alone analyze, recent developments in mathematics—it's good historiographical advice to focus on mathematical changes that occurred a century or more before the historian's birth. Second, I am sympathetic to Lakatos's thought that the methodology of mathematics should start by trying to understand the emergence of the body of knowledge the would-be foundationalists intended to bring to order.

During the roughly four millennia separating the Babylonian techniques for what we would see as solving equations (simple equations and quadratic equations) from late nineteenth-century algebra, analysis, geometry, topology, and probability theory, changes occurred at a number of different scales. The smallest ones are the most philosophically familiar. A mathematician uses received ideas to prove a new theorem. The largest are those on which I shall primarily concentrate. They are the ones revealing the most dramatic enrichment of mathematical language and accepted methods of mathematical reasoning. New notation is introduced, sometimes to express concepts that have not previously figured in mathematics,

---

[1] I take mathematical methodology to be a normative discipline, one that tries to characterize the ways in which mathematics should grow. Like Lakatos, I inherit the historicist spirit of philosophers like Kuhn and Feyerabend, and believe that we understand how a form of inquiry should grow by examining its past successes—and perhaps its failures, as well.

on other occasions to provide a more perspicuous way of working with familiar concepts, one that enlarges the class of permissible methods. The replacement of Roman numerals with the language we have inherited from Arab mathematicians of the tenth century is an obvious instance of the latter. The (different) notations offered by Newton and Leibniz for their versions of the calculus exemplify the former.

Lakatos's most famous example, the career of the Descartes-Euler conjecture, is a mid-scale transition. It does not leave the language of mathematics unchanged—new concepts are introduced to classify polyhedra—but the novel vocabulary is applicable only to a restricted class of questions. That is: of questions that provoke mathematical interest. The principal differences between the approach to mathematical methodology I take here and that pursued by Lakatos stem from my beginning from a question I take to be crucial: What makes a proposed change worthy of mathematical interest? As we shall see, the sources of mathematical interest are quite diverse. Typically, they lie in a sense of some deficiency or incompleteness with the status quo. Mathematical advances clear up the puzzles and dissatisfactions that prompted mathematical inquiry. They do so in diverse ways. Hence, my mathematical methodology breaks with the narrow Popperian framework Lakatos employs. It retains, however, the spirit of the enterprise he began.

2.3 A Broad-Brush History

To begin addressing the crucial question, it will help to have a broad-brush treatment of the history of mathematics that focuses on some of the large-scale changes generating the mathematics of the late nineteenth century. The roots of elementary mathematics, the simplest parts of arithmetic and geometry, are buried deep in prehistory, and any attempt to uncover them must be conjectural. At some point in our deep past, probably long before the invention of writing, our ancestors introduced into their language words for numbers, for the basic arithmetical operations, for shapes, distances and areas. *Perhaps* they did so in order to avoid quarrels that arose from recognizably unequal division of resources, or to facilitate exchanges of goods. Trade among groups dates back at least twenty thousand years (McBrearty & Brooks, 2000; Renfrew & Shennan, 1982). *Perhaps* the Babylonian interest in equations results from complex regulations about the shares inherited by relatives of different degrees (or sexes). *Perhaps* geometrical problems have their origin in efforts to assign portions of land so as to satisfy different claimants. If we want an account of how basic parts of arithmetic, geometry, and algebra might reasonably have been adopted by people in Mesopotamia, India, and Egypt, possibly independently, possibly through cultural transmission, speculations of these kinds furnish how-possibly explanations.

By the beginning of the common era, mathematical practice already outran the practical applications I have gestured towards as providing the initial rationales for introducing arithmetical and geometrical concepts. The integration of mathematics into a wide array of ventures surely exceeded its original, more limited roles.

Not only surveying and trade, but engineering, finance, and astronomy called for people with developed arithmetical and geometrical skills. But Euclid had already systematized geometry, introducing the idea of proving new theorems, whether or not they served any useful purpose. Mathematicians had formulated concepts for special types of numbers—prime numbers are only the most obvious example. Locus problems, equations of several degrees, and Diophantine equations all exercised the mathematical community. It is likely that these explorations were spinoffs from the more immediately practical techniques of arithmetic and geometry. Yet, as we shall see, they provided growing points for major expansion.

Fast forward to the early Renaissance. Although the work of using the mathematical framework inherited from the ancients to solve practical problems has intensified, generating further techniques and results within that framework, little conceptual expansion has occurred. One large achievement of the interval I've skipped over, is the provision of a notation that enables easier and more systematic methods for applying the fundamental arithmetical operations: doing sums—adding, subtracting, multiplying, and dividing. Try dividing MDXCI by XLIII—without translating. The traditional parts of mathematics—arithmetic, geometry, and the algebra of polynomial equations—have been further developed, both for practical purposes and with respect to solving "theoretical" problems of the kinds that have been recognized since antiquity: new Euclidean theorems, a few new solutions to locus problems and Diophantine equations, more results about prime and perfect numbers. But not much conceptually new.

Apart, perhaps, from one step whose full significance took centuries to manifest itself. Given that mathematicians' primary function seemed to be to use established techniques to help the bankers and the bridge-builders, it's hardly surprising that they were not held in high esteem within the academy. It's worth recalling that Galileo sought the title of 'philosopher' not 'mathematician'. Indeed, some mathematicians who devoted themselves to the abstract "impractical" problems, derived a significant portion of their income from performing for the high-born. Long before the age of television, an evening's entertainment for the privileged might feature a trip to the Pitti Palace to watch Niccoló Tartaglia or Gerolamo Cardano tackle problems that require a method for solving cubic equations.

For, possibly on the basis of ideas of earlier mathematicians (Scipione del Ferro is the leading candidate), Tartaglia had formulated a technique for attacking such problems. Unwisely, he let Cardano in on the secret, and was enraged when Cardano published it. His reaction is easy to understand. No entertainer wants a trade secret divulged.

Some modern students don't find the method easy to apply, and very few would have the skill to devise it.[2] Admiration for these mathematicians, and for their

---

[2] A clear explanation of the mathematical details underlying this historical episode can be found in (Cooke, 2008). With respect to other historical examples I employ, either this work or (Cooke, 2013) will provide helpful discussions.

contemporaries in the same line of work should be increased by knowing that they operated, as all their predecessors had done, without the aid of algebraic notation.

The introduction of a perspicuous way of formulating equations comes at the end of the sixteenth century (shortly after the careers of Cardano and Tartaglia), and begins the accelerated expansion giving rise to mathematics as Frege knew it. In 1591, François Viète, originally trained as a lawyer, published a book in which he offered a new notation (close to the one still employed) for representing algebraic equations. This enabled him to formulate explicitly the formulae underlying the solutions to quadratic equations (already achieved by the Babylonians), and to cubic equations (the methods of Tartaglia, Cardano, and other Italian mathematicians). Viète also recognized the relations between the sums of the roots of polynomial equations and their coefficients, as well as devising geometrical methods for tackling such equations.

His work paved the way for one of the most fruitful transitions in the history of mathematics. In 1637, Descartes made public his most important intellectual accomplishment, linking geometry to the new algebra in coordinate geometry. Rightly proud of the power of his method—as he frequently points out, the examples he provides illustrate how to solve infinitely many similar problems—he contrasts the scatter of results painfully generated by previous geometers with the systematic success he can offer the mathematical world (Descartes, 1954).

Armed with these tools, mathematicians of the mid-seventeenth century tried to extend them to classical geometrical questions that resisted solution: finding the lengths of segments of curves, constructing tangents and normals, computing areas, and discovering maxima and minima. Descartes himself, Fermat, Roberval, and Cavalieri all achieved some partial advances with these problems. A fully general approach came only at the end of the century, with the techniques of the calculus, independently (and differently) offered by Newton and Leibniz (Kitcher, 1983, pp. 230–241).

Though Leibniz won the notational contest, and his more uninhibited tolerance for infinitesimals eventually triumphed, Newton's conservative preference for tying the calculus close to geometry played a decisive role in the subsequent transformation of mathematics. For Newton proposed an approach to geometry connecting that branch of mathematics with the study of motion. He writes:

> In ye description of any Mechanicall line what ever, there may bee found two such motions wch compound or make up ye motion of ye point describeing it, whose motion being by them found by ye Lemma, its determinacon shall bee in a tangent to ye mechanicall line (Newton, 1967, p. 377).

By adopting this kinematic approach, Newton gains the ability to move to and fro between geometry and the theory of motion. Tools crafted for one domain can be applied in the other. In particular, the successes of the calculus in solving geometrical problems can be mirrored in kinematics, and form the platform on which Newton will erect his strikingly successful dynamics.

The subsequent history of the calculus is a tale of free-swinging success, in which most of those who participate don't share the qualms of Newton and his successors about "infinitely little" quantities that can be "blotted out" at the appropriate moments. Only when the liberated appeals to thinking about terms that are sometimes (opportunistically) positive and sometimes (opportunistically) treated as zero start to interfere with problem-solving, or when clever substitutions in infinite series generate odd-looking conclusions, do mathematicians turn their attention to mopping up what's become an annoying mess (Kitcher, 1983, pp. 264–268). So, through the eighteenth and nineteenth centuries, clearer conceptions of limit, continuity, convergence, differentiation, and integration emerge to allow mathematicians to deal with the full array of functions they want to consider. The end result is the real analysis of Weierstrass and his school—although some mathematicians, Kronecker and Dedekind for example, contend that further steps are required. Frege, of course, has a far more tender conscience than any of them (Kitcher, 1986).

Between Newton and the late nineteenth century, there's an explosion of mathematical developments, and a concomitant change in the status of the mathematician. As the role of differential equations in physics becomes recognized, the idle games of mathematicians appear in a new light. The "useless" problems with which they toy generate concepts and methods for doing serious investigation of the natural world. Cardano and Tartaglia are no longer court entertainers. People like them deserve prestigious chairs in prestigious academies, and bountiful rewards for their ingenious play.

Let's pick up one thread in a rich and complex tapestry. Mathematicians quickly discovered, that applying the Tartaglia-Cardano method to some cubic equations results in a bizarre designation of the roots. So, for instance, if the initial equation is

$$x^3 - 15x - 4 = 0$$

the method yields the result.

$$x = \left(2 + \sqrt{(-121)}\right)^{1/3} + \left(2 - \sqrt{(-121)}\right)^{1/3}.$$

Recognizing 4 as one root, Rafaello Bombelli was inspired to extend the usual arithmetic operations to expressions containing designations of the "nonexistent square roots of negative numbers." So, using modern notation, he defined:

$$(a + ib)(c + id) = (ac - bd) + i(ad + bc) \text{ where } i^2 = -1$$

Is it then possible to choose a value of k so that

$$(2 + ik)^3 = \left(2 + \sqrt{(-121)}\right) \text{ and } (2 - ik)^3 = \left(2 - \sqrt{(-121)}\right)?$$

If so, the offensive terms can be eliminated, and the root will be given as $2 + 2$. Bombelli saw that taking $k = 1$ would do the trick. But, as he also recognized, his ability to choose an appropriate value depended on his already knowing the value of the root: because of that, he knew that $8 - 6k^2$ would have to be 2. In general, without knowing the root, you'd just have to guess at the value for the pertinent parameter.

So, an apparent curiosity. Most contemporaries and immediate successors accepted Bombelli's own verdict on the new expressions: "subtle and useless." A few, however, continued to explore. As the new Leibnizian analysis played unrestrainedly with infinite series representations of functions and unchecked tolerance for substitutions, many unproblematic results emerged.

$$e^x = 1 + x + x^2/2! + x^3/3! + \ldots$$
$$\sin x = x - x^3/3! + x^5/5! - \ldots$$
$$\cos x = 1 - x^2/2! + x^4/4! - \ldots$$

As Euler saw, if you extend these functions to allow complex numbers as arguments, you can obtain a remarkable identity:

$$e^{iz} = 1 + iz - z^2/2! - iz^3/3! + z^4/4! + \ldots$$
$$= \left(1 - z^2/2! + z^4/4! - \ldots\right) + i\left(z - z^3/3! + \ldots\right)$$
$$= \cos z + i \sin z$$

which yields as a special instance (when $z = \pi$)

$$e^{i\pi} = -1$$

Euler's "beautiful identity."

Play with the cubic has thus started a line of development that leads to an extraordinary connection among trigonometric and exponential functions, mediated by numbers most mathematicians had dismissed. Yet this is only one part of Tartaglia's and Cardano's legacy. Systematic attempts to solve polynomial equations of higher degree yielded success with the quartic, but a frustrating sequence of failures with the quintic. These developments prompted Lagrange to seek an understanding of why particular idiosyncratic substitutions of variables transformed the original equation into an equation of higher degree (a polynomial of degree six in the case of the cubic) that could then be reduced by some formula used for lower degree equations: the sextic is a quadratic in the cube of the artfully chosen variable (Cooke, 2008, pp. 82–85, 92–101). Given the understanding of the relations between functions of the roots and the coefficients of the original equation (extended and deepened since Viète's pioneering work), he focused on permutations of the roots, considering those functions that were invariant under permutations. His work made it clear why the techniques for cubic and quartic were successful, and it inspired early nineteenth-century mathematicians, Abel and Galois in particular, to introduce the concept of a group. The night before the duel in which he would be killed, Galois wrote out a synopsis of his ideas, revealing why the quest for a method to solve the quintic was doomed.

One final episode from nineteenth century mathematics will complete my review of some important transitions. Late eighteenth century studies of complex numbers introduced the notion of the complex plane as an analog of the real number line. They inspired William Rowan Hamilton to ask if there were higher-dimensional numbers. Hamilton first sought a three-dimensional generalization. Relatively quickly he convinced himself of the impossibility of generalizing in ways that would preserve the features of the elementary arithmetic operations he took to be important. The four-dimensional case appeared much more promising, although he encountered recurrent difficulties in defining multiplication. (Over a period of many years, he would retreat to his study to work on the problem. According to legend, when he emerged his wife would ask "Have you discovered quaternions yet?"—and he would ruefully shake his head.) Hamilton filled many waste-paper baskets with potential multiplication tables, all of which failed. Finally, on a walk around Dublin, inspiration came, and he carved the multiplication table into the stonework of a bridge. The breakthrough was to abandon commutativity.

We no longer talk of Hamilton's numbers—"quaternions" as he called them. Instead, his work is absorbed, like that of Abel and Galois, in the abstract algebra, already well-developed in Frege's time, which is central to the mathematics of the twentieth century.

2.4 Mathematical Progress: Significant Questions

A whirlwind tour of a few of the major transitions giving rise to the body of mathematics for which a few mathematicians and a whole long philosophical tradition have wanted to find foundations. If we reject Lakatos's satirical suggestion that there was no mathematical knowledge until the foundationalists provided it—or, since it's not clear that they've yet succeeded in providing what's needed, maybe there's no mathematical knowledge at all—we are faced with two obvious, and connected questions. What made the transitions out of which modern mathematics grew *reasonable*? What made them deserve the title of *advances*? We require accounts of mathematical reasonableness and mathematical progress that accommodate the history I have handled so roughly and crudely.

Preliminaries: I should explain how I think about progress, in mathematics and in the natural sciences, and defend my preference for talking about reasonableness in this context, rather than adhering to the familiar idiom of rationality. Many people adopt a narrow view of progress, keyed to salient examples. Unless you are Don Quixote, your travels aim at a destination, and your progress is measured by the decreasing distance to your goal. Some kinds of progress are like that. Many are not. Children learning to play musical instruments make progress by overcoming their technical problems and expanding the limits of their interpretive skills. The technology of computers and smartphones makes progress by eliminating the glitches of the devices and increasing the range of things they can do. Teleological progress is progress *to*. Pragmatic progress is progress *from*. In mathematics and in science, progress is pragmatic progress. Moreover, although professional communities are often reasonably viewed as pursuing significant problems, ultimate authority on that issue lies with the broader human population. Solving the problems experts identify as significant ought to contribute to human progress in the broadest sense—another form of pragmatic progress.

If all goes well with a community of inquirers, that community will select and resolve questions contributing to the progress of the discipline and to enhancing human lives and improving human societies. Investigators do not always need to ask whether, and how, what they propose to do will bear on human interests. Much of the time, but not always, they can take it for granted that pursuing the kinds of questions they and their fellows single out as significant will do no harm, and may even have positive consequences (possibly quite remote) for human projects. As in much ethical life, they can operate by habit, not constantly interrogating themselves about the worth of their enterprises. Perhaps, from time to time, it would be good for them to reflect on that issue. As I would put it (with thanks to Rudner, 1953), the scientist *qua* scientist, or the mathematician *qua* mathematician, is an ethical agent.

Following the ideas about significance that have been inculcated during your apprenticeship is usually a reasonable strategy. Since the notion of rationality oscillates between two unsatisfactory senses, it is better to talk about reasonableness here. One sense is far too thin and puny to serve the methodology of mathematics or of any natural science. Nobody should be interested in guidelines for not acting madly. The other, an artefact of much work in philosophy of science, is embedded in a technical formalism, often elegant, but inapplicable to any number of different contexts of inquiry. The difficulties of specifying appropriate constraints on assignments of probabilities and utilities are all too well known. Methodology does better by seeking informal canons of good judgment.

That, I hope, clears the decks for approaching the kinds of historical episodes I have very briefly described, with an eye to eliciting characterizations of the problems and solutions through which mathematics expanded from Babylon to late-nineteenth-century Jena.

2.5 Arts of Significant Extension

Here are some very obvious features of those episodes. Start with Hamilton's search for quaternions. It's clearly motivated by the urge to generalize. Bombelli tentatively proposed to extend arithmetic to encompass new numbers, and his successors eventually recognized the fertility of doing so. Recognizing that complex numbers can be identified with ordered pairs of reals, Hamilton sees Bombelli's initial move as extending the arithmetic operations to cover ordered pairs of reals. Is further extension possible? Can you do it for ordered triples? No. For ordered quadruples? Yes, but it takes some effort, and the abandonment of a principle that holds for multiplying real numbers and complex numbers alike: $xy = yx$. Commutativity has to go. But there's a systematic relation that obtains when commutativity lapses: for some of Hamilton's new numbers, $xy = -yx$.

Let's stop for a moment and ask why Hamilton judged the case of triples to be insoluble. The answer: familiar constraints on the arithmetical operations would have to be amended, without offering any regular way of doing so. Extending the system of mathematics with which you are working to introduce expressions that satisfy a previously unsatisfiable requirement can typically be done, but, most of the time, it leads nowhere. A very simple example will make the point. Imagine yourself before the introduction of negative numbers. As things now stand, $m - n$ is undefined when $m < n$. I am an ambitious young mathematician, eager to make my mark. I decide to introduce a single new expression—'N' to denote The Negative (I have Heideggerian sympathies). For any values of m and n such that $m < n$, $m - n = N$. Have I succeeded in my goal?

"No," you reply, pointing out that I have now trivialized arithmetic by making all numbers identical. $1 - 2 = N = 1 - 3$. Switching terms with change of sign, $1 = 1 + 2 - 3 = 0$, and we're now off to the races. I've anticipated this rejoinder, and point out that standard arithmetic practices are not allowed in equations where N figures. So my extension preserves all of positive whole number arithmetic (Kitcher, 1983, p. 208).

I hope you find this response exasperating. You would be entirely justified in deriding my proposal on the grounds that it goes nowhere. Let's not worry for the present about what "going nowhere" means, but simply refer to extensions that go nowhere as "dumb extensions." Hamilton judges that all ways of defining the arithmetical operations for triplet numbers are dumb extensions, but that abandoning commutativity for quaternions leads to an extension that isn't dumb.

When Bombelli originally announced his proposals for the arithmetic of complex numbers, he worried that he was proposing a dumb extension—that's what lies behind his description of them as "subtle and useless." Just modesty? I think not. Operating in the context of a method for solving cubic equations, assumed to be fully general, he wanted to be able to extract cube roots of complex numbers—to specify the value of $(a + bi)^{1/3}$ and $(a - bi)^{1/3}$ for any values of a and b, without guessing in advance the roots of the equation to be solved. In effect, he's recognizing his extension of the arithmetic operations to enable the supposed algorithm to work helps in just the cases where you don't need it. That's one specific way in which a candidate extension can "go nowhere."

By the time Euler celebrates the "beautiful identity", complex numbers have been embedded in a number of different contexts, and, most pertinently, they expose a connection between exponential functions and trigonometric functions—thus fostering the definition of the hyperbolic analogues of the trigonometric functions (that then turn out to have interesting physical applications—e.g. in studying hanging chains). The "beautiful identity" condenses all this, the aesthetic tribute fully deserved by the fact that when three interesting (and mysterious) numbers are connected in a mysterious way (raising numbers to powers starts with squares and cubes—who'd have taken raising the base of natural logarithms to the power $i\pi$ to be a paradigm for exponentiation?) you get an everyday negative integer.

Underlying Hamilton's judgment, then, is a long history. Practical problems lead Babylonians to investigate equations, and inspire a search for methods to solve them. They succeed with two important classes—simple first-degree equations and quadratic equations—even though they lack any perspicuous notation for making the methods explicit. Few cubic equations have the same practical payoff, but the game of solving them is fun to play (for mathematicians) and apparently fun to watch (the aristocratic audiences), and Cardano publishes (plagiarizes?) a method. Difficulties in understanding that method in a whole range of instances provoke Bombelli to generalize real arithmetic. As that generalization proves fruitful in a wide number of unanticipated contexts, "imaginary" numbers become accepted. Hamilton tries to generalize further. The three-dimensional attempt forces him to modify without providing any systematic understanding of multiplication: the extension is unlikely to bear much fruit. On the other hand, the multiplication table for quaternions offers a pleasing symmetry. In the instances where commutativity fails, we find anti-commutativity (ij = k; ji = −k). The algebra for these "numbers" appears worth exploring further.

	1	i	j	k
1	1	i	j	k
i	i	-1	k	-j
j	j	-k	-1	i
k	k	j	-i	-1

Hamilton's reasonableness consists in his emulation of a mathematical practice that has proved useful in the past. Pragmatic concerns enter into the judgments of all the major characters in this story. Tartaglia and Cardano play mathematical games, because they and their fellows enjoy the games, the games are harmless, and outsiders find them entertaining. Bombelli's modesty is grounded in recognizing that his extension won't do the work for which he undertook it. Euler's enthusiasm rests on seeing that the extension is useful for all sorts of questions that interest mathematicians, that some of the answers to those questions can play a role in investigations of natural phenomena, and that it provides aesthetic satisfaction (Edna St. Vincent Millay got it wrong: Euler as well as Euclid "has looked on Beauty bare.") Hamilton sees his own enterprise as potentially having all these virtues—and, despite the fact that we no longer think of quaternions as special numbers, history has justified his confidence.

Sometimes, the grounds for supposing an extension to be worthwhile are more straightforward than in the history I have reviewed. Consider the streamlining afforded by introducing Arabic numerals, or by Viète's notation. In the first instance, the change eases the daily work of all those who do arithmetic, whether they are mathematicians pursuing projects of no obvious practical significance or are accountants or shopkeepers or engineers. In the second, the benefits for extra-mathematical practice are less evident, but the new notation helps anyone who has to solve an algebraic equation, and, within mathematics, it aids the search for methods for solving the quartic (and, until Galois, ventures beyond.) Moreover, after Viète, the power of this style of notation is revealed in Descartes's coordinate geometry, and, even more spectacularly, in the emergence of the calculus, and in its scientific payoffs: first, in Newtonian dynamics, and then through the growing incorporation of differential equations in physics.

When mathematicians provide tools for addressing scientific problems, their discipline can no longer be disparaged as mere game-playing. It's no accident that Newton's work is done in the middle of a century during which the status of the mathematician is dramatically elevated. A Galileo active in the mid-eighteenth century would not have been so anxious to be known as a philosopher rather than as a mathematician. Moreover, Newton's paradigmatic achievement is bracketed by two other practically fruitful expansions of mathematics. Pascal considers how to divide up the antecedent stakes in unfinished games of chance, and establishes the theory of probability as a new mathematical discipline. Euler muses on the difficulty of traversing the bridges of Königsberg without retracing your steps, and takes the first steps towards topology. In all three instances, new mathematics is, we might say, *purpose built*, growing out of efforts to tackle practical problems outsiders can recognize.[3]

The extent to which pragmatic goals dominate can be appreciated by considering the career of the calculus from the 1680s to the late nineteenth century. For a very long time, the obvious difficulties with the methods practitioners employ—obvious enough to be lucidly pointed out by an Anglican bishop—were mostly ignored. For Newton and Leibniz, both dead by the time Berkeley's *Analyst* was published in 1734, the characterization of the central method would not have been news: both knew they were treating small quantities sometimes as positive and sometimes as zero. Yet, as even Berkeley acknowledged, the calculus was immensely successful. The successes prompted *one* of the major traditions of eighteenth century analysis simply to go on and to pile up results. Indeed, to couple the central method to another dubious technique—substituting freely in infinite series, oblivious to the odd results that sometimes emerged. After all, mathematicians knew enough to toss aside the "result" that.

$$-1 = 1 + 2 + 4 + 8 + \ldots$$

---

[3] As the referee helpfully pointed out, Pascal and Euler seem to have clear goals: they want a mathematical apparatus for settling questions that arise for them. These are short-term goals, "ends-in-view". They do not define any long-term goal for mathematics (e.g. the complete mapping of mathematical reality), but are set by the problems arising in a particular historical context. So, we can see their pragmatic progress—solving a particular problem—in teleological terms, without supposing that the overall progress of mathematics is teleological.

and, although Leibniz offered a quasi-theological explanation of why.

$$1 - 1 + 1 - 1 + \cdots = \frac{1}{2}$$

most (though not all) of his followers consigned it to the scrapheap.

Conservative Newton proceeded differently, preferring, from the beginning, to develop a complex geometrical account that would undergird the manoeuvres he allowed himself. The result was a mathematical tradition, prominent in Britain but very much a minority in the rest of Europe, that proceeded far more slowly, by cautious steps. In the hands of the Leibnizians, the Bernouillis, Euler, and an increasingly large majority of mathematicians, a deluge of results poured in from functional analysis, making the post-Newtonian approach appear a quaint irrelevance. Eighteenth century analysis reveals the triumph of the pragmatic rationale for extending mathematics.

But, to summarize a story I have told at length elsewhere, eventually the piper had to be paid. For pragmatic reasons. To solve the problems the analysts wanted to tackle, they had to become clearer about just when their free-swinging methods would let them down. This occurred piecemeal, as different mathematical interests exposed difficulties with different aspects of mathematical practice—with convergence or continuity or differentiation or integration. They needed explanations of when a method worked and when it didn't, explanations that could guide them when they couldn't just "see" that a potential "result" was absurd.

The pragmatic virtue that sometimes prompts and warrants acceptance of mathematical extension is sometimes an increase in understanding—effected by clearing up a previous source of perplexity.

2.6 Varieties of Misunderstanding

Two episodes crudely outlined in my history can be approached in terms of explanation and understanding. To his mathematical contemporaries, Descartes offered new ways of understanding traditional geometrical problems. Consider, for example, locus problems. The ancients discovered answers to a few questions of the form "What is the locus of a point which ... ? " where the blank is filled in by a characterization of the distance or distances from some fixed points or fixed lines. The locus of a point whose distance from some fixed point remains constant is a circle; the locus of a point such that the sum of its distances from two fixed points is constant is an ellipse. Ancient mathematicians had gone far beyond such elementary instances, but, as was well known in the early seventeenth century, large numbers of general classes of locus problems existed, that could only be solved for the (simplest) cases. The Pappus problem, involving fixed lines and fixed angles where the products of specified distances must stand in a given ratio, was a parade case (indeed one, many historians believe, that may have spurred the development of Cartesian coordinate geometry.) As the *Géometrie* proudly declares, the algebraic reformulations enable mathematicians not only to solve a vast number of previously intractable problems, but also to explain why the limited techniques they replace worked when they did.

Similarly, Lagrange attempted to understand why the available techniques for solving cubic and quartic equations succeed. It looks like magic. Consider the cubic: $x^3 + ax + b = 0$ (any cubic can be brought into this form by a prior substitution to eliminate the quadratic term). You make what appears to be, at first sight, a completely unmotivated substitution: let $x = y - a/3y$; and, hey presto, the equation transforms to.

$$y^6 + by^3 - a^3/27 = 0$$

a nice, soluble, quadratic equation in y3. Lagrange wants to know why this cunning trick works—indeed why *any* substitution should transform the equation to one that would succumb to methods devised for equations of lower degree. To this end, he introduces a new idea—thinking about functions of the roots that remain invariant under permutations.

Now it's tempting to assimilate both cases to a pattern of change much heralded in discussions about science, the idea that scientific theories grow by developing deeper and deeper explanations, and, since the kind of explanation provided isn't readily assimilated to causal explanation, to seek some other general account of explanation that will apply across the board, underlying mathematical explanation, high-level theoretical explanation in the natural sciences, and (ultimately) causal explanation. An incautious philosopher who went down this road might be led to propose that explanation is fundamentally unification.

I hope that, as I have grown older, I have become wiser. Also that a sign of wisdom is abandoning the search for any general theory of explanation *tout court*. I continue to believe that, on occasion, in both mathematics and the natural sciences, investigators seek increased understanding, that they adopt new concepts and new principles because doing so helps relieve their perplexities, and that they reasonably hail the new framework as explaining what previously puzzled them. But I no longer hope for a general understanding of explanation (or should it be a general explanation of understanding?). The explanation-seeking questions people pose are diverse, and the virtues of the successful answers they find (when they are lucky) are a motley collection.

Philosophy is beguiled, I think, by the idea of understanding as a general ideal, something to be achieved completely in particular instances and (for utopians) to be achieved completely with respect to everything. Here, too, it's easy to equate progress—progress in understanding—with approximating a long-range goal. I suggest that the progress of our understanding is another species of pragmatic progress. Not only in the sense that the goal is an impossible one—there are far too many questions, and, however long we lived, we'd never run out of potential puzzles. Also because the questions are diverse, and the ideal of understanding, *even in specific cases*, could never be realized. Pragmatists should start in a different place, with *mis*understanding, and appreciate that misunderstanding comes in many guises.

Descartes and Lagrange want to know *how* to do more generally things they can do partially. They also would like to see *why* the available methods succeed where they do. Euler responds to the bizarre results obtained by the technique of substituting in infinite series, and asks *what* has gone wrong—*what* differentiates these from the many instances in which the technique delivers a new result whose correctness can be recognized by doing a bit of arithmetic. Bolzano wants a *rein analytischer Beweis* of the theorem that a continuous function that takes positive values for some arguments and negative values for others must have a zero; he wonders what the relation between analysis and geometry is, and why we need the detour of thinking about curves drawn on paper without lifting the pencil (Kitcher, 1975). Galileo, Bolzano, and Dedekind all want to know how to handle a "paradox of the infinite": they want to know *whether* there are more natural numbers than even numbers—does the inclusion criterion settle relative size, or should we appeal to one-one correspondence?[4] Cauchy and Abel want to know if there can be Fourier series representations of discontinuous functions—and which one of them has made an error. Bombelli wants to know the relation between the odd expression involving square roots of negative numbers and the roots of the cubic equation. Gauss, Bolyai, and Lobachevsky are puzzled by the difficulties in trying to prove Euclid's fifth postulate, and wonder if they can evade them by using *reduction ad absurdum*— and are led to non-Euclidean geometry.

Forty years ago, Bas van Fraassen offered a "pragmatic theory of explanation" (van Fraassen, 1980, Chap. 5). To explain, he suggested, is to answer a why-question (distinguished by a topic and a contrast class). Explanation is achieved by giving a sentence that stands in a relevance relation to the topic and the contrast class. Van Fraassen placed no constraints on relevance, and, as Wesley Salmon and I showed, this led to a trivialization of his account: anything can explain anything (Kitcher & Salmon, 1987). Salmon and I assumed that an adequate theory of explanation must supply a characterization of the relevance relation. We differed on what that account should be. Salmon opted for a particular causal relation; I proposed that relevance relations were those generated from the patterns figuring in the best overall unification of our beliefs. We were both mistaken. Salmon appreciated van Fraassen's insight far earlier than I did (Salmon, 1998). Even though a vast number of relations are not relevance relations, the class of relevance relations is large and diverse. That's the core of a "pragmatist non-theory of explanation": explanation-seeking questions come in many varieties, and with respect to each of the varieties there are diverse appropriate relevance relations, apt in different contexts. To go beyond this non-theory is a matter of exploring the diversity, and characterizing the relevance relations in ways that show how they meet the needs of the context in which the explanation-seeking question arises.

Hence, I suggest, we do best to begin with the particular perplexities of particular people, with some understanding of the varieties of misunderstanding. In the case of mathematics, I hope my sampling from history reveals how interestingly diverse the forms of misunderstanding are.

---

[4] Paolo Mancosu's brilliant (Mancosu, 2016) shows how it is possible to develop an alternative to the familiar Dedekind-Cantor approach that gives priority to the one-one correspondence criterion.

2.7 Games Worth Playing

Let me try to pull some of the threads of this discussion together. During the long period I have surveyed, mathematics expands and thereby makes progress. It does so by building on, and generalizing, parts of the accepted mathematics of the past. Some expansions allow mathematicians to address questions they have previously been unable to tackle, or only to answer partially. Others are directed towards understanding the limited successes of methods previously employed. Or resolving other kinds of misunderstanding. Sometimes, mathematics is directed towards tackling some particularly puzzling natural phenomenon (the purpose built mathematics of Pascal and Newton and Euler). At others, it issues in new systems mathematicians take pleasure in exploring—new games that are fun to play (and even attractive to those who watch.) The pleasures enjoyed can be aesthetic.

By assimilating mathematics to game-playing, I may seem to be insulting its dignity. Not so. Renaissance patrons may have condescended to the mathematical performers they invited in for the evening, according them the status appropriate to the conjurer brought in for the children's party or the touring grandmaster who comes to play simultaneously against forty local chess players. Our attitude should be different. We should endorse the social shift that occurred between the sixteenth century and the eighteenth. It has turned out to be a very good idea to provide mathematicians with a license to pursue the games that interest them. Not only has it offered them and a wider public plenty of amusement (Martin Gardner's *Mathematical Games* column in *Scientific American* was eagerly devoured by many readers), and even glimpses of "Beauty Bare." It has also furnished large numbers of tools for use in investigating the natural and social worlds.

Many philosophers would respond to this last fact by characterizing mathematics in ways designed to celebrate its dignity. The Book of Nature is written, Galileo told us, in the language of mathematics. Dignity-obsessed philosophers take the language to refers to denizens of an abstract universe—of special objects? special structures?—that underlies reality. Mathematics isn't mere game-playing, but the description of the deep structure of the cosmos. Despite the mysteries this conception brings in its train, a host of philosophers and some mathematicians subscribe to it. Why else is mathematics so "unreasonably effective" in theorizing about nature?

Because, quite reasonably, we attempt to write the book of nature in the language of mathematics. (Galileo's dictum needs rewriting.) From the very beginning, thinking in terms of shapes and numbers helped our ancestors in their practical activities, and, to this day, measurement of vast numbers of kinds pervades our lives. New chapters in the book are facilitated by building up a store of tools from which investigators may draw. In response to Wigner's celebrated question, it's appropriate to ask just how effective we might have expected mathematics to be for its successes to strike us as "unreasonable." Or, more pointedly, as Sidney Morgenbesser once wittily remarked to me, "What do you expect scientists to use? Mathematics that hasn't been invented yet?" The question needn't be merely rhetorical, for there are occasions on which great mathematicians—Pascal, Newton, Euler –fashion a new mathematical language to inquire into natural phenomena.

The perspective I have offered is supported by recognizing the opportunistic ways in which different types of inquiry borrow from one another. Borrowing occurs within mathematics: Andrew Wiles' famous proof of Fermat's last theorem (the "proof of the century") built on previous efforts to connect previously separate branches of mathematics (Singh, 1998). It also happens across independent provinces of scientific investigation. The game theory, originally introduced to study human interactions, inspired evolutionary biologists to elaborate it in a particular way, and the results of their extensions have fed back, both into the original socio-political investigations, and into mathematics. The history of mathematics is intertwined with that of the natural and social sciences, in which ideas travel and grow in criss-crossing paths—to meet the needs of current inquiry.

The games mathematicians play have turned out to be wonderful sources of concepts and techniques for many people to use. No wonder mathematicians are held in high regard. Even when they only seem to be playing.

2.8 Homage to Lakatos

Play seems to be going on in Lakatos' classroom—is it a cooperative or a competitive game? Lakatos provides us with a detailed, but historically streamlined, account of a particular series of progressive steps in mathematics. The initial question is prompted by a long-recognized piece of elementary geometry: the number of angles in a polygon is equal to the number of sides. Is there a similar relation in the three-dimensional case, for polyhedra? The discussion begins after decisions have already been made about how to specify similarity: the class asks after the relation among the number of faces, vertices, and edges.

The question has no obvious practical significance. It is spurred, as so many questions in the history of mathematics seem to be, by an interest in generalizing from an existing part of mathematics (think of Bombelli, Descartes, and Hamilton). As the question is explored, difficulties are generated, and the class seeks ways to circumvent them (just as Euler did with respect to bizarre series summations, and Cauchy and Abel did when faced with a puzzle about sums of continuous functions). Lakatos' account of the steps is easily liberated from his reliance on Popperian ideas, and it will fit well within the larger framework I have sketched.

So I end as I began, with acknowledgement of the brilliance of his work in this area of philosophy. The philosophy of mathematics ought to be developed by investigating the methodology behind mathematical practice, using a general historiographical framework—not necessarily the one I have outlined here—to reveal the reasonableness and progressiveness of particular transitions, the virtues sought and attained. *Proofs and Refutations* is a paradigm of how that can be done. I hope a future philosophy of mathematics, liberated from the scholastic debates that have dominated professional activity in this area, will imitate Lakatos—and look back on him as the founder of the subject as it should be pursued.

References

Arabatzis, T. (2017). What's in it for the historian of science? Reflections on the value of philosophy of science for history of science. *International Studies in the Philosophy of Science, 31*, 69–82.
Cooke, R. (2008). *Classical algebra. Its nature, origins, and uses*. John Wiley.
Cooke, R. (2013). *History of mathematics: A brief course*. John Wiley.
Descartes, R. (1954). *The geometry of René Descartes* (trans: David Eugene Smith, Marcia L. L.). Dover.
Kitcher, P. (1975). Bolzano's ideal of algebraic analysis. *Studies in the History and Philosophy of Science, 6*, 229–269.
Kitcher, P. (1977). "On the uses of rigorous proof", (review of Lakatos 1976). *Science, 196*(4291), 782–783.
Kitcher, P. (1983). *The nature of mathematical knowledge*. Oxford University Press.
Kitcher, P. (1986). Frege, Dedekind and the philosophy of mathematics. In L. Haaparanta & J. Hintikka (Eds.), *Frege synthesized* (pp. 299–343). Reidel.
Kitcher, P., & Salmon, W. (1987). van Fraassen on Explanation. *Journal of Philosophy, 84*, 315–330.
Lakatos, I. (1963–1964). "Proofs and refutations" Parts I–IV, *British Journal for the Philosophy of Science*, 14, 1–25, 120–139, 221–243, 296–342.
Lakatos, I. (1976). *Proofs and refutations*. Cambridge University Press.
Lakatos, I. (1978). *Philosophical papers, volume II, mathematics, science, and epistemology*. Cambridge University Press.
Mancosu, P. (2016). *Abstraction and infinity*. Oxford University Press.
McBrearty, S., & Brooks, A. (2000). The revolution that Wasn't: A new interpretation of the evolution of modern human behavior. *Journal of Human Evolution, 39*, 453–563.
Newton, I. (1967). In D. T. Whiteside (Ed.), *The mathematical papers of Isaac Newton* (Vol. 1). Cambridge University Press.
Renfrew, C., & Shennan, S. (1982). *Ranking, resource and exchange*. Cambridge University Press.
Rudner, R. (1953). The scientist *qua* scientist makes value judgments. *Philosophy of Science, 20*, 1–6.
Salmon, W. (1998). Scientific explanation: Causality *and* unification. In *Chapter 4 of Salmon causality and explanation*. Oxford University Press.
Singh, S. (1998). *Fermat's enigma*. Anchor.
van Fraassen, B. (1980). *The scientific image*. Oxford University Press.

Open Access This chapter is licensed under the terms of the Creative Commons Attribution-NonCommercial-NoDerivatives 4.0 International License (http://creativecommons.org/licenses/by-nc-nd/4.0/), which permits any noncommercial use, sharing, distribution and reproduction in any medium or format, as long as you give appropriate credit to the original author(s) and the source, provide a link to the Creative Commons license and indicate if you modified the licensed material. You do not have permission under this license to share adapted material derived from this chapter or parts of it.

The images or other third party material in this chapter are included in the chapter's Creative Commons license, unless indicated otherwise in a credit line to the material. If material is not included in the chapter's Creative Commons license and your intended use is not permitted by statutory regulation or exceeds the permitted use, you will need to obtain permission directly from the copyright holder.

Chapter 3
Proofs as Dialogues: The Enduring Significance of Lakatos for the Philosophy of Mathematical Practice

Catarina Dutilh Novaes

Abstract This paper discusses the enduring significance of Lakatos' account of mathematical knowledge in in *Proofs and Refutations* for the philosophy of mathematical practice. While the account has been criticized for being historically inaccurate and for relying on contentious (idealist) assumptions, I argue that it remains an insightful source for philosophers of mathematical practice. In particular, I spell out how Lakatosian proofs and refutations have inspired the formulation of a dialogical account of deduction and mathematical proof, as presented in my book *The Dialogical Roots of Deduction*. In particular, Lakatos' account provides the conceptual tools for an analysis of the intricate interplay between adversariality and cooperation in practices of mathematical proof. I conclude by highlighting some of the differences between Lakatos' Hegelian idealism and the pragmatism that underpins my dialogical account of deduction.

Keywords Mathematical proof · Dialogical account of deduction · Adversariality and cooperation · Proofs and refutations

3.1 Introduction

Lakatos' account of mathematical knowledge in terms of a 'proofs and refutations' dialectic, first presented in his dissertation (1959), subsequently developed in various articles, and then posthumously published as the influential book *Proofs and Refutations* (Lakatos, 1976) (henceforth, P&R), has had phenomenal impact on the philosophy of mathematics. In particular, his ideas have been influential for the emergence of the *philosophy of mathematical practice* tradition (Kitcher & Aspray, 1988; Tanswell, 2016)—even if, perhaps ironically, Lakatos himself did not view

---

C. Dutilh Novaes (✉)
Department of Philosophy, VU Amsterdam, Amsterdam, Netherlands
e-mail: c.dutilhnovaes@vu.nl

© The Author(s) 2025
R. Frigg et al. (eds.), *Proofs and Research Programmes: Lakatos at 100*, Synthese Library 498, https://doi.org/10.1007/978-3-031-88213-5_3

the analysis of mathematical practices as such as belonging to the realm of the philosophy of mathematics.

His critique of the main strands in philosophy of mathematics of his time—logicism, intuitionism, and formalism—was based on the observation that they did not correctly represent the history of mathematics. Moreover, for Lakatos, these strands were overly focused on the axiomatic, formal structures of mathematical theories, thus neglecting their conceptual, contentual aspects. Lakatos criticized what he called the *deductivist* style in mathematics, which focuses on the quasi-mechanical deduction of theorems from axioms in proofs. Rather, by attending to the interplay between proofs and refutations, Lakatos was interested in the *processes* through which new mathematical concepts and findings come into being, and in the relevant *heuristic* background.

There has also been much critique of Lakatos' proofs and refutations model. In particular, it has been argued that it is actually not historically accurate (Koetsier, 1991; Musgrave & Pigden, 2016) (something that Lakatos himself also recognizes in a much discussed footnote of P&R),[1] applying at best only to a handful of examples from the history of mathematics. It has also been argued that, besides not being historically accurate, it does not provide a compelling *philosophical* account of mathematics either (Larvor, 2001; Gillies, 2014). In particular, it is debatable whether the Hegelian assumptions underpinning the account (Larvor, 1999) are compatible with a wide-ranging understanding of mathematics in all its facets.

These criticisms may seem to imply that Lakatos' ideas no longer have anything to offer to contemporary discussions in the philosophy of mathematics and mathematical practice. In what follows, I argue against this pessimistic conclusion. In particular, I show how P&R has provided significant inspiration for the development of a dialogical conception of mathematical proof, which I presented and defended in (Dutilh Novaes, 2020).[2]

The paper proceeds as follows: Sect. 3.2 introduces the main elements of P&R for the present purposes. In Sect. 3.3, I present the Prover-Skeptic dialogues that are the conceptual spine of the dialogical account presented in (Dutilh Novaes, 2020). In Sect. 3.4, I reflect on the interplay between cooperation and adversariality, both in the Lakatosian account and in Prover-Skeptic dialogues. Finally, in Sect. 3.5 I discuss some Hegelian features of the Lakatosian account, which leads me to examine the main philosophical difference between his account and my dialogical

---

[1] "Pi's statement, although heuristically correct (i.e. true in a rational history of mathematics) is historically false. (This should not worry us: actual history is frequently a caricature of its rational reconstructions.)" (P&R, p. 84).

[2] Lakatos' work did not provide direct initial inspiration for the project, which was originally mostly inspired by the historical emergence of deduction in ancient Greece (both in logic and in mathematics) (Netz, 1999). But sustained engagement with Lakatos' ideas was essential for its subsequent development. The fact that the book has received the 2022 Lakatos Award is particularly rewarding, in view of the significance of his ideas for the overall project. I have also argued elsewhere that the dialectic of proofs and refutations provides important insights for practices of *philosophical* argumentation as well, *mutatis mutandis* (Dutilh Novaes, 2022).

account of deduction. In a nutshell, while Lakatos is primarily interested in *concepts* and their development, my dialogical account focuses primarily on the relevant *human agents*, thus having a markedly *pragmatic* orientation which contrasts with the Hegelian idealism of P&R.

3.2 *Proofs and Refutations* in a Nutshell

The main text of P&R consists in a classroom dialogue between a teacher and students, named after letters of the Greek alphabet; they discuss various (attempted) proofs of Euler's conjecture for polyhedra, which links vertices (V), edges (E) and faces (F) through the formula $V - E + F = 2$. The dialogue is presented as a rational reconstruction of the historical development of (attempted) proofs for the conjecture and their refutations; the different students are portrayed as representing various positions and reactions.[3]

The dialogue starts with the teacher presenting an argument (due to the nineteenth century mathematician Cauchy) supporting the conjecture, which the students then go on to scrutinize and criticize for various reasons. At each objection, the proof is modified so as to withstand the force of the objection, for example by restricting the relevant definition so that the counterexample produced is now excluded from the range of the (new) hypothesis/conjecture; this is what Lakatos describes as 'lemma-incorporation' (P&R, p. 36). Other strategies to deal with objections include 'monster-barring', which consists in questioning the legitimacy of the purported counterexample ("It is a monster, a pathological case..." (P&R, p. 14)), 'exception-barring' (a conjecture is only valid in a certain domain excluding exceptions (P&R, p. 24)), and 'monster-adjustment' (a change in the interpretation of the counterexample (P&R, p. 31)).

Through this process, it becomes clear that many of the key concepts involved (e.g., the concept of a polyhedron itself) were in fact vague and poorly understood at the starting point, and through the dialectic of proofs and refutations these concepts are clarified (Tanswell, 2016, Sect. 3.3) and transformed (Tanswell, 2018). As the dialogue progresses, there is a shift from treating polyhedra as solids to treating them as closed surfaces (the Cauchy argument discussed assumes that a polyhedron is not a solid), and later as networks and vector spaces. Given these changes, one may wonder if the participants are still talking about the same 'things,' or whether the subject has changed along the way—which may or may not be a problem. It isn't a problem for Lakatos, as he is interested primarily in the emergence of new mathematical concepts (see Sect. 3.5); as Alpha puts it, "behind surreptitious changes in the meaning of the terms [hides] an essential improvement" (P&R, p. 41).

---

[3] A formal reconstruction of Lakatosian dialogues is to be found in (Pease et al., 2017).

In fact, the very notion of 'mathematical proof' seems to be taken in different senses throughout P&R. Lakatos does not draw a sharp distinction between correct and incorrect proofs; the different arguments pertaining to Euler's conjecture discussed along the way are variously referred to as 'proofs', 'thought experiments', 'proof-ideas', or 'proof-analysis'. Cauchy's initial argument, presenting polyhedra as hollow objects with a surface of thin rubber which can be stretched, is described as a proof-idea (P&R, fn. 1, p. 8). As the dialogue progresses, different proof-analyses are proposed to unpack the initial proof-idea. One may be tempted to conclude that only a fully worked-out, detailed proof-analysis, where all assumptions and inferential steps are made explicit, can properly count as a proof, but Alpha notes that "a mature mathematician understands the entire proof from a brief outline" (P&R, p. 51). More generally, and in the context of his criticism of the deductivist style, Lakatos rejects the idea that there is a unique, absolute standard of rigor that must be satisfied for an argument to merit the honorific 'mathematical proof' (Tanswell, 2016).

As the discussion progresses, the method of proofs and refutations emerges, and is described in the following terms (these are not rules in a strict sense, but rather fallible strategies, heuristics):

> *Rule 1.* If you have a conjecture, set out to prove it and to refute it. Inspect the proof carefully to prepare a list of non-trivial lemmas (proof-analysis); find counterexamples both to the conjecture (global counterexamples) and to the suspect lemmas (local counterexamples). (P&R, p. 50)
>
> *Rule 2.* If you have a global counterexample discard your conjecture, add to your proof-analysis a suitable lemma that will be refuted by the counterexample, and replace the discarded conjecture by an improved one that incorporates that lemma as a condition. Do not allow a refutation to be dismissed as a monster. Try to make all hidden lemmas explicit. (P&R, p. 50)
>
> *Rule 3.* If you have a local counterexample, check to see whether it is not also a global counterexample. If it is, you can easily apply Rule 2. (P&R, p. 50)
>
> *Rule 4.* If you have a counterexample which is local but not global, try to improve your proof-analysis by replacing the refuted lemma by an unfalsified one. (P&R, p. 58)
>
> *Rule 5.* If you have counterexamples of any type, try to find, by deductive guessing, a deeper theorem to which they are counterexamples no longer. (P&R, p. 76)

As made evident by these rules, counterexamples play a key role in the dialectic of proofs and refutations. They can be of two sorts: local counterexamples to specific steps in the proof (lemmas), or global counterexamples that affect the whole conjecture. In the case of local counterexamples, the conjecture itself may still hold up, but cannot be shown to be true by that very proof. Global counterexamples, in turn, require a more serious modification of the initial conjecture, as described in Rule 2.

While Lakatos himself does not speak of specific characters carrying out proofs and refutations, we may think of these two sides as two *functional* roles, which can be associated to participants in dialogical games, as is done in comparable accounts such as Lorenzen's dialogical logic (Lorenzen & Lorenz, 1978) (proponent and opponent), and Hintikka's game-theoretical semantics (Hintikka & Sandu, 1997) (verifier and falsifier). Importantly, the functions of challenging and defending can be switched among participants as the dialogue progresses: for example, in dialogical logic, the negation operator corresponds to such a switch (if a participant asserts ~p, then the other participant must then *defend p* in order to attack ~p).

In this vein, we may want to introduce two functional characters, *Prover* and *Refuter*, each performing one of the two crucial functions in the dialectic of proofs and refutations.[4] Importantly, unlike in dialogical logic and game-theoretical semantics, these Lakatosian dialogues do not correspond to typical zero-sum games with losers and winners, and there does not seem to be a counterpart for the notion of a *winning strategy* in these dialogues. Indeed, Lakatos defends the controversial view that there is no end-point for a proof. He does seem to suggest at times that interlocutors are 'opponents' to each other, and debates in P&R can get quite heated, but the ultimate goal of the 'game' is the higher-order (cooperative?) purpose of improving the conjecture-proof pair. (More on adversariality and cooperation in Sect. 3.4).

Thus described, however, the role of Refuter, restricted to the search for counterexamples, appears to be rather limited. In reality, someone at the receiving end of a mathematical proof is likely to be interested in a number of other features of the proof, such as: does it start from plausible, potentially fruitful premises?[5] Is the argument as a whole illuminating, i.e., does it improve one's understanding of the issue? Are the individual inferential steps compelling? In other words, there is much more to being at the receiving end of a mathematical proof than merely looking for counterexamples. Indeed, as suggested in (Pease et al., 2017), these dialogues appear to be essentially of the *persuasive* kind (following the taxonomy introduced in (Walton & Krabbe, 1995)):

> Lakatos describes the interactions of mathematicians, when aiming to prove the conjecture, in the way that is the most closely related to the concept of persuasion dialogue, and in particular its "conflict resolution" subtype. Persuasion dialogue is triggered by a difference of opinion between participants, each of them aims to persuade each other and the main goal of the conversation is to achieve the resolution of the conflict. (Pease et al., 2017, p. 187)

I suggest, however, that the putative audience for a mathematical argument does *not* start out with the conviction that it is incorrect, or that there must be a mistake in it (which would count as a more straightforward 'difference of opinion'). Instead, it seems more plausible that the audience initially adopts an agnostic, neutral position, but that the bar for persuasion is set quite high. Walton and Krabbe recognize this type of dialogue, and classify it as a subclass of persuasive dialogues:

---

[4] Prover-Refuter games are widely used and studied in the computer science literature on theorem-proving. Trafford (Trafford, 2017, pp. 88–94) explicitly introduces these two 'players' when discussing Lakatos. But a different reading of P&R suggests that the characters in fact represent different *philosophical positions*, not being particularly concerned with Euler's theorem as such except insofar as it allows them to expound their philosophical views. On this reading, it makes no sense to think of them as Provers or Refuters. (I owe this remark to Brendan Larvor.)

[5] "We may still reasonably enquire how the premises for proofs are chosen. Why should mathematicians explore the deductive closure of *this* set of axioms rather than *that?*" (Larvor, 2001, p. 214).

> Every persuasion dialogue is based on an initial conflict of opinions, but these conflicts can be of various types. In a simple conflict, only one party has a positive thesis to defend, the other party being merely a critical doubter or questioner, with no expressed positive viewpoint of his own. (Walton & Krabbe, 1995, p. 118)

This is, I submit, a more accurate description of how the interlocutors position themselves in dialogues involving mathematical arguments. The receiver of a mathematical proof will engage critically with it, searching for counterexamples and other imperfections, but not with the goal of refuting it at all costs. The audience functions rather as an 'examiner', who will not allow Prover to produce sloppy arguments, but who is not in the business of knocking down the proof no matter what. In fact, the audience may even *help* Prover to improve the initial proof-idea. Thus, interventions in mathematical dialogues seem to comprise more than just 'proofs and refutations', which suggests the possibility of an expansion of these Lakatosian dialogues. A good proof is one that convinces a fair but critical opponent; as remarked by mathematician Mark Kac, "the beauty of a mathematical proof is that it convinces even a stubborn proponent" (Fisher, 1989, p. 50).[6]

3.3 Prover-Skeptic Dialogues

In view of these considerations, I submit that the 'stubborn proponent' (or better put, stubborn interlocutor) is better described as a *skeptic* than as a refuter. A skeptic is simply someone who is not easily convinced—a "critical doubter or questioner"—not someone who from the beginning has a different opinion and who will dogmatically hold on to it. What characterizes a skeptic is a general questioning attitude, and a tendency to withold judgment unless there is strong evidence that a given belief is well-founded. Indeed, the Greek terms at the origin of the term 'skeptic', namely σκεπτικός and σκέψις, mean 'inquirer' and 'inquiry': the skeptic is the one who inquires, the one who asks for (further) clarification. When presented with sufficiently persuasive argumentation, a skeptic may well become convinced of a conclusion, but this will not constitute any kind of 'loss' for him.[7] In other words, a skeptic is not actively trying to disprove the arguer, but he will only become convinced of the conclusion if the argument is strongly persuasive.[8]

Therefore, I submit that, instead of Prover-Refuter games, practices of mathematical proofs viewed from a dialogical perspective are best approached through the lenses of *Prover-Skeptic dialogues*. As presented in (Dutilh Novaes, 2020),

---

[6] Of course, some proposed proofs fail to be convincing, such as the purported proof of the ABC conjecture by Mochizuki, which still divides the mathematical community (Dutilh Novaes, 2020, Chap. 11).

[7] I use feminine pronouns for Prover, and masculine pronouns for Skeptic.

[8] The notion of 'organized skepticism' is presented as one of the pillars of modern science in R. Menton's influential 1942 book *The Sociology of Science* (Merton, 1942).

Prover-Skeptic dialogues are intended as a *rational reconstruction* of practices of mathematical proofs. This means that, while the model systematizes, simplifies and idealizes various features of these diverse practices,[9] it remains accountable towards their main structural features (see Sect. 3.5 below). Chapter 11 of (Dutilh Novaes, 2020) consists in an application of the model to mathematical practices of proof, including detailed discussions of four case studies, thus showcasing its explanatory value vis-à-vis its target phenomenon. Hence, an important difference between the rational reconstruction presented in P&R and the Prover-Skeptic dialogues is that they have related but importantly different target phenomena: respectively, the historical development of arguments pertaining to Euler's conjecture vs. practices of mathematical proof more generally, both historical (Chap. 5) and contemporary (Chap. 11) (see Sect. 3.5 for the contrast between Lakatos' idealist orientation and my pragmatist orientation).

Prover-Skeptic games have been studied in the theoretical computer science and mathematics literature (Moore & Mertens, 2011; Chap. 8), including games of information hiding (Kolata, 1986; Wayner, 2009). The presentation of Prover-Skeptic games that comes closest to what we are after here is to be found in (Sørensen & Urzyczyn, 2006), where these games are described as a dialogue between a Prover, who should produce a construction (proof) for a formula, and a Skeptic, who doubts that the construction (proof) exists and thus must be persuaded that it does (Sørensen & Urzyczyn, 2006, p. 89). Notice though that, while Sørensen and Urzyczyn often use zero-sum concepts such as that of a winning strategy, it is not clear that such dialogues are indeed best viewed as games with winners and losers (see below).

Back to Pease et al.'s characterization of Lakatosian games as dialogical games of persuasion mentioned above, the framing of such games in terms of Prover and Skeptic allows us to see that these are *asymmetric* persuasion games: Prover wants to persuade Skeptic that there is a proof for a given conclusion from given premises, but Skeptic does not have a particular stake in convincing Prover that there is no such proof (or more precisely, that the particular proof proposed by Prover is not valid). The game is asymmetric also in terms of the moves available to each player, thus being a variant of what Walton and Krabbe describe as 'Rigorous Persuasion Dialogue' (RPD):

> Whereas PPD [Permissive Persuasion Dialogue] was symmetric in that both players made the same kinds of moves, RPD is asymmetric. One player plays a positive role of proponent, while the other plays a negative or questioning role of opponent.[10] (Walton & Krabbe, 1995, p. 154)

---

[9] As Wittgenstein puts it, "mathematics is a motley of techniques of proof" (RFM III, §46) (Wittgenstein, 1978).

[10] I do not agree with the use of the terms 'positive' and 'negative' to describe the moves by each player, as Skeptic is not only making a negative contribution, so to speak. But otherwise, the general idea is clear, and in the spirit of the present proposal.

Prover (proponent, in Walton and Krabbe's terminology) seeks to persuade Skeptic (opponent, in Walton and Krabbe's terminology) of the correctness and cogency of her proof, while Skeptic raises objections but also specifies what it would take for him to become convinced—for example by pointing out inferential steps that are not sufficiently compelling to him, even if he does not have a knock-down counterexample. After all, absence of counterexamples is not sufficient for a proof to count as genuinely persuasive; witness proofs that are considered correct by the mathematical community, but which somehow fail to be persuasive (such as Snale's proof of the eversion of the sphere, discussed in (Dutilh Novaes, 2018)).

What does it take for Skeptic to become persuaded by a proof? Formulating an adequate answer to this question at a suitable level of abstraction and generality is not an easy task, given that persuasion is arguably an agent-relative, contextual notion, even in mathematics. However, we may formulate a few minimal conditions:

- To become convinced of the conclusion of a proof, Skeptic must accept its premises. Alternatively, for conditional persuasion (*if* the premises are true, *then* the conclusion must be true), he must at the very least be persuaded that the particular set of premises in question is relevant and interesting enough such that establishing what follows from it is not a futile endeavor.
- He must not be in possession of (easily fixable) counterexamples, either global or local.
- He must also deem each step in the proof/argument to be individually perspicuous and convincing.[11] As already noted, an inferential step may be unconvincing even if Skeptic cannot immediately find a counterexample; in such cases, he is still entitled to request for further clarification, for example that the step be broken down into smaller, more perspicuous steps.

These three components seem to be necessary (though they may not be sufficient, for example in very long, unsurveyable proofs) for Skeptic to become persuaded by the proof. Consequently, for each of these components, there must be moves available to Skeptic to ensure that persuasion (or refutation, as the case may be) occurs.

Thus seen, a Prover-Skeptic dialogue begins with Prover stating explicitly the conjecture to be proved and then requesting Skeptic's endorsement of certain premises. At this point, Skeptic has the option of refusing to endorse one or more of the premises, perhaps because he thinks they are problematic or unwarranted. In this case, and if Prover cannot provide alternative premises from which she can also derive the desired conclusion and which would meet with Skeptic's approval, then the dialogue halts. Another way in which the dialogue can be aborted early on is if Skeptic can provide a global counterexample, namely a situation where the premises requested hold but the conjecture just stated does not.

---

[11] But notice that 'real life' Skeptics, e.g., referees, do not always engage in line-by-line checking of a proof; rather, they typically focus on overall plausibility and check more carefully only the more 'suspicious' parts of the proof (Andersen, 2018).

But assuming that Skeptic does grant the premises proposed by Prover and no global counterexample is found, then Prover proceeds to put forward a sequence of other statements that she claims follow necessarily from what Skeptic has already granted. If these statements follow from what has been granted, then Skeptic is indeed committed to them, and thus may not refuse to grant them. Prover may also perform *imperative* speech-acts, whereby Skeptic is instructed to execute a given operation or construction that is relevant for him to become convinced of the conclusion, given the premises (Weber & Tanswell, 2022).

At this point, there are three possible moves for Skeptic: (i) the 'null move', which means that he does nothing explicitly (perhaps a silent nod of approval), simply because he accepts the inferential move proposed by Prover as legitimate; (ii) he proposes a local counterexample to a specific inferential step; (iii) he asks for further clarification if a particular inferential step is not sufficiently convincing and perspicuous; he may simply ask, 'why does this follow?' In response to (ii), Prover may withdraw or modify that particular inferential step, but may still want to pursue the proof as a whole by taking a different route (for example, by means of lemma-incorporation). In response to (iii), Prover may maintain the same route, but she must provide further clarification for the particular step just questioned, for example by a process of interpolation.[12]

If the intended conclusion is eventually reached through successive inferential steps that were not questioned or refuted by Skeptic, and if Skeptic's objections and requests for clarification have been dealt with satisfactorily, then Prover will have succeeded in her goal of persuading Skeptic of the conclusion. Otherwise, if pending challenges have not been dealt with, and in particular if Skeptic has provided compelling counterexamples,[13] then Prover will not have succeeded to persuade Skeptic. (See Fig. 3.1 representing a tree for these dialogues.)

Notice however that the second outcome does not necessarily count as a 'win' for Skeptic, as his goal was not to block the establishment of the conclusion at all costs; his goal was merely to ensure that Prover's attempted proof is indeed valid and persuasive, and if not, to show what is wrong with it (which is in fact a way for Skeptic to *help* Prover). We turn to the cooperative and adversarial components of these dialogues in more detail in the next section, but for now it is important to note that the concepts of 'winning and losing conditions', in particular the idea that a win for one participant will necessarily entail a loss for the other, do not apply here.

There are a number of differences between the Prover-Skeptic model and Lakatos' account. An important one seems to be that, on Lakatos' account, Prover

---

[12] That is, to explicate A —>B, Prover may clarify the path from A to B in terms of intermediate steps: A —>A_1, A_1 —>A_2, ..., A_n —>B. Notice however that there is no upper bound for this process of interpolation, as noted in (Rav, 1999).

[13] Of course, there is always the possibility of the putative counterexamples proposed being dismissed as illegitimate counterexamples, in what Lakatos describes as 'monster-barring': "It is the 'criticism' that should retreat. It is a fake criticism. The pair of nested cubes is not a polyhedron at all. It is a monster, a pathological case, not a counterexample." (P&R, 14). See (Dutilh Novaes, 2022) for further discussion on monster-barring.

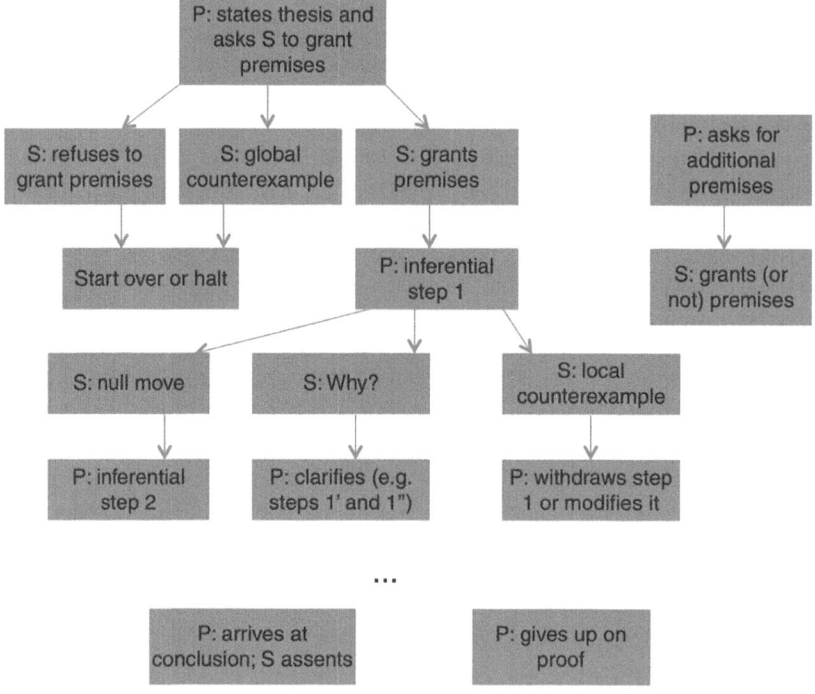

Fig. 3.1 Tree representing the Prover-Skeptic dialogues

first fully formulates a proof in all its steps (or puts forward a proof-idea), which is then scrutinized by Refuter who may come up with counterexamples (be they global or local), thus forcing Prover to revise the original proof. In the Prover-Skeptic model, by contrast, Prover and Skeptic interact during the very production/presentation of the proof; even when Skeptic has no objections to raise, he is tacitly agreeing with and endorsing the steps of the proof. This is thus a component of Prover-Skeptic dialogues that they share with Lorenzen's dialogical logic and Hintikka's GTS: participants take turns in making moves at each inferential step of the proof. Moreover, notice that such dialogues do not correspond straightforwardly either to a Popperian 'context of justification' or 'context of discovery'[14] of a proof: when starting one such dialogue, Prover already has a plan for how to persuade Skeptic that the conclusion follows from

---

[14] As is well known, Lakatos was heavily influenced by Popper, and the very title of P&R is a reference to *Conjectures and Refutations* (Popper, 1989). See (Musgrave & Pigden, 2016) on the complex ways in which Lakatos was both influenced by but also came to disagree with Popper, and Sect. 3.5 below on one specific point of influence.

the premises.[15] Skeptic, in turn, is acting as a proof checker, but along the way the proof may change in view of Skeptic's challenges in the form of requests for further clarification or local counterexamples, which force revisions of the original proof-concept.

What about the 'change of subject' that occurs in the dialogue of P&R? Does it have a counterpart in the Prover-Skeptic model?[16] At first this might appear to be a problem for the model, as the dialogue should lead to a conclusion specifically for the theorem initially proposed. However, on further reflection, it seems perfectly compatible with the model that, throughout the dialogue, Prover and Skeptic in fact *construe new concepts together*, as long as they continue to be sufficiently aligned in how they understand key concepts at each stage. (This is in fact something that happens in 'regular' conversations, theorized for example in the literature on *metalinguistic negotiations* (Plunkett, 2015)). The 'change of subject' phenomenon only becomes concerning when Prover and Skeptic begin to talk *past* each other, each adopting different conceptions of the key terms being used in the dialogue. But conceptual change as such is not a problem for the Prover-Skeptic model, which can also account for the co-creation of new concepts through dialogue, especially as various inferential relations are spelled out. (After all, as Wittgensteinians would say, 'meaning is use'.)

3.4 Cooperation and Adversariality

As mentioned above, both Lorenzen's dialogical logic (Lorenzen & Lorenz, 1978) and Hintikka's GTS (Hintikka & Sandu, 1997) are based on adversarial, zero-sum games. Of course, a minimal amount of cooperation among the players is required for them to accept entering the game and playing it by the agreed-upon rules. But otherwise, the games are presented as adversarial in that players have opposite goals, and a win for one of them entails a loss for the other.

However, both in Lorenzen's dialogical logic and in Hintikka's GTS, the adversarial component of these games quickly becomes strained (Hodges, 2001; Marion, 2009; Dutilh Novaes, 2020; Chap. 3). Many of the moves as defined by the rules of the game do not neatly fit into the categories of 'attack' and 'defense',

---

[15] In other words, how a mathematician comes up with the idea for a proof (its 'context of discovery') is not registered in these dialogues, given that their main goal is to produce persuasion. However, once the mathematician has had the idea in question, she still needs to formulate an argument for public consumption, and at that point she will be entertaining a dialogue with her own inner Skeptic (see (Dutilh Novaes, 2020), Chaps. 4 and 9, for the idea of *internalization* of Skeptic). Moreover, insofar as the original proof is modified in response to Skeptic's objections (for example, peer reviewers in actual mathematical practice), there is a sense in which the dialogue is also a process of *joint* discovery involving Prover and Skeptic (Andersen, 2018; Dutilh Novaes, 2020; Chap. 11).

[16] Thanks to Brendan Larvor for raising this question.

and in fact hint at a hidden layer of cooperation. This suggests that Lorenzen and Hintikka do not sufficiently accentuate the potential *cooperative* components of these dialogues, while overstating their *adversarial* components. Prover-Skeptic dialogues have adversarial as well as cooperative components; they are not zero-sum games. This hybrid status can be cashed out in different ways:

1. Prover and Skeptic have a common goal, that of establishing the validity or invalidity of a proof, and no (conflicting) individual goals (they either win or lose together). They each perform a different task, but in view of a common interest (or converging individual interests). This is a purely cooperative, division-of-labor game, where neither player can 'win' alone; both players will benefit from achieving the overall goal of correctly establishing the (in)validity of the proof.
2. Prover wants her proof to go through no matter what (as this counts as a 'win' for her), regardless of whether it is a valid proof or not. Skeptic, by contrast, wants valid proofs to go through and invalid ones to be refuted, and is neutral with respect to 'pay-offs' of the game for him (no win or loss). Here, Prover can win or lose the game, and Skeptic can neither win nor lose (the outcome is neutral for him, as long as validity is correctly established).
3. Skeptic wants to block (refute) the proof no matter what (as this counts as a win for him), regardless of whether it is a valid proof or not. Prover, by contrast, wants valid proofs to go through and invalid ones to be refuted, and is neutral with respect to 'pay-offs' of the game for her (no win or loss). Here, Skeptic can win or lose the game, and Prover can neither win nor lose (the outcome is neutral for her, as long as validity is correctly established).
4. At a lower level, the game is a classical adversarial, zero-sum game: Prover wins if the proof goes through, Skeptic wins if the proof is refuted or otherwise blocked. But at a higher level, they are in fact cooperating to establish whether a proof is valid or not.[17]

Games such as those in (2) and (3) are known as *partial-conflict-of-interest games* (Zollman et al., 2013). These two kinds of games can be described as semi-adversarial, non-zero-sum games, given that a legitimate win for Prover (the proof is valid) does not necessarily constitute a loss for Skeptic in (2), and a legitimate win for Skeptic (the proof is invalid) does not necessarily constitute a loss for Prover in (3). The game described in (1) is purely cooperative, with no adversarial component at all. The other three games feature different combinations of cooperation and adversariality: unilateral cooperation/adversariality (games (2) and (3)) or lower-level adversariality combined with higher-level cooperation (game (4)).[18]

---

[17] An analogy with adversarial justice systems may be helpful here: at a lower level, defense and prosecution are competing, and only one of them can 'win', but at a higher level they are ideally both pursuing the ultimate goal of achieving justice.

[18] Another way in which the complex interplay between cooperation and adversariality in these games may be spelled out is by distinguishing explicitly the play level from the strategy level (Rahman et al., 2018). To be sure, there are many ways in which cooperation and competition can

Lakatos' dialectic of proofs and refutations does not seem to recognize a motivational asymmetry, and thus arguably falls under scenario (4)[19] (though it could also plausibly be construed as falling under scenario (1)). The different participants are simultaneously enemies *and* allies of each other, as somewhat dramatically put by one of the characters in the dialogue, Sigma:

> Then not only do refutations act as fermenting agents for proof-analysis, but proof-analysis may act as a fermenting agent for refutations! What an unholy alliance between seeming enemies! (P&R, p. 48)

While all four scenarios are prima facie plausible, I submit that scenario (2) comes closest to the actual 'game of proof' in mathematical practice, for the following reason: a mathematician's reputation is very strongly connected to her ability to produce interesting proofs (Jaffe & Quinn, 1993). By contrast, rewards for finding mistakes in proofs are considerably less significant (though some counterexamples and refutations become famous in themselves).[20] For this reason, scenario (3) is in fact not very plausible. In other words, there seems to be a motivational asymmetry between the roles of Prover and Skeptic (something not clearly recognized by Lakatos), even if we optimistically prefer to believe that mathematicians, Provers and Skeptics alike, are ultimately moved by the pursuit of mathematical knowledge (scenario (1)).

Following the account of adversariality in terms of (mis)alignment of interests presented in (Dutilh Novaes, 2021), we may say that there is a modicum of adversariality between Prover and Skeptic. This is because there is a slight discrepancy of interests insofar as Prover prefers for the proof to go through, while Skeptic is indifferent with respect to Prover's goal; instead, he is interested in ensuring that the proof is correct/compelling/persuasive. This slight discrepancy is consistent with the observation that these dialogues are by and large cooperative—Skeptic is *helping* Prover to produce a better proof (Dutilh Novaes & French, 2018)—despite a modest but non-negligeable misalignment of interests (Prover prefers for the proof to go through; Skeptic prefers for the proof to be 'good').

At this point, a few clarifications concerning potential *abuses* of the rules of the game by different participants are called for. This becomes a significant issue in particular with respect to games where participants have asymmetric motivations, such as (2) and (3), as the participant with a higher stake in the game is more likely to engage in deviant behavior. (In games such as (1), none of the participants is

---

interact in games (see (Brandenburger & Nalebuff, 1996) for an influential account), so this list does not purport to exhaust the possible combinations.

[19] The dialogical accounts by Lorenzen and Hintikka can also be seen as falling under scenario (4).

[20] Here I am considering the 'game of proof' primarily in terms of the author-reviewer dyad: mathematicians producing proofs that are then scrutinized by peers, either formally (in the context of peer review for journals) or informally. However, the 'game of proof' is multifaceted, including other kinds of interactions: those between teachers and learners, between co-authors, or between rivals. For these, perhaps scenarios other than (2) make more sense (for rivals, for example, perhaps scenario (4) makes most sense insofar as they are really competing with each other but, at a higher level, producing mathematics 'together'). (I owe this point to Colin Rittberg.)

likely to 'cheat', and in (4) they should keep each other in check since they both want to win.)

In games such as (2), Prover may want to deploy a number of strategies to confuse Skeptic, which nevertheless technically do not represent infringements of the rules of the game (Krabbe, 2008). For example, in its minimal version, the game does not feature restrictions on relevance regarding the premises that Prover asks Skeptic to grant. For example, Prover may ask for premises she will in fact not use in the deduction, in order to produce an 'information overload' effect in Skeptic (who needs to keep track of all concessions he has made so far), which may hinder his ability to survey the correctness of the proof.[21]

In a similar vein, in games such as (3), where Skeptic seeks to block the proof at all costs, there are a few 'tricks' he may turn to. He may for example refuse to grant any premise and thus obstruct the dialogue from the get-go. Alternatively, he may make excessive use of the 'why does it follow?' move, obtusely refusing to be convinced by even the most obvious, self-evident inferential step. This uncooperative Skeptic is exquisitely personified in the figure of the Tortoise in Lewis Carroll's famous fable "What the Tortoise Said to Achilles" (Carroll, 1895). The Tortoise refuses to be persuaded by simple instantiations of modus ponens, and forces Achilles to write down an endless number of additional premises in his notebook, thereby blocking the argument.[22]

Does the possibility of such abuses count as an argument against the present account? I do not think so, given that virtually every interesting enough game is liable to some form of misuse or abuse of its rules. What these observations suggest, however, is that the game is not deterministic, and this plasticity leaves room for extreme uses of the moves that may not be conducive to a fruitful unfolding of the dialogue (even if not explicitly banned by its rules). More restrictive versions of Prover-Skeptic games may include provisions to avoid such pitfalls, such as restrictions related to relevance imposed on Prover's requested premises, or a foundation of agreed-upon valid inferential moves that Skeptic cannot legitimately question (i.e., an underlying formal inferential system that both participants agree upon). But at this stage, we want to describe the game in its most general formulation, and different restrictions will give rise to sub-species of the Prover-Skeptic template for deductive dialogues (which in turn has implications for the issue of logical pluralism—see (French, 2019; Dutilh Novaes, 2020; Chap. 4)).

Thus, while both recognize that mathematical dialogues have cooperative and adversarial components, the Lakatosian dialectic of proofs and refutations differs from Prover-Skeptic dialogues with respect to the exact details of the interplay

---

[21] See (Dutilh Novaes & French, 2018; Dutilh Novaes, 2020; Chap. 4) on relevance as a cooperative desideratum, which suggests that relevant logics are more cooperative than classical logics (left-weakening being a feature that can be exploited by Prover to confuse Skeptic). Interestingly, and at odds with the previous observation, in *Logik und Agon* (Lorenzen, 1960) Lorenzen suggests that intuitionistic logic is more adversarial than classical logic, which he says is largely cooperative. (I owe this reference to an anonymous referee.)

[22] See (Scotto di Luzio, 2000) on this problem.

between cooperation and adversariality. While the former seems to fall under scenario (4) as described above, the latter falls under scenario (2), where there is a motivational asymmetry between the participants. In the next section, I discuss other aspects in which these two accounts differ.

3.5 Lakatosian Hegelianism and Dialogical Pragmatism

Lakatos openly acknowledged three 'ideological' sources for his proofs-and-refutations account: Popper's critical philosophy and the dynamic of conjectures and refutations (Popper, 1989), Pólya's mathematical heuristic (before moving to the UK, Lakatos produced a translation into Hungarian of *How to Solve It* (Pólya, 2014)), and Hegelian dialectic (Larvor, 1999).[23] Lakatos had grand philosophical ambitions when developing the dialectic of proofs and refutations: he sought to establish the superiority of dialectical philosophy of mathematics over other philosophical accounts of mathematics, especially those that emphasized formality and infallibility in mathematical proofs (Larvor, 2001; Tanswell, 2016). According to Lakatos, mathematical concepts develop through repeated cycles of conjectures, attempted proofs, refutations and proof-analysis; such processes are open-ended (in keeping with his fallibilist conception of mathematical knowledge) and follow their own intrinsic rationality. They are presented as inherently dialectical in the Hegelian sense of a clash between contraries that gives rise to a new entity, integrating aspects from both (Larvor, 1999).

As is well known, while at first a 'disciple' of Popper, in later years Lakatos came to diverge intellectually as well as personally from Popper (Musgrave & Pigden, 2016). As for Hegel's influence, there is much discussion among interpreters of Lakatos on the extent to which he really was and remained a Hegelian over the years, in particular with respect to his philosophy of mathematics, where his presumed Hegelianism is more plausible than in other aspects of his thought (Musgrave & Pigden, 2016). This is not the place to discuss the minutiae of these debates; for the present purposes, what is relevant is Lakatos' focus on mathematical *concepts* and the rationality of the development of mathematics as a process largely detached from the *humans* involved.

This general stance draws not only from Hegel's idealism and the primacy of *Geist* (Hegel et al., 2018), but also from Popper's conception of World 3 as the partially autonomous realm of the products of thought. Popper distinguished between World 1 (material reality), World 2 (mental states and processes) and World 3 (quasi-autonomous concepts) (Popper, 1968). Lakatos' dialectical philosophy of mathematics focused primarily on World 3 entities (even if Lakatos did not seem to have a fully worked-out mathematical ontology). Indeed, he had little interest in the

---

[23] Importantly, his knowledge of Hegelian dialectic was mediated by the Hungarian Marxist tradition, of which György Lukács was the main exponent.

individual humans involved in these processes, as revealed by his refusal to include anything remotely 'sociological' in his philosophy of mathematics (Gillies, 2014). (He also nurtured a vivid antipathy for the later Wittgenstein.)

It is telling that, despite the fact that P&R is written in dialogical form, Lakatos' dialectical philosophy of mathematics is not particularly *dialogical*.[24] The protagonists of P&R are in fact not the different characters (the teacher and the students), but rather the mathematical *concepts* (especially the different instantiations of the concept of 'polyhedron') that undergo a continuous process of refinement through the dialectic of proofs and refutations. This feature seems to be connected to a Hegelian disregard for specific agents, given that the objective *Geist* should transcend subjective minds; the development of these concepts follows its course quasi-autonomously (though Lakatos rejected Hegel's teleological, fatalistic views on the development of concepts (Larvor, 1999)). For Lakatos, the human activities that produce mathematics are epiphenomenal; at the core is the "wonderful dialectic of mathematical ideas".

> Mathematics, this product of human activity, 'alienates itself' from the human activity which has been producing it. It becomes a living, growing organism, that *acquires a certain autonomy* from the activity which has produced it; it develops its own autonomous laws of growth, its own dialectic. The genuine creative mathematician is just a personification, an incarnation of these laws which can only realise themselves in human action. Their incarnation, however, is rarely perfect. The activity of human mathematicians, as it appears in history, is only a fumbling realization of the wonderful dialectic of mathematical ideas. But any mathematician, if he has talent, spark, genius, communicates with, feels the sweep of, and obeys this dialectic of ideas. (P&R, p. 146)

(In a footnote immediately following this passage, he describes the idea of autonomy and alienation from human activity as 'Hegelian'.) In this respect, the Lakatosian account differs quite substantially from the dialogical account in terms of Prover-Skeptic dialogues presented here and in (Dutilh Novaes, 2020). The latter resolutely prioritizes the 'human factor' (Dutilh Novaes, 2019): it seeks to understand logic, mathematics and deductive reasoning more generally as human phenomena embedded in broader social, cultural practices and contexts. The focus on human practices is perhaps best described as a form of *pragmatism*, as pointed out by Robert Brandom in his commentary on *The Dialogical Roots of Deduction* (Brandom, 2020). The dialogical analysis starts *and remains* at the level of the human, social activity of producing mathematical knowledge, rejecting the idea that mathematics 'alienates itself' from these practices and acquires a quasi-autonomous ontological status.[25]

---

[24] Notice that the rules of the method of proofs and refutations (quoted above) do not make explicit reference to multi-agent situations at all.

[25] A potential challenge to this 'human' conception of mathematics is the increasing use of computational methods in mathematical research; one might view the 'digitalization' of mathematics as a process of 'alienation' from human activity. (I owe this point to Donald Gillies.) However, insofar as different kinds of technologies have arguably always been part of mathematical practices (e.g., notations can be viewed as 'cognitive technologies' (Dutilh Novaes, 2012)), it is not clear whether

The choice to remain at the level of praxis is primarily methodological, but it also reveals an ontological commitment to the primacy of human experiences and social practices for epistemic phenomena; the latter supervene upon the former. The main object of analysis are the dialogical, discursive practices embodying instances of deductive reasoning in various domains, primarily in mathematics: how mathematicians make discoveries, how they communicate and justify their findings, the social structures underlying mathematical practices etc. In short, the account presented in (Dutilh Novaes, 2020) is best described as a form of *dialogical pragmatism*, as I argue in (Dutilh Novaes, Forthcoming).

In particular, in Chapter 11 of (Dutilh Novaes, 2020) I present a detailed study of the practices of mathematicians as instantiations of real-life Provers and Skeptics, examining their patterns of interaction such as peer review, online cooperation, and 'adversarial collaboration,' and discussing a number of case studies analyzed from a Prover–Skeptic perspective. Indeed, the role of Skeptic is performed by peer reviewers and the mathematical community as a whole (Andersen, 2018), and this role goes beyond (but includes) Lakatosian attempts to find counterexamples and refutations.

A vivid illustration of a Prover-Skeptic dialogue that happened in real life is the peer-review process for Andrew Wiles' celebrated proof of Fermat's last theorem (FLT) (Taylor & Wiles, 1995). When he submitted his 200-page proof of FLT to the journal *Annals of Mathematics*, the editor split up the proof among six referees; one of them was mathematician Nick Katz. What ensued was a typical instance of Prover-Skeptic adversarial/collaborative interaction:

> For two months, Katz and a French colleague, Luc Illusie, scrutinized every logical step in Katz's section of the proof. From time to time, they would come across a line of reasoning they couldn't follow. Katz would email Wiles, who would provide a fix. But in late August, Wiles offered an explanation that didn't satisfy the two reviewers. And when Wiles took a closer look, he saw that Katz had found a crack in the mathematical scaffolding. At first, a repair seemed straightforward. But as Wiles picked at the crack, pieces of the structure began falling away. (Brown, 2015)

As we now know, this turned out to be a serious issue, which took Wiles a full year to fix (in collaboration with mathematician Richard Taylor). He had not identified the problem while working on the proof by himself; it took the external perspective of two Skeptics, Katz and Illusie, for the issue to be spotted. Interestingly, they did not at first come up with a straightforward counterexample; instead, they asked a pointed 'why?' question when one of the steps of the proof did not seem truly convincing. When Wiles took a closer look at the step himself, he realized that there was a real problem there.

Thus, what Lakatos views as the abstract 'dialectic of mathematical ideas', I view as the concrete co-production of mathematical knowledge by Provers and Skeptics

---

we should speak of an 'alienation' from human activities in these recent developments. Perhaps they are best described as a *transformation* of these human activities, understood as human-technology couplings of various kinds.

in a mathematical community: an inescapably social process where the protagonists are mathematicians and their collective practices.[26] True enough, the Prover-Skeptic model is also a rational reconstruction, but it is a rational reconstruction having dialogical practices as their object, not the 'disembodied' historical development of mathematical ideas and concepts as is the case in Lakatos' rational reconstruction.

3.6 Conclusions

In this chapter, I defended the enduring significance of Lakatos' work, in particular the dialectic of proofs and refutations, for contemporary philosophy of mathematical practice. In particular, Lakatos was a significant influence in the development of the dialogical account of deduction and mathematical proof presented in (Dutilh Novaes, 2020). This observation is compatible with a critical engagement with Lakatos' work, where possible objections to his account are discussed and perhaps 'improved upon', as it were—incidentally, very much in the dialectical spirit of proofs and refutations.

Acknowledgments This research was supported by the Dutch Research Council (NWO) with grant 276-20-014 for the project 'The Roots of Deduction' and by the European Research Council (ERC) with grant ERC-2017-CoG 771074 for the project 'The Social Epistemology of Argumentation'. Thanks to Donald Gillies, Brendan Larvor, Colin Rittberg and an anonymous referee for helpful feedback on earlier versions of the paper.

References

Andersen, L. E. (2018). Acceptable gaps in mathematical proofs. *Synthese, Online First, 1*–15.
Brandenburger, A. M., & Nalebuff, B. J. (1996). *Co-Opetition: A revolution mindset that combines competition and cooperation*. Crown.
Brandom, R. (2020). *The pragmatist roots and some expressivist extensions of the dialogical roots of deduction*. https://sites.pitt.edu/~rbrandom/Texts/Pragmatist%20Roots%20and%20 Expressivist%20Extensions%20of%20the%20Dialogical% 20Roots%20of%20Deduction.pdf
Brown, P. (2015). How math's most famous proof nearly broke. *Nautilus*.
Carroll, L. (1895). What the tortoise said to Achilles. *Mind, 104,* 691–693.
Dutilh Novaes, C. (2012). *Formal languages in logic—A philosophical and cognitive analysis*. Cambridge University Press.
Dutilh Novaes, C. (2018). A dialogical conception of explanation in mathematical proofs. In P. Ernest (Ed.), *Philosophy of mathematics education today*. Springer.

---

[26] Given its social, dialogical focus, my position bears some resemblance to that of Feyerabend, who argued against overly abstract approaches in the philosophy of science (Feyerabend, 2001), and who had a particular fondness for dialogical practices (citing frequently in particular Galileo's *Dialogue Concerning the Two Chief World Systems*). (I owe this point to Mauricio Suárez).

Dutilh Novaes, C. (2019). *The human factor: Doing philosophy in a messy world by asking inconvenient questions*. https://research.vu.nl/en/publications/the-human-factor-doing-philosophy-in-a-messy-world-by-asking-inco

Dutilh Novaes, C. (2020). *The dialogical roots of deduction*. Cambridge University Press.

Dutilh Novaes, C. (2021). Who's afraid of Adversariality? Conflict and cooperation in argumentation. *Topoi, 40*(5), 873–886. https://doi.org/10.1007/s11245-020-09736-9

Dutilh Novaes, C. (2022). Two types of refutation in philosophical argumentation. *Argumentation, 36*(4), 493–510. https://doi.org/10.1007/s10503-022-09583-5

Dutilh Novaes, C. (Forthcoming). Dialogical pragmatism and the justification of deduction. In K. Tanaka (Ed.), *Learning from Buddhist logic*. Springer.

Dutilh Novaes, C., & French, R. (2018). Paradoxes and structural rules from a dialogical perspective. *Philosophical Issues, 28*, 29–158.

Feyerabend, P. (2001). In B. Terpstra (Ed.), paperback ed *Conquest of abundance: A tale of abstraction versus the richness of being*. University of Chicago Press.

Fisher, M. (1989). Phases and phase diagrams: Gibbs' legacy today. In G. D. Mostow & D. G. Caldi (Eds.), *Proceedings of the Gibbs symposium: Yale University, May 15–17*. American Mathematical Society.

French, R. (2019). A dialogical route to logical pluralism. *Synthese, Online First*, 1–21.

Gillies, D. (2014). Should philosophers of mathematics make use of sociology? *Philosophia Mathematica, 22*(1), 12–34. https://doi.org/10.1093/philmat/nkt021

Hegel, G. W. F., Pinkard, T. P., & Baur, M. (2018). *Georg Wilhelm Friedrich Hegel: The phenomenology of spirit*. Cambridge University Press.

Hintikka, J., & Sandu, G. (1997). Game-theoretical semantics. In J. van Benthem & A. ter Meulen (Eds.), *Handbook of logic and language* (pp. 361–410). Elsevier.

Hodges, W. (2001). Dialogue foundations. A sceptical look. *Proceedings of the Aristotelian Society, Supplementary, LXXV*, 17–32.

Jaffe, A., & Quinn, F. (1993). Theoretical mathematics: Towards a cultural synthesis of mathematics and theoretical physics. *Bulletin of the American Mathematical Society, 29*, 1–13.

Kitcher, P., & Aspray, W. (1988). An opinionated introduction. In *History and philosophy of modern mathematics* (pp. 3–57). University of Minnesota Press.

Koetsier, T. (1991). *Lakatos' philosophy of mathematics: A historical approach*. Distributors for the U.S. and Canada, Elsevier Science Pub. Co.

Kolata, G. (1986). Prime tests and keeping proofs secret. *Science, 233*, 938–939.

Krabbe, E. C. W. (2008). *Strategic maneuvering in mathematical proofs* (Vol. 22, pp. 453–468). Springer.

Lakatos, I. (1976). *Proofs and refutations*. Cambridge University Press.

Larvor, B. (1999). Lakatos's mathematical Hegelianism. *Owl of Minerva, 31*(1), 23–44. https://doi.org/10.5840/owl199931119

Larvor, B. (2001). What is dialectical philosophy of mathematics? *Philosophia Mathematica, 9*, 212–229.

Lorenzen, P. (1960). Logik und Agon. In *Atti del XII Congresso Internazionale di Filosofia* (pp. 187–194). https://doi.org/10.5840/wcp1219604110

Lorenzen, P., & Lorenz, K. (1978). *Dialogische Logik*. Wissenschaftliche Buchgesellschaft.

Marion, M. (2009). Why play logical games? In I. O. Majer, A. V. Pietarinen, & T. Tulenheimo (Eds.), *Games: Unifying logic, language, and philosophy* (pp. 3–26). Springer.

Merton, R. (1942). *The sociology of science: Theoretical and empirical investigations*. University of Chicago Press.

Moore, C., & Mertens, S. (2011). *The nature of computation*. Oxford University Press.

Musgrave, A., & Pigden, C. (2016). *Imre Lakatos*. Stanford encyclopedia of philosophy. https://plato.stanford.edu/archives/win2016/entries/lakatos/

Netz, R. (1999). *The shaping of deduction in Greek mathematics: A study in cognitive history*. Cambridge University Press.

Pease, A., Lawrence, J., Budzynska, K., Corneli, J., & Reed, C. (2017). Lakatos-style collaborative mathematics through dialectical, structured and abstract argumentation. *Artificial Intelligence, 246*, 181–219.

Plunkett, D. (2015). Which concepts should we use?: Metalinguistic negotiations and the methodology of philosophy. *Inquiry, 58*(7–8), 828–874. https://doi.org/10.1080/0020174X.2015.1080184

Pólya, G. (2014). *How to solve it: A new aspect of mathematical method*. Princeton University Press.

Popper, K. R. (1968). Epistemology without a knowing subject. In *Studies in logic and the foundations of mathematics* (Vol. 52, pp. 333–373). Elsevier. https://doi.org/10.1016/S0049-237X(08)71204-7

Popper, K. R. (1989). *Conjectures and refutations: The growth of scientific knowledge* (5th ed.). Routledge.

Rahman, S., Klev, A., McConaughey, Z., & Clerbout, N. (2018). *Immanent reasoning or equality in action*. Springer.

Rav, Y. (1999). Why do we prove theorems? *Philosophia Mathematica, 7*, 5–41.

Scotto di Luzio, P. (2000). Logical systems and formality. In M. Anderson, P. Cheng, & V. Haarslev (Eds.), *Theory and application of diagrams. Diagrams 2000* (pp. 117–132). Springer.

Sørensen, M. H., & Urzyczyn, P. (2006). *Lectures on the curry-Howard isomorphism*. Elsevier.

Tanswell, F. (2016). *Proof, Rigour & Informality: A Virtue Account of Mathematical Knowledge*. PhD Thesis. University of St. Andrews.

Tanswell, F. S. (2018). Conceptual engineering for mathematical concepts. *Inquiry, 61*, 881–913.

Taylor, R., & Wiles, A. (1995). Ring theoretic properties of certain Hecke algebras. *Annals of Mathematics, 141*, 553–572.

Trafford, J. (2017). *Meaning in dialogue*. Springer.

Walton, D., & Krabbe, E. (1995). *Commitment in dialogue*. State University of New York Press.

Wayner, P. (2009). *Disappearing cryptography: Information hiding: Steganography and watermarking*. Morgan Kaufmann.

Weber, K., & Tanswell, F. S. (2022). Instructions and recipes in mathematical proofs. *Educational Studies in Mathematics, 111*(1), 73–87. https://doi.org/10.1007/s10649-022-10156-2

Wittgenstein, L. (1978). *Remarks on the foundations of mathematics* (3rd ed.). Blackwell.

Zollman, K., Bergstrom, C., & Huttegger, S. (2013). Between cheap and costly signals: The evolution of partially honest communication. *Proceedings of the Royal Society B, 280*, 1–8.

Open Access This chapter is licensed under the terms of the Creative Commons Attribution-NonCommercial-NoDerivatives 4.0 International License (http://creativecommons.org/licenses/by-nc-nd/4.0/), which permits any noncommercial use, sharing, distribution and reproduction in any medium or format, as long as you give appropriate credit to the original author(s) and the source, provide a link to the Creative Commons license and indicate if you modified the licensed material. You do not have permission under this license to share adapted material derived from this chapter or parts of it.

The images or other third party material in this chapter are included in the chapter's Creative Commons license, unless indicated otherwise in a credit line to the material. If material is not included in the chapter's Creative Commons license and your intended use is not permitted by statutory regulation or exceeds the permitted use, you will need to obtain permission directly from the copyright holder.

Chapter 4
Lakatos and the Euclidean Programme

A. C. Paseau ⓘ and Wesley Wrigley ⓘ

Abstract Euclid's *Elements* inspired a number of foundationalist accounts of mathematics, which dominated the epistemology of the discipline for many centuries in the West. Yet surprisingly little has been written by recent philosophers about this conception of mathematical knowledge. The great exception is Imre Lakatos, whose characterisation of the Euclidean Programme in the philosophy of mathematics counts as one of his central contributions. In this essay, we examine Lakatos's account of the Euclidean Programme with a critical eye, and suggest an alternative picture that builds on his while differing from it in a number of important ways.

Keywords Euclideanism · Epistemological foundationalism · Lakatos · Mathematical knowledge · Axioms

In the *Elements*, Euclid begins by laying down definitions, postulates and common notions. From these, he methodically solves problems and proves theorems, deriving the geometry of his time step by step. Euclid himself offers no philosophical gloss on his method; as a mathematician rather than a philosopher, that is to be expected. Others, however, have not shied away from doing so. They have read into the *Elements* a methodological ideal to be emulated throughout mathematics and elsewhere. The *Elements* has inspired a foundationalist vision of mathematical knowledge and of knowledge in general.

We are grateful to audience members at the Imre Lakatos Centenary Conference in November 2022 for many helpful questions and comments, and to Brendan Larvor and an anonymous referee for comments on a later draft.

A. C. Paseau
Wadham College, Oxford, UK
e-mail: alexander.paseau@philosophy.ox.ac.uk

W. Wrigley (✉)
London School of Economics and Political Science, London, UK
e-mail: w.wrigley@lse.ac.uk

© The Author(s) 2025
R. Frigg et al. (eds.), *Proofs and Research Programmes: Lakatos at 100*, Synthese Library 498, https://doi.org/10.1007/978-3-031-88213-5_4

Although Euclidean foundationalism has been much praised over the centuries, it has been little analysed by recent philosophers. An important exception is Imre Lakatos, whom the present volume honours. Lakatos was no Euclidean; quite the contrary. But he believed in knowing his enemy, so was careful to describe the Euclidean picture in some detail. With Euclideanism as a foil, he developed his own 'quasi-empiricist' and fallibilist epistemology of mathematics.

Following in Lakatos's footsteps, we will take a closer look at Euclideanism. Our main motivation is that although the picture is commonly referred to, it is not entirely clear what it is. Contemporary philosophers are superficially familiar with 'Euclidean foundationalism' in the philosophy of mathematics; but dig down, and the details are fuzzy. Euclidean foundationalism is like a great-aunt who has always been around and seems very familiar, though you have never bothered to get to know her. When you finally have a long conversation with her, you realise quite how interesting she is, even if you don't necessarily agree with her. Lakatos would have concurred: as he put it, '[t]he fascinating story of the Euclidean programme and of its breakdown has not yet been written' (1962, p. 6). The first half of that story, before the breakdown, must start with what the programme actually is.

Our essay is devoted to drilling down into the details of Euclideanism, with Lakatos as our guide. It falls into two parts. The first and principal part (Sect. 4.1) outlines Lakatos's views about Euclidean foundationalism, which we follow him in calling the Euclidean Programme, or EP for short. Along the way, we analyse his account of it, noting where we part company with him. In Sect. 4.2, using Lakatos's discussion as a springboard but moving beyond it, we characterise the EP in our preferred way by means of seven principles. Our own assessment of where the EP stands today is too lengthy and unrelated to Lakatos for inclusion in this volume; it may be found in our recent book *The Euclidean Programme* (2024). In the present essay, we compare and contrast Lakatos's account of the EP with our own.

4.1 Lakatos on the EP

Lakatos wrote about the EP in several places. It crops up in his writings as something to be opposed, attacked and rejected, sometimes head-on, sometimes glancingly. An article in which the focus is squarely on the EP is the relatively early piece 'Infinite Regress and Foundations of Mathematics' (Lakatos, 1962). In this article, Lakatos considers several ways of organising knowledge in a deductive system. Here is how he describes the Euclidean mode of organisation:

> I call a deductive system a '*Euclidean theory*' if the propositions at the top (*axioms*) consist of perfectly well-known terms (*primitive terms*), and if there are *infallible truth-value-injections* at this top of the truth-value *True*, which flows downwards through the deductive channels of truth-transmission (*proofs*) and inundates the whole system. (If the truth-value at the top was *False*, there would of course be no current of truth-value in the system.) Since the Euclidean programme implies that all knowledge can be deduced from a finite set of trivially true propositions consisting only of terms with a trivial meaning-load, I

shall call it also the *Programme of Trivialization of Knowledge*. Since a Euclidean theory contains only indubitably true propositions, it operates neither with conjectures nor with refutations. In a fully-fledged Euclidean theory meaning, like truth, is injected at the top and it flows down safely through meaning-preserving channels of nominal definitions from the primitive terms to the (abbreviatory and therefore theoretically superfluous) defined terms. A Euclidean theory is *eo ipso* consistent, for all the propositions occurring in it are true, and a set of true propositions is certainly consistent.[1] (1962, pp. 4–5)

In the rest of this section, we'll dissect this passage, and others, to extract some of the key features that Lakatos ascribes to the EP. We should clarify at the outset that we aren't concerned with Lakatos's criticisms of the EP, or with his criticism of dogmatic epistemology more generally or formalism more specifically. We will not, for example, consider how Lakatos's criticism of formalism stands up in today's age of computer proof. Our chief concern is his *description* of the EP, and our question is whether he got this target right. Of course, the EP is a rational reconstruction, not a historically attested manifesto, so there is some leeway in how to describe it. Nevertheless, given its history, there is something to get right here. And in our opinion, Lakatos gets some aspects of the EP right, but not others. This will be our concern in the rest of this section.

4.1.1 Truth

Both in the core passage above and elsewhere, Lakatos emphasises that the axioms of a Euclidean theory are true, or at least aspire to be true. He is clearly right that the axioms of a Euclidean theory are (supposed to be) true, and that this is an essential aspect of the EP.

Lakatos is not particularly clear about precisely *why* the truth of the axioms is an essential feature of the EP, and appears to take this for granted. An obvious point is that the EP is a foundationalist account of mathematical knowledge, and knowledge implies truth. It also chimes with how mathematicians down the ages have thought of, say, geometry, unhesitatingly taking its axioms to be correct.

A further point is that the major figures in the history of mathematics and its philosophy that one would want to identify as Lakatos's targets are all explicit that mathematics is a body of truths, starting from true axioms or first principles. For

---

[1] There is a footnote accompanying a sentence in this passage (ending with the words '*Programme of Trivialization of Knowledge*'). The footnote refers to Pascal's *De l'esprit géométrique* ('On the Geometrical Mind'), which Lakatos calls the EP's *locus classicus*. Never published in Pascal's lifetime, the *Esprit* is a short work that influenced the *Port-Royal Logic*. In his one-sentence footnote, Lakatos refers to it as 'Pascal [1657–8]', but more recent scholarship has tended to settle on 1655 as the date of its composition, following Jean Mesnard, the editor of Pascal's works. Lakatos's reference to it is interesting in that the work is largely unknown to English-speaking philosophers; indeed, the philosophical literature on the *Esprit* in the analytic tradition is almost non-existent, even today. For discussion of the *Esprit* by us, see Sect. 5.2 of *The Euclidean Programme*.

example, Aristotle gives an account of how *episteme* ('understanding' or 'scientific knowledge', which he identifies as the highest epistemic state) can be gained from demonstrations, in terms that resemble Lakatos's characterisation of proof in the EP. And Aristotelian demonstrations all start from true principles (*Posterior Analytics* I.2 71b17–25).[2] Centuries later, Descartes gives a Euclidean account of *scientia* (the epistemic ideal of the scholastic and early modern periods) where first principles are truths which are understood so clearly and distinctly as to be rationally indubitable (1637, pp. 16–17; AT 6, p. 19).[3] And Pascal, who Lakatos paints as the arch-Euclidean (see footnote 1, above) unambiguously describes the axioms of geometry as 'vérités' ('truths').

So on the truth of the axioms, we are in complete agreement with Lakatos. No commitment is thereby made to any particular analysis or philosophical account of truth. As we shall see below, however, we do not entirely share Lakatos's view about the role that truth plays in a Euclidean theory.

4.1.2 Flow

One of the most interesting features of the passage above is Lakatos's metaphor of the *flow* of truth in a Euclidean system, downward from axioms at the 'top' of the theory to theorems at the 'bottom'. To illustrate the significance of this point, Lakatos contrasts the Euclidean Programme with the 'Empiricist Programme':

> The Euclidean programme proposes to build up Euclidean theories with foundations in meaning and truth-value at the top, lit by the *natural light of Reason*, specifically by arithmetical, geometrical, metaphysical, moral, etc. intuition. The Empiricist programme proposes to build up Empiricist theories with foundations in meaning and truth-value at the bottom, lit by the *natural light of Experience*. Both programmes however rely on Reason (specifically on logical intuition) for the safe transmission of meaning and truth-value. (1962, p. 5)

It is important to appreciate that empiricist theories do not need to be strictly empirical. In particular, one could have an empiricist account of mathematics, in Lakatos's sense. The salient epistemological point is that in an empiricist theory, the relevant flow is not *downward*, from axioms to theorems, but rather *upward* from 'basic statements' (perhaps observations, or elementary arithmetical sentences) to higher-level statements (perhaps theoretical scientific principles or mathematical axioms).

---

[2] Although Aristotle predates Euclid by a few decades, one way to read him is as an early exponent of the EP. There are several important commonalities between Aristotle's account of method and the EP as described by Lakatos. And naturally, Aristotle's account of geometric method was influenced by the geometry of his time, likely very similar to that of Euclid's time. For more details, see Sect. 4 of *The Euclidean Programme*.

[3] The first citation is to Olscamp's English translation of the *Discourse on Method* (2001). The second is to Adam and Tannery's *Oeuvres de Descartes* (volume, page).

Lakatos returns to Euclideanism in a later work, 'A Renaissance of Empiricism in the Recent Philosophy of Mathematics'.[4] In this article, he clarifies that empiricist theories are *quasi-empirical*. The relevant direction of flow is upward, and what is transmitted is typically falsity, rather than truth. If a theory, empirical or otherwise, implies a false basic statement, this 'inundates' the system, which is refuted. To think that truth, in addition to falsity, can be transmitted upward is to indulge in what Lakatos refers to in a Popperian vein as the *inductivist delusion* (1976a, p. 41).

We do not wish to dwell on this point, since we are not concerned here with quasi-empirical theories. We simply note, against Lakatos, that the inductivist idea of basic statements retransmitting truth to the axioms which imply them is not obviously a delusion, as he characterises it. It is common enough, of course, to take a scientific theory to be confirmed to some degree when its observational predictions are correct. And in mathematics as well, many philosophers have thought that axioms are given some degree of confirmation when they imply elementary truths which we already take ourselves to know (such as that $2 + 2 = 4$, for instance).[5] But Lakatos is clearly correct about the direction of flow in the Euclidean account of mathematics. Indeed, so deeply embedded is this idea that it has seemed obvious to many that the direction of flow is top-down in mathematics, an idea just as obvious as that the direction of flow in the empirical sciences is bottom-up.

But Lakatos does more than just identify this commitment of the EP. In the above passages, and in others to be quoted later, he consistently talks of truth-value, and of meaning, as flowing through the channels of the system, a point to which we return with a more critical eye later in this section. His focus on meaning and truth notwithstanding, he also offers what we see as an absolutely crucial insight. This is the observation that the EP is less about *what* flows from axioms to theorems and more about *how* it flows. As Lakatos puts it:

> We can get a long way merely by discussing *how* anything flows in a deductive system without discussing the problem of *what in fact flows* there, infallible truth or only, say, Russellian 'psychologically incorrigible' truth, Braithwaitian 'logically incorrigible' truth, Wittgensteinian 'linguistically incorrigible' truth or Popperian corrigible falsity and 'verisimilitude', Carnapian probability. (1962, p. 6)

What makes the EP distinctive as a methodological account of mathematics, therefore, is its emphasis on mathematicians' prior access to the axioms from which they establish theorems by means of proof. Different Euclideans might mean different things by the terms 'access' and 'establish'. With his key observation that what flows is of lesser interest than how it flows, Lakatos is a locksmith who has opened the way to a proper understanding of the EP.

Years later, in the 'Renaissance' article, Lakatos re-iterated the flow idea: truth is injected at the top and flows down to the bottom. Indeed, he draws the very

---

[4] This essay, posthumously published as Lakatos (1976a), is an expanded version of an earlier 1967 paper.

[5] For an influential expression of this idea, see Russell (1907).

distinction between Euclidean and quasi-empirical theories in these terms: as he says, '[i]t is the *how* of the flow that is decisive' (1976a, p. 29).

The insight we extract from Lakatos is, as we put it elsewhere, that the EP is all about Euclidean hydraulics (2024, p. 4). For comparison, consider the Phillips Machine, an analogue computer developed in 1949. In this machine, coloured water moves through a series of clear pipes in order to model the flow of money in the economy. By analogy, in a Euclidean theory, some theoretical good corresponds to the coloured water. This for Lakatos is truth, but for us (see Sect. 4.2) it will be something epistemic, such as (rational) certainty, knowledge, or justification. Much as the colour of the water is inessential to the modelling process in the Phillips Machine, the choice of a particular theoretical good is immaterial to the Euclidean explanation of mathematics, at least in broad outline. What matters is that the theoretical good is injected with the axioms, and flows downward through the logical structure of the theory to the theorems.

4.1.3 What Is Injected?

Although we agree with Lakatos that the axioms of a Euclidean theory are supposed to be true, we part ways with him in a crucial respect on the point of truth. In the passage cited above from the 'Foundations' paper, Lakatos speaks of an injection of truth and meaning. This idea that truth is injected into the theory via the axioms persists into the 'Renaissance' paper, where he writes:

> Classical epistemology has for two thousand years modelled its ideal of a theory, whether scientific or mathematical, on its conception of Euclidean geometry. The ideal theory is a deductive system with an indubitable truth-injection at the top (a finite conjunction of axioms) – so that truth, flowing down from the top through the safe truth-preserving channels of valid inferences, inundates the whole system. (1976a, p. 28)

We find this talk at best misleading, at worst confused. What would it even mean for truth itself to flow from axioms to theorems? In mathematics at least, truth is not tensed: mathematical propositions are either eternally true or eternally false. The theorems of geometry are all eternally true, and there is no literal sense in which the truth of one proposition is transmitted to another. Of course, logicians like to speak of rules being 'truth-preserving', but that image is more easily literalised than the flow or transmission idea: it simply means that if the rule's premises are true then so is the conclusion. It's possible, of course, that Lakatos meant no more than this. A similar point applies to meaning: the meanings of the theorems do not depend on the meanings of the axioms. Although perhaps that view is more tenable than the analogous one about truth, especially if the axioms are consciously stipulated at the start of the practice rather than extracted from it.

We highlighted above that the crucial point about flow in a Euclidean theory is its direction. We think the flow metaphor is best construed as transmission of an *epistemic* good of some sort. What this good is exactly will vary from one Euclidean theorist to another. But the EP as we see it represents an epistemological conception. It is best thought of as a form of epistemic foundationalism, in which the axioms enjoy an initial justification (for instance), which flows to the theorems when these are proved.[6] The channel along which the epistemic good flows, from axioms to theorems, is an epistemic path the mathematician themself follows, or could follow.[7]

As mentioned, it's quite possible that Lakatos appreciated this point but wrote misleadingly. (Or to be fairer to him, that he wrote in a way that two philosophers in the 2020s taking him very literally find misleading.) He seems to recognise as much in passages such as the following:

> Whether a deductive system is Euclidean or quasi-empirical is decided by the pattern of truth value flow in the system. The system is Euclidean if the characteristic flow is the transmission of truth from the set of axioms 'downwards' to the rest of the system – logic here is an *organon of proof*; it is quasi-empirical if the characteristic flow is retransmission of falsity from the false basic statements 'upwards' towards the 'hypothesis' – logic here is an *organon of criticism*. (1976a, p. 29)

The focus on the role of proof in Euclidean theories and criticism in quasi-empirical ones is most welcome. But despite that, Lakatos seems to think epistemic facts enable—in the best case, *guarantee*—truth-injection, rather than constitute the injection itself. This is plain in the talk of truth value being transmitted from axioms to theorems. And as he put it much earlier, empiricists 'criticized the guarantee of the intuitive Euclidean truth-injection: self-evidence' (1962, p. 9). We shall clarify our way of putting things in the next section. For now, we simply note that our characterisation of the EP as involving the transmission of an epistemic good from axioms to theorems is sufficiently general to encompass the main historical figures that one would wish to characterise as Euclideans.

4.1.4 Finitude

In the long quotation from 'Foundations' at the start of Sect. 4.1, Lakatos describes the axioms of a Euclidean theory as a 'finite set of trivially true propositions'. In the 'Renaissance' passage cited at the start of Sect. 4.1.3, he characterises the 'top' of a Euclidean theory as 'a finite conjunction of axioms'. No clear justification for this is given by Lakatos; he claims only that the finitude of the axioms is implied by the EP (1962, p. 4).

That sets of axioms should be finite is a broadly, but not entirely, correct historical observation. Axiomatic theories prior to the twentieth century, including

---

[6] However, some epistemic goods possessed by the axioms will not flow from axioms to theorems. For instance, *self*-evidence will not be transferred via deduction.

[7] Since deduction is truth-preserving, and the axioms of a Euclidean theory are true, so are the theorems. This point stands even if we think of *Flow* as a primarily epistemic, rather than alethic or semantic, principle.

Euclid's geometry, are finite. Even today, most axiomatic theories of mainstream mathematical interest are finitary, in an important sense. This makes Lakatos's inclusion of this point defensible, though it needs to be made more precise. In particular, we must take due notice of axiom schemata. First-order Peano Arithmetic (PA), for example, cannot be finitely axiomatised, and hence cannot be presented as a finite conjunction. But PA *can* be finitely formulated as long as schemata are allowed, the usual way of doing so being to adopt a schematic form of the induction axiom. Euclideans should allow this sort of latitude.

Moreover, there are important Euclidean thinkers who contradict this point, or remain silent on it. For example, Aristotle, who can be identified as a forerunner of the EP, is explicit in the *Posterior Analytics* that science as a whole requires an infinity of axioms (I.32 88b6). Prominent advocates of the EP in the seventeenth century do not share this commitment as far as we know, but nor are we aware of an active commitment to the finitude of the axioms in these writers. No clear endorsement is discernible in the relevant works of Descartes (*The Discourse on Method,* including *The Geometry*) or Pascal (*On the Geometric Mind*), for instance.

So on this point, we broadly agree with Lakatos, but insist that more care be taken over its formulation, and that the principle is not central to the EP. We return to it in the next section.

4.1.5 Triviality

As we saw in the quotation from 'Foundations', Lakatos takes the axioms of a Euclidean theory to be 'trivially true' and says they bear a 'trivial meaning-load' (1962, pp. 4–5). What does Lakatos mean by 'trivial'? We confess we're not entirely sure.

One understanding of 'trivial' is logical. But this cannot be the sense of triviality that Lakatos has in mind. If the axioms were logical truths they would be redundant, as they would be deducible from anything. And no Euclidean theory could go beyond logic.

Another way to understand triviality is along logical empiricist lines: mathematical statements are void of content, in virtue of being analytically true. If this is what Lakatos intends, we disagree in the strongest terms. The central philosophical commitments of the EP, as both a general epistemology of mathematics and as a historical phenomenon, are entirely consistent with the axioms being substantive truths about a 'third realm' of abstract objects, or as being apprehended by a non-linguistic faculty of mathematical intuition. But given Lakatos's general disparagement of logical empiricism, and his explicit mention of intuition in 'Foundations', it is unlikely that this is his intended sense.

A third possibility is that triviality in 'Foundations' is related to non-explanatoriness in the 'Renaissance' article, where the term 'trivial' does not appear. The earlier Lakatos describes proof as *giving way* to explanation (1962, p. 14), as

Euclidean theories are replaced by their Empiricist successors. And the later Lakatos draws the contrast in the following way:

> [I]n a Euclidean theory the true basic statements at the 'top' of the deductive system (usually called 'axioms') *prove*, as it were, the rest of the system; in a quasi-empirical theory the (true) basic statements are *explained* by the rest of the system. (1976a, p. 29)

So, perhaps the intended sense of *trivial* in the early paper is simply that the axioms are non-explanatory. They represent fossilised truisms rather than theoretically hard-working explanatory principles that have a role to play in making bold theoretical conjectures.

Lakatos believes that in quasi-empirical theories of the type he favours, basic statements are explained by the rest of the system, but (by implicit contrast) that this is not the case in a Euclidean theory. We can agree on at least one point: that axioms are explanatory of theorems need not be built into a Euclidean theory. But we do not see why explanatoriness has to be ruled out either. Perhaps a Euclidean could think of the axioms as explaining the theorems. Indeed, this seems to be Aristotle's position (*Posterior Analytics* I.2 71b17–25), and his thought bears a strong resemblance to the EP, according to at least one major interpretative school.[8] So the thought that axioms explain theorems is not *per se* un-Euclidean or un-foundational. Perhaps there is good reason to think that the axioms cannot *in fact* explain the theorems in a Euclidean theory. But in so far as we are undertaking a rational reconstruction of the EP, we see no reason to include non-explanatoriness as one of its features.

There is a fourth way to interpret the word 'trivial'. This is the idea that axioms are self-evident; that their discovery, as opposed to their content, is trivial. This reading is suggested by Lakatos' insistence that the injection of truth value in a Euclidean theory is supposed to be infallible. In the development of his quasi-empirical account of mathematics, Lakatos sets himself fiercely against the axioms' alleged self-evidence and the indubitability of mathematics, so he clearly meant to bake this idea into the EP. If this is the intended sense of 'trivial', then we agree. Indeed, we take it to be central to the EP that axioms are thought of as self-evident (more on this in Sect. 4.2).

4.1.6 Primitive Terms

Related to the issue of triviality, Lakatos requires that the primitive terms of a Euclidean theory be *perfectly* well-known (1962, p. 4). What is significantly less clear to us, however, is why he insists on this.

In the context of the 1962 paper, the reason seems to be scepticism. Here, Lakatos is concerned with two regressive sceptical arguments that aim to show that

---

[8] See chapter 4 of *The Euclidean Programme*.

meaning and truth cannot be conclusively established (1962, p. 3). Meaning cannot be established because when defining an expression E, for instance, one must use at least one expression, call it F. Presumably F itself requires a definition, and if circularity is to be avoided, this definition will use expressions that are not defined in terms of E or F. Rather, a new term, G, must be introduced. G apparently needs its own non-circular definition, and so the regress goes on *ad infinitum*. There is also the more familiar regress in terms of proof and knowledge. If one claims to know some mathematical theorem by giving a proof, that proof will have premises, which in turn require their own proofs, and so on *ad infinitum* again.

Lakatos paints the EP as a response to *both* problems. The regress in proof is blocked by the axioms; these truths are known indubitably and are not in need of proof at all. If the EP is to block the semantic regress also, it is natural to think of the Euclidean theorist as making a similar pronouncement on the primitive terms of the theory: they are understood *perfectly*, and so are not in need of a definition or elucidation, and the regress is blocked. In short, Lakatos sees the EP as both an epistemological and a semantic manifestation of foundationalism.

It is questionable, however, whether Lakatos's requirement is justifiable in historical terms. It is not clear (to us, at any rate) that the history of the EP is so closely connected to semantic scepticism. In the *Posterior Analytics*, Aristotle discusses a relevant issue. He claims that '[a]ll teaching and all learning of an intellectual kind proceed from pre-existent knowledge', clarifying that '[t]here are two ways in which we must already have knowledge: of some things we must already believe that they are, of others we must grasp what the items spoken about are (and of some things both).' The requirement that we *must grasp what the items spoken about are* is something like a requirement that the terms of a theory be previously understood, and Aristotle gives the example 'of the triangle, that it means *this*' as something we need to know in order to learn about triangles (I.1 71a1–16).[9] Now of course one must know the meaning of the term 'triangle' *in some sense* to have knowledge of triangles. But there is nothing here to suggest that such understanding must be perfect. Rather, the use of the demonstrative seems to suggest that Aristotle requires simply that one be able to identify triangles when confronted with them, which on its own seems to fall far short of understanding 'triangle' perfectly. And, at least at this juncture, Aristotle is not even responding to the sceptical regress about meaning that Lakatos had in mind. Rather, he is responding to 'the puzzle in the *Meno*' (I.1 71a29–31), that one cannot enquire after what one is ignorant of, since one will not know what to enquire after, nor will one recognize the correct answer to the query if one comes across it.

Reading the great Euclideans of the seventeenth century also casts Lakatos's claims in a dubious light. Descartes, of course, is extremely concerned to respond to scepticism. But he is most naturally read as responding to *epistemological* scepticism about the possibility of knowledge (*scientia*) rather than to semantic scepticism about the meaning of terms in mathematics or elsewhere.[10]

---

[9] The translations here are from Barnes's edition (Aristotle 1993).

[10] See *The Meditations* (AT 7) for instance.

The situation is even worse when we turn to Pascal, Lakatos's paradigmatic Euclidean. Pascal is clearly alive to both of Lakatos's regresses. He claims that, ideally at least, we would like to define *all* the geometric terminology, and give a proof of *all* the propositions of geometry. But neither is possible in practice, thanks to the threat of infinite regress. We must, therefore, use primitive terms that are so clear that we cannot explain them in clearer terms, and unproved principles that are so obvious as to admit of no proof from principles more obvious still. But he goes on to assert that trying to further explain the geometrical primitives would cause more confusion than it would resolve.[11] This makes it clear that Pascal also requires primitive terms to be understood, but suggests that the understanding may be imperfect, otherwise there would be no confusion to even try to resolve. Pascal is also clear that geometric knowledge is in good standing when it is obtained by the method he outlines. Hence failing to live up to our initial ideals is not to the detriment of geometry, contrary to what Lakatos asserts about the Euclidean position on semantics.

In short, we see the primitive terms requirement as a convenience that Lakatos adds to his characterisation of the EP in order to present it as a broader form of foundationalism than the textual evidence allows for. Lakatos's thoughts on primitive terms and meaning in mathematics are certainly interesting. Yet we do not find in practice that historical Euclideans address these semantic issues in the way Lakatos describes, if indeed they address them at all.

4.1.7 *Formality*

In the earlier 'Foundations', Lakatos also comments parenthetically that deductions in a Euclidean system need not be formal. Changing the clause's italicisation to emphasise this aspect of it:

> The basic definitional characteristic of a (*not necessarily formal*) deductive system is the principle of retransmission of falsity from the 'bottom' to the 'top' ... (1962, p. 4)

Lakatos is of course famous for denigrating 'formalist' philosophies of mathematics, or at least insisting that there is a lot more to the philosophy of mathematics than 'formalist' approaches. As he explains in the introduction to *Proofs and Refutations* (1976b), formalists identify mathematics with its formalised axiomatic version; Carnap, Church, Peano, Russell and Whitehead are examples of formalists in this sense. As far as the EP goes, however, Lakatos builds no requirement of formality into it. Formalism, in the hands of some of its advocates, is a late nineteenth/early twentieth-century incarnation of Euclideanism. Yet the two should not be identified

---

[11] Pascal (1655, p. 396).

more generally, on pain of being blind to all pre-nineteenth-century forms of Euclideanism, which were non-formal.

In this, Lakatos is surely correct. Logicians today conceive of a theory as a set of sentences in a formal language that is closed under the deductive apparatus of some formal logic. Yet Euclidean theories can be formulated in natural languages, and until the late nineteenth century that is how they were formulated. So it would be entirely anachronistic to insist that a Euclidean theory must be formal.[12] To do so would rule out huge swathes of mathematics, and even reconstructions of mathematics, by definition.

Lakatos stresses a related point. Not only can entailment in a Euclidean theory be informal, it can also be non-logical (1962, p. 13). Subject-specific inferential rules and construction techniques may be taken to be perfectly legitimate components of informal deductions if certain conditions are met; for instance, if they are licensed by spatial intuition in geometry. If such modes of inference are ineliminably used in a proof, the conclusion is implied by the premises but does not follow from them in the deductive sense of modern formal logic. In order to cast a suitably wide historical net, this sort of implication should count as well on our reconstruction of the EP.

Just as significant is the fact that which entailments we count as logical depends on which background logic we use. However, it would be anachronistic to insist that all Euclidean theories employ the same background logic, since different Euclideans may disagree on the legitimacy of particular inference rules. For example, a more modern Euclidean would likely contest Aristotle's rule that from *All As are Bs* one can infer *Some A is B*. In short, we must not insist that a Euclidean theory is a set of *formal* sentences, nor that it is closed under a *formal* deduction relation.

4.1.8 Lakatos and the EP

Although we will not dwell on Lakatos's own assessment of the Euclidean picture, it is clear which side he is on. He thinks that '[f]rom the seventeenth to the twentieth century Euclideanism has been on a great retreat', and that rearguard attempts to 'break through beyond the hypotheses, towards the peaks of *first principles*' have all failed. The upshot: '[t]he fallible sophistication of the empiricist programme has won, the infallible triviality of Euclideans has lost'. That said, the 'four hundred years of retreat seems to have by-passed mathematics', and Lakatos clearly sees his own role as being to wield the axe in this subject too.[13] The point of his most famous work in the philosophy of mathematics, *Proofs and Refutations*, is to show 'that informal, quasi-empirical, mathematics does not grow through a monotonous

---

[12] As Barnes highlights, ancient logicians did not even have the concept of a formal language (2005, p. 512).

[13] The quotations in this paragraph so far are from his (1962, p. 10). The theme of Euclidean theories' decline, especially outside mathematics, is repeated in his (1976a, p. 30).

4 Lakatos and the Euclidean Programme

increase of the number of indubitably established theorems [the Euclidean picture] but through the incessant improvement of guesses by speculation and criticism, by the logic of proofs and refutations' (1976b, p. 5). In the words of Pi, one of that great book's Greek-alphabet-named characters: '[h]euristic is concerned with language-dynamics, while logic is concerned with language-statics' (1976b, p. 93). The latter could equally well apply to the Euclidean picture, which is static rather than dynamic. To present mathematical knowledge in static fashion, as an unchanging pyramidal-shaped system of immutable truths, is to belie it. Towards the end of *Proofs and Refutations*, its author comments:

> In deductivist style, all propositions are true and all inferences are valid. Mathematics is presented as an ever-increasing set of eternal, immutable truths. Counterexamples, refutations, criticism cannot possibly enter. An authoritarian air is secured for the subject by beginning with disguised monster-barring and proof-generated definitions and with the fully-fledged theorem, and by suppressing the primitive conjecture, the refutations, and the criticism of the proof. Deductivist style hides the struggle, hides the adventure. The whole story vanishes, the successive tentative formulations of the theorem in the course of the proof-procedure are doomed to oblivion while the end result is exalted into sacred infallibility. (1976b, p. 142)

Like Lakatos, we are also critical of the Euclidean Programme, though we do not have the space to discuss our criticisms of it here. These may be found in *The Euclidean Programme*.

We conclude this section by raising a more general sort of worry. One may criticise Lakatos's ambition to even discuss the EP in the relatively ahistorical way that he does. Perhaps it is historically insensitive to throw a single critical blanket over a great swath of the past.[14] Perhaps one should not try to capture the essence of Euclideanism in the way Lakatos tried to. Perhaps there is no such thing as Euclideanism, only a series of authors inspired by *The Elements* in different ways.

We have some sympathy with this complaint, but only up to a point. Clearly, different authors tempted by Euclideanism have stressed different points and added their individual imprint to its expression. Indeed, the body of work attributed to Euclid has varied across time and place, so that different Euclideans may have even drawn their inspiration from varied sources. That said, we believe there is an identifiable body of doctrine reasonably called 'The Euclidean Programme' that runs through the ages, even if it is not precisely defined and differs from writer to writer. As philosophers, we see our role as trying to identify these doctrines and, once identified, to assess them. Euclideanism is not at bottom different from the many other 'isms' philosophers blithely engage with in fairly ahistorical fashion—just within epistemology, think of coherentism, foundationalism, internalism, externalism, etc. If there is room for discussion of these 'isms' in a relatively abstract way, so should there be for Euclideanism.

A more sensitive approach might be to compare a rational reconstruction that tries to capture the centre of gravity of a body of thought—Euclideanism—within

---

[14] We owe this phrase to Brendan Larvor.

individual historical writers' conceptions. Although there is no space for the latter here, we have attempted it in our book, which compares the EP as an abstract methodological ideal with some historical authors: Aristotle, Euclid, Descartes, Pascal and other more recent ones. This leads on to the next section: having seen what Lakatos thinks the EP is, it is high time we say what we take it to be.

4.2 The Euclidean Programme in Seven Principles

Lakatos's discussion of the EP was instructive. Let's try to capture the general picture and the lessons learnt in a more structured way, and set aside the historical scruples just mentioned.

In *The Euclidean Programme*, we argue that the EP is characterised by seven principles. Three of these are core principles, which we take to be present in any historical manifestation of the Euclidean Programme worthy of the name (2024, p. 8). The other four principles are peripheral to the programme, manifesting themselves in many, but not all, occurrences of the EP throughout history (2024, p. 9).

4.2.1 Core Principles

The first core principle is that the axioms of a Euclidean theory are supposed to be true. We agree with Lakatos that this is essential to the EP, and see its inclusion as mandatory. All the major figures in the Euclidean tradition subscribe to some version of it: Aristotle (on one common interpretation), Descartes, and Pascal, to mention just a few.[15]

The second core principle of the EP in our reconstruction is that the truth of the axioms should be *self-evident*. While this is not something Lakatos focused on,[16] all the major Euclidean thinkers subscribe to this principle, or something very similar. As a distinctly foundationalist epistemology, the EP requires knowledge of the axioms to be completely secure and unmediated by inference. Given that this aspect of the programme is so historically well-attested, we shall not dwell on it here. Suffice it to say that we are sympathetic with Lakatos' assessment that our assurance of the truth of the axioms should be infallible in the context of a Euclidean theory.

---

[15] A difficult question is how Euclidean Euclid himself was. As we see it, the EP is a programme inspired by the methodology of the *Elements*, whether or not its author was what we would now call a Euclidean foundationalist. For a little more detail, see chapter 3 of *The Euclidean Programme*.

[16] Unless self-evidence is how we are supposed to understand his idea of 'triviality'.

Given our interpretation of the Euclidean Programme as a foundationalist epistemology of mathematical propositions, rather than an account of mathematical terms, we have no parallel to Lakatos' requirement that the primitive terms of a Euclidean theory be perfectly well-understood. At best, some requirement on the understanding of the terms can be inferred from the requirement that the axioms be self-evident. For if we cannot even understand what proposition is expressed by an axiom, then the truth of the proposition expressed can hardly be evident to us. But *perfect* understanding of the terms is a far stronger requirement than merely understanding them well enough to grasp the self-evidence of the axioms in which they appear.

The easiest way to appreciate this point is to think of simple logical propositions involving imperfectly understood terms. For example, it should be as obvious as can be that 'all democracies are democracies' is true. It is equally obvious that if 'all horses are ungulates' and 'all ungulates are mammals' are both true, then 'all horses are mammals' is true. One can appreciate this even if the terms 'democracies' and 'ungulates' (or even 'horses' and 'mammals') are not understood with perfect clarity. But we can go beyond logical truths. It is self-evident, for example, that a gallon of water is less than a gallon and a fluid ounce of water, even to those with only the haziest understanding of American units of measurement.

So on our picture, the mathematician must have an understanding of the primitive terms which is developed enough to allow them to understand the axioms. But their grasp of the terms may be less than perfect. If, consequently, their understanding of the axioms is less than perfect, this is permissible so long as the mathematician is still able to appreciate their self-evidence. To return to our hydraulic metaphor, there is no requirement that the water in the Phillips Machine be chemically pure.

This brings us to the third core Euclidean principle. In one respect, we take it to be Lakatos' greatest contribution to the study of the EP that he identifies the flow idea as an essential and defining principle of it, a principle understood in distinct ways by distinct philosophers and mathematicians working in the tradition. However, we part company with Lakatos in one key respect. In his reconstruction of the EP, the flow from axioms to theorems is of semantic content, such as truth and meaning. We think this is an unfaithful representation of the historical Euclidean ideal, where the emphasis has been squarely on epistemological issues.

In our reconstruction, the direction of flow is indeed downwards, from axioms to theorems. But what is inherited is an epistemological good, which one exactly varying from one manifestation of the EP to the next. In addition to an account of the relevant epistemic good, a particular manifestation of the EP must include a principle governing the flow or transmission of said good. In a strong version, the epistemic good (such as justification) is perfectly preserved from premises to conclusion; in a weaker version, it is more or less preserved. Since different Euclidean thinkers have had different ideas about the relation that the mathematician bears to the axioms, and to the theorems, of a theory, our reconstruction simply uses the placeholder relation E to represent a mathematician's having this epistemic good with respect to a proposition. We allow for degrees as well, as this is relevant to the epistemic good in some cases. So for example, $E(x, p, d)$ might mean that x has a justified belief to degree d in p.[17]

Our account of the core principles of the EP is therefore as follows:

EP-Truth   All axioms and theorems are true.

EP-Self-Evidence All axioms are self-evident. If a subject clearly grasps a self-evident proposition then she bears E to it to the maximal degree.

EP-Flow   If a conclusion is deducible from some premises, and the subject clearly grasps this, and bears E to these premises to a high degree, she thereby bears E to the conclusion to the same, or a similarly high, degree.

The point of this reconstruction is to enable the systematic comparison of a range of historical views, both to the abstract prototype of the Euclidean view, and to one another. This has two important implications. The first is that the actual views of historical figures will inevitably diverge to some extent from the reconstructed ideal we have presented. That ideal is supposed to represent the views of multiple philosophers and mathematicians, which of course differ from one another. Consequently, in any historical manifestation of the EP, there will be extra details that do not appear in our account, details that will be fleshed out in slightly different ways, and numerous other small discrepancies. But we hope to have avoided the potential charge of anachronism or caricature, which, as Lakatos warns us, is often levelled at these kinds of projects by '[r]espectable historians' (1962, p. 4).

The second implication is that, since the reconstructed ideal of the EP is indeed supposed to represent these actual historical views (albeit imperfectly), we do expect our rational reconstruction to bear significant similarities with the specific historical manifestations of the programme. The reason we have identified the above principles as core is that, despite the disagreements on detail between the Euclidean thinkers of the past, all of them subscribe to some version of these three. We take them to be an accurate, if not exhaustive, characterisation of any species of the EP worth its salt.

4.2.2 Peripheral Principles

In order to try and account for popular, though less significant, currents in the history of the EP, we also give the following four peripheral principles:

EP-Finite   The axioms are finitely many.

---

[17] Chapters 3–6 of *The Euclidean Programme* discuss several more specific accounts of this relation and how it might be taken to flow in the accounts of particular thinkers.

EP-General   All axioms are general propositions.

EP-Independence   Each axiom is independent of the others.

EP-Completeness   All truths of a certain kind can be deduced from the axioms.

Most Euclideans have historically subscribed to at least some of these, although we do not take them to characterise an essential aspect of Euclidean foundationalism. We now explain these peripheral principles, some of which are familiar from Lakatos, some of which are not.

We are happy to follow Lakatos in including *EP-Finite* as part of the EP, but only as a peripheral principle and with the caveat mentioned in the previous section. We do not take the relevant sense of finitude quite so literally as Lakatos, since schematic theories can have a finite presentation, despite having an infinite number of axioms. When we are dealing with historical theories that were formulated prior to the development of modern formal logic, it is anachronistic to ask whether the theory is 'really' a first-order theory with axiom schemata, or a second order theory, where a principle which would be schematic in the first-order context is formulated as a single axiom (e.g. the induction principle in arithmetic). But it seems to us that, for the purposes of Euclidean epistemology, a theory with a finite number of schemata is on a par with a truly finite theory. So we understand this principle as requiring only that a theory have a finite presentation. And the principle is only peripheral, since important Euclideans do not endorse it (see Sect. 4.1.4).

EP-General is also a popular principle amongst historical Euclideans. Although it is not a feature of the Euclidean Programme as Lakatos reconstructs it, we include it in our characterisation due to its prevalence in the history of the EP. It passes muster only as a peripheral principle, however, due to the existence of prevalent Euclideans who appear to deny it.[18]

We confess that it is not immediately clear what the requirement of generality actually amounts to. The logical form of a statement is not a particularly helpful guide here, since a (logically) singular statement, such as 'London is a city' can be made general in a narrow sense simply by being prefixed with a redundant universal quantifier. Nonetheless, the axioms of standard mathematical theories are general in a recognizable sense. For example, the standard axioms of arithmetic, such as that the natural numbers are closed under the successor relation, and that distinct numbers have distinct successors, clearly do not concern particular individuals. And even when an individual is mentioned, for instance 0, the singular term can be eliminated in favour of a definition using only general vocabulary (in this case, "the only natural number x such that $x+x=x$"). So we understand *EP-General* as requiring that the axioms include only general vocabulary, and expressions that are definable in terms of it.[19] This certainly accords with axiomatic theories as

---

[18] For example, see pp. 29–30 of *The Euclidean Programme* for our argument that Descartes does not subscribe to this principle.

[19] Of course, we have not actually defined the notion of general vocabulary. But that is a job for particular Euclideans, not for us.

they are usually found. A Euclidean-style presentation of geometry in which one of the axioms is about the properties of a triangle of sides 3, 4 and 5 units is not inconceivable; but it would be eccentric, and quite different from the usual historical practice.

EP-Independence requires that each axiom of the theory is not logically redundant. (In other words, no axiom is derivable from the rest of the axioms.) Like *EP-General*, this principle is not one that Lakatos builds into his reconstruction of the EP, so it is worth saying something to justify its inclusion. The most historically significant episode relating to independence concerns the status of Euclid's Parallel Postulate.[20] From ancient times until modern, a number of mathematicians attempted to prove this postulate from the other four. An interesting feature of these attempts is that the mathematicians were *already* convinced that the postulate was true. Moreover, they do not, in the main, seem to have thought that the Parallel Postulate wasn't evident or obvious. The issue was simply that they thought that a proof *could* be, and so *should* be given. As Proclus (one of Euclid's prominent commentators) puts it, '[the Parallel Postulate's] obvious character does not appear independently of demonstration but is turned by proof into a matter of knowledge' (1970, p. 151). There is evidence, therefore, of a historical current in thinking about mathematics which requires that no provable proposition is included as an axiom.[21] Much as with Lakatos's finitude requirement, independence is not discussed or endorsed by all the Euclidean theorists we consider. And independence problems have been of significant interest outside the Euclidean tradition too, for example in Hilbert's work, simply for their purely mathematical (as opposed to epistemological) significance. Thus we take *EP-Independence* to be a merely peripheral component of the programme.

EP-Completeness says that *all* truths in some important class can be deduced from the axioms. Although the issue is not prominent in Lakatos's characterisation of the EP, he is clearly aware of its presence in Euclidean thought generally; for example, he highlights that the Euclidean believes '*all knowledge* can be deduced from a finite set of trivially true propositions' (1962, p. 4, our emphasis). This characterisation is ambiguous (as we'll see below) and Lakatos does not return to it in the later 'Renaissance' article. There he writes only that the axioms prove 'the rest of the system' (1976a, p. 29), though he does discuss completeness in (what he sees as) some specific manifestations of the EP, such as logicist foundations for mathematics and Hilbert's finitist programme.

Completeness is properly seen as a schematic requirement. It can be understood in various ways corresponding to different understandings of which class of truths the axioms must be complete with respect to. A weak, though still mathematically

---

[20] In Heath's translation, the postulate is: 'That, if a straight line falling on two straight lines make the interior angles on the same side less than two right angles, the two straight lines, if produced indefinitely, meet on that side on which are the angles less than the two right angles' (1968, p. 155).

[21] We discuss the case in of the Parallel Postulate in more detail on pp. 10–12 of *The Euclidean Programme* (2024).

significant, understanding of this requirement is that the axioms should be complete with respect to the known truths in the relevant area of mathematics. Completeness in this sense facilitates the achievement of what Russell described as 'an organization of our knowledge, making it more manageable and more interesting' (1907, p. 580). A stronger principle demands completeness with respect to the knowable truths of the relevant mathematical discipline (although the notion of *knowability* here stands in need of clarification). The strongest completeness requirement is that every truth in the relevant mathematical domain is derivable from the axioms.

We include the (schematic) completeness principle in our reconstruction of the EP because of its historical prominence, particularly amongst canonical Euclideans such as Descartes.[22] Despite this, not all significant Euclideans subscribe to this principle, and indeed it is not a distinctively Euclidean principle at all. Even its strongest version is subscribed to by figures whose classification as Euclideans is best resisted, for example Hilbert and Kant.[23] Hence we include it only as a subsidiary principle.[24]

Our reconstruction of the EP does not include a principle stating that theorems are *dependent* on the axioms from which they are derived. It isn't clear to us whether Lakatos intended this principle to be included in his reconstruction of the EP, though something like it is perhaps suggested by his talk of truth flowing from the axioms to the theorems. And the idea is of course a historically prominent one. Frege famously claimed that the aim of proof was 'to afford us insight into the dependence of truths upon one another' (1884, Sect. 2). This relationship of dependency (*Abhängigkeit*) is supposed to be an objective matter, and similar views can be found in earlier writers as far back as Aristotle. But this relationship of dependency is clearly metaphysical,[25] concerning as it does a relation that holds between truths independently of our epistemological stance toward them. However, the EP is an epistemology, rather a metaphysics, of mathematics. Just as we resist Lakatos' attempt to substitute semantic notions for those that, in the EP, should be epistemological, so too we resist the importation of anything metaphysical into our picture, to whatever extent that is possible. While the dependency of theorems on axioms may have appealed to a number of Euclidean thinkers in the past, we do not reflect this with the inclusion of a relevant EP-principle, not even a peripheral one.

---

[22] See p. 29 of *The Euclidean Programme* (2024).

[23] See, for example, (1787, A480/B508) for remarks by Kant, and (1902, p. 445) for remarks by Hilbert.

[24] Evaluating the plausibility of *EP-Completeness* would take us too far afield here, but it is tackled in Sect. 8.4 of *The Euclidean Programme* (2024).

[25] As emphasised by Shapiro (2009, p. 183), for instance.

4.3 Conclusion

With Lakatos's help, we reconstructed the EP. His idea that what matters is *how* some theoretical good flows from axioms to theorems, not *what* flows, was key to this reconstruction. In other ways, we parted company with Lakatos, for the reasons given. The next thing to do would be to assess the EP in light of developments in contemporary epistemology and contemporary mathematics. We take that next step in *The Euclidean Programme*, and also compare and contrast the ahistorical EP with some flesh-and-blood authors.

Lakatos, as we have had occasion to mention, was strongly opposed to the EP. But he was clear-sighted enough to recognise it as a formidable opponent. Well-versed in Popper's philosophy, he knew how hard existential claims are to refute. We will let him have the last word:

> A Euclidean never *has* to admit defeat: his programme is irrefutable. One can never refute the pure existential statement that there exists a set of trivial first principles from which all truth follows. Thus science may be haunted for ever by the Euclidean programme as a regulative principle, 'influential metaphysics'. A Euclidean can always deny that the Euclidean programme as a whole has broken down when a particular candidate for a Euclidean theory is tottering. (1962, pp. 6–7)

References

Adam, C., & Tannery, P. (Eds.). (1897–1910). *Oeuvres de Descartes* (Vols. 1–12). Leopold Cerf.
Aristotle. (1960). *Posterior analytics, topica* (H. Tredennick & E. S. Forster, Trans.). Harvard University Press: Loeb Classical Library.
Aristotle. (1993). *Posterior analytics* (J. Barnes, Trans.) (2nd ed.). Oxford University Press.
Barnes, J. (2005). What is a disjunction? In D. Frede & B. Inwood (Eds.), *Language and learning*. Cambridge University Press. Reprinted in J. Barnes (2012), *Logical matters* (pp. 512–537). Oxford University Press.
Descartes, R. (1637). *Discourse on method, optics, geometry, and meteorology* (P. Olscamp, Trans.) (revised ed.) (2001). Hackett.
Euclid. (1968). *The thirteen books of the elements* (T. Heath, Trans.) (2nd ed.). Cambridge University Press.
Frege, G. (1884). *The foundations of arithmetic*. (J. L. Austin, Trans.) (2nd ed.) (1953). Blackwell.
Hilbert, D. (1902). Mathematical problems. *Bulletin of the American Mathematical Society*, 8(10), 437–479.
Kant, I. (1787). *Critique of pure reason* (2nd ed.). (P. Guyer, & A. Wood, Trans.) (1998). Cambridge University Press.
Lakatos, I. (1962). Infinite regress and foundations of mathematics. In J. Worrall & G. Currie (Eds.), *Mathematics, science and epistemology* (pp. 3–23). Cambridge University Press.
Lakatos, I. (1976a). A renaissance of empiricism in the recent philosophy of mathematics. In J. Worrall & G. Currie (Eds.), *Mathematics, science and epistemology* (pp. 24–42). Cambridge University Press.
Lakatos, I. (1976b). *Proofs and refutations*. Cambridge University Press.
Pascal, B. (1655). De l'esprit géométrique. In J. Mesnard (Ed.) (1991), *Pascal: OEuvres Complètes, Vol. III: OEuvres Diverses (1654–1657)* (including commentary by Mesnard) (pp. 360–428). Desclée de Brouwer.

Paseau, A. C., & Wrigley, W. (2024). *The Euclidean programme*. Cambridge University Press.
Proclus. (1970). *A commentary on the first book of Euclid's Elements* (G. R. Morrow, Trans.). Princeton University Press.
Russell, B. (1907). The regressive method of discovering the premises of mathematics. In G. Moore (Ed.) (2014), *The collected papers of Bertrand Russell* (Vol. 5, pp. 571–580). Oxford University Press.
Shapiro, S. (2009). We hold these truths to be self-evident: But what do we mean by that? *Review of Symbolic Logic,* 2(1), 175–207.

Open Access This chapter is licensed under the terms of the Creative Commons Attribution-NonCommercial-NoDerivatives 4.0 International License (http://creativecommons.org/licenses/by-nc-nd/4.0/), which permits any noncommercial use, sharing, distribution and reproduction in any medium or format, as long as you give appropriate credit to the original author(s) and the source, provide a link to the Creative Commons license and indicate if you modified the licensed material. You do not have permission under this license to share adapted material derived from this chapter or parts of it.

The images or other third party material in this chapter are included in the chapter's Creative Commons license, unless indicated otherwise in a credit line to the material. If material is not included in the chapter's Creative Commons license and your intended use is not permitted by statutory regulation or exceeds the permitted use, you will need to obtain permission directly from the copyright holder.

Chapter 5
Proofs and Refutations, Non-classically and Game Theoretically

Can Başkent

Abstract Lakatos's seminal work *Proofs and Refutations* depends heavily on counter-examples and refutations. In this work, I argue that the said dependancy goes further than anticipated, rendering *Proofs and Refutations* a working example of paraconsistent reasoning in mathematical methodology. I also maintain that *Proofs and Refutations* is an example of paraconsistent reasoning with strategies, making it an example of game theoretical and strategic reasoning in mathematical methodology.

Keywords Proofs and refutations · Lakatosian methodology · Paraconsistency · Game theory

5.1 Introduction

In Lakatosian epistemology contradictions promote knowledge growth. Lakatos, particularly in *Proofs and Refutations*, suggests various methods to eliminate inconsistencies: *monsters* help us to revise the given theory by following a dialectic heuristic, *proofs that do not prove* allow us to revise mathematical theorems or conjectures (Lakatos, 2015, 1979).

However, Lakatos seems to have missed one thing: Whilst the method of proofs and refutations carries out the aforementioned procedures to *maintain* a consistent theory of scientific inquiry, it still needs to *work with inconsistencies* at the object level. Proof attempts, inquiries and lemmas may turn out inconsistent, despite the fact that the meta-theory is committed to maintain the consistency of the theory. Yet, not "everything goes" once such inconsistencies or contradictions are identified. The method of proofs and refutations, like many revisionist methodologies, has some particular rational procedures to follow to revise the theory at hand.

C. Başkent (✉)
Department of Computer Science, Middlesex University, London, UK
e-mail: c.baskent@mdx.ac.uk, https://www.canbaskent.net/logic

This is what I will address in this paper, and argue that, intrinsically, the Lakatosian method of proofs and refutations, exemplified in his masterpiece *Proofs and Refutations*, is an inconsistency-friendly system.[1] Moreover, I will show that the inconsistency-friendliness of *Proofs and Refutations can* and *should* be approached game theoretically as it resorts to strategic and rational reasoning with inconsistencies. In short, I argue that the Lakatosian method is *both* paraconsistent and game-theoretical.

There are a number of reasons to think so. First, the Lakatosian method is dialectical (Larvor, 1998).[2] Furthermore, dialectic is perhaps one of the major points of intersection between paraconsistency (or dialetheism) and the Lakatosian thought. As discussed by Ficara, Hegelian dialectic has some strong dialetheic tones (Ficara, 2013).[3] Similarly, it is widely argued that the Lakatosian method converges to Hegelian dialectic (Kvasz, 2002; Musgrave & Pidgen, 2021). Musgrave and Pidgen maintain that

> [T]hese water-tight deductions from well-defined premises are the (perhaps temporary) *endpoints* of an evolutionary, and indeed a *dialectical*, process in which the constituent concepts are initially ill-defined, open-ended or ambiguous but become sharper and more precise in the context of a protracted debate.
> (Musgrave & Pidgen, 2021, their emphasis)

Therefore, through the common point of a dialectical approach, the Lakatosian method of proofs and refutations has some dialetheic and inconsistency-friendly aspects.

---

[1] It is important to notice that the Lakatosian revisionism is not the only method that may benefit from an inconsistency-friendly approach. The Lakatosian method is similar to Hintikka's interrogative models of inquiry in some ways, and it may be helpful to draw some analogies between the two (Başkent, 2015a, 2015b; Başkent, 2017).

[2] Larvor indicates that "(...) Lakatos expressed a desire to become the founder of a dialectical school in the philosophy of mathematics" in a letter written to Larvor (1998, p. 9).

[3] The debate on the Law of Contradiction and Hegel is an illuminating one, as Ficara notes:

> "Classically, interpretations of Hegel's dialectics either take Hegel's claims against the law of non-contradiction (LNC) as a serious logical argument, and therefore do not take Hegel's philosophy seriously, or consider Hegel's philosophy as a serious enterprise, and therefore deny that his critique of LNC should be taken seriously.
> According to a widespread view, whose most authoritative exponent is probably Karl Popper, Hegel's dialectic is unscientific because it implies a refusal of LNC. Popper writes:
>
>> [Hegel's idea of the fertility of contradictions] amounts to an attack upon the 'law of contradiction' [...] of traditional logic, a law which asserts that two contradictory statements can never be true together, or that a statement consisting of the conjunction of two contradictory statements must always be rejected as false on purely logical grounds
>
> For this reason: 'If we are prepared [like Hegel] to put up with contradictions, criticism, and with it all intellectual progress, must come to an end'. And on a similar line, Charles Sanders Peirce observes: 'As far as I know, Hegelians profess to be self-contradictory'."

(See Ficara's paper for the full reference information for the quotes.)

Second, the Lakatosian method *needs* and therefore justifies the existence of inconsistencies. What follows from an inconsistency is certainly not *everything* unlike in classical logic. Particularly in *Proofs and Refutations* (PR, for short), Lakatos offers a wide variety of tools to guide what *should* follow from an inconsistency. Such methods include *monster-barring*, *exception-barring*, *method of surrender* and *lemma-incorporation* to name a few (Başkent & Bağçe, 2009). What is common in all these is that they rely and depend on the existence of inconsistencies. For Lakatos, the existence of inconsistencies in a rational theory is not surprising, his method of proofs and refutations relies on their existence. In common with many other revisionists, Lakatos offers certain methods to *fix* the theory. Yet, whilst doing so, the theory works with the very inconsistencies.[4]

Third, in PR, Lakatos identifies certain methods to deal with inconsistencies that show up throughout mathematical practice. These methods are perhaps first there to describe and maintain a classical and consistent theory. Yet, they also serve an important strategic goal, which makes them game-theoretical. This is certainly evident in PR which was written as a sometimes competitive, sometimes cooperative game. Students (that are players) have various strategies that Lakatos identifies, their strategies depend on what they know or learn from other players' moves and strategies. The game is also evolutionary, especially after the Teacher's interventions who keeps introducing new *signals* to the game. Classically, the game identifies rationality with consistency and proceeds as such. However, in due time, I will argue that this is a mistake as it is an unnecessary restriction on the Lakatosian method.

The current paper is organised as follows. First, I briefly discuss the Lakatosian method of proofs and refutations from a paraconsistent point of view. Following, I offer two inter-connected solutions: Lakatosian paraconsistency *and* paraconsistent games for PR. Granted, such solutions are relatively high level, I will conclude with a discussion.

5.2 Lakatos's Method of Proofs and Refutations: Briefly

A brief review of the method of proofs and refutations is a good starting point. Corfield summarises the steps of the Lakatosian method as follows (Corfield, 1997):

---

[4] An analysis of the nature inconsistencies in the Lakatosian philosophy of mathematics falls outside the scope of the current paper. Many non-classical logicians carefully distinguish inconsistencies and contradictions, and certainly, this approach has some merits (Carnielli & Coniglio, 2016). What is left is to apply it to a particular philosophy of mathematics, such as the Lakatosian method of proofs and refutations.

Moreover, some inconsistencies can be classified as local and global, where the former refutes a local lemma and the latter an overarching theorem. Formal logical approach to such distinctions between inconsistencies, with a direct application to the Lakatosian philosophy of mathematics, remains a challenging future work opportunity.

1. Primitive conjecture.
2. Proof (a rough thought experiment or argument, decomposing the primitive conjecture into subconjectures and lemmas).
3. Global counterexamples.
4. Proof re-examined. The guilty lemma is spotted. The guilty lemma may have previously remained hidden or may have been misidentified.
5. Proofs of the other theorems are examined to see if the newly found lemma occurs in them.
6. Hitherto accepted consequences of the original and now refuted conjecture are checked.
7. Counterexamples are turned into new examples, and new fields of inquiry open up.

This algorithm allows us to make many "searches" and, as such, gives us some room to control the parameters. Searching for counterexamples, re-examining proofs and the methods that are developed to turn them into examples are all strategic moves. Moreover, the existence of inconsistencies is embedded in the algorithm: the algorithm reasons with them rather than "exploding" under their existence. Consequently, the above algorithm makes it clear that the reasoning in PR is a game with inconsistencies—a game with "proofs that do not proof", a game with "guilty lemmas".[5]

If PR enjoys paraconsistent and game theoretical reasoning, then we can use this idea to further our discussion of rationality. What distinguishes game theoretical agents from, say, automata or probabilistic and randomised guesses, is that first and foremost they are rational. As such, they aim at increasing their own pay-offs and maximising their utilities, and consequently winning the game. In order to reach that goal, players need to be allowed to enjoy inconsistent reasoning in their rational strategies. This practice is more common than it seems (Ariely, 2008, 2010; Kahneman, 2011). *Homo economicus* is assumed to be rational yet makes emotional decisions based on inconsistencies in a *systematic* and *predictable* way. Seen as a game, the *game* of mathematical discovery and practice has the potential to share this approach. One can, therefore, imagine a dialogue similar to PR where the players adopt not a revisionist, but a game theoretical approach to mathematical discovery. Instead of discussing how to revise a proof that does not prove, they take turns and make moves to develop a proof that works—even paraconsistently. They may discuss their preferences, pay-offs and strategies to reach an equilibrium in

---

[5] The way the Lakatosian method resolves inconsistencies shows some notable similarities to Hintikkan method of interrogative inquiry (IMI, for short). IMI is a well-known example of an epistemic method that may result in knowledge increase (Başkent, 2016b). It excludes inconsistencies by *bracketing* them—that is some pieces of information are excluded from the epistemic reasoning as they may lead to contradictions. The decision to choose what to bracket needs to be strategic and rational. This renders IMI also as a game with inconsistencies—a game with "bracketing".

discovering mathematical knowledge. They may talk about "cheaper" proofs which use less resources.[6]

If PR is paraconsistent,[7] then we have a problem. And, this problem requires a solution. I will offer two. First, the theory must allow us to work with inconsistencies with whatever meta-theoretical commitments one might have. Second, the way we progress or resolve the inconsistencies must be strategic in the sense that the players (or rational agents) must be able to *compute* their responses under uncertainty or imperfect information. Furthermore, players must know when the game reaches a solution, or an equilibrium—a state of balance where all players involved are satisfied enough not to make a further move.

In conclusion, the problem of having inconsistencies in PR requires a solution that is *friendly* to contradictions and strategic reasoning as it is the way that the dialogue (the method of proofs and refutations) is presented in PR. In what follows, I will explain how.

5.3 Solution: Paraconsistency for the Lakatosian Method

I have argued earlier that Lakatos's theory of PR is dialetheic. This means that the theory admits true contradictions. Moreover, I claim that the formal system in which PR seems to be operating in is paraconsistent.

First, as we explained earlier, the Lakatosian method is dialectical (Musgrave & Pidgen, 2021). This idea can be supported by various historical and even political arguments, as many maintained (Corfield, 1997; Koetsier, 1991; Kvasz, 2002; Larvor, 1998). For example, Corfield is critical of Lakatos's dialectic,[8] Koetsier argues that PR is a rational reconstruction of history,[9] Kvasz criticises Lakatos for

---

[6] Considering the computational cost of a proof is a well-known approach in computer science. Such costs may include the time it takes to develop a proof, the memory space or the processor power that it requires to compute a proof. Therefore, an agent may have a strict preference towards "cheaper" proofs.

[7] It needs to be noted that arguing that PR is paraconsistent does not suggest that Lakatos himself is a paraconsistent logician nor a dialetheic thinker. The current paper focuses only on PR and leaves it for future work how the Lakatosian philosophy may benefit from dialetheic and paraconsistent approaches.

[8] "[...] an important part of the dialectical process is being missed in that good intuitive ideas, which are often the material for the most fruitful variety of rigorous exploration, are being drowned in a sea of conjectures from which they may only be extracted by great effort." (Corfield, 1997).

[9] "(...) there is no doubt that *Proofs and Refutations* contains a highly counterfactual rational reconstruction" (Koetsier, 1991).

his "confusion" of logic and dialectic,[10] and Larvor's approach is more historical.[11] Such inconsistencies exist in PR, arguably for the purpose of reconstructing the history of Euler's theorem on polyhedra in order to construct a dialectic theory.

Furthermore, Larvor underlines the difference between a logical contradiction and a Hegelian one within the context of the Lakatosian methodology of mathematics.

> For something to contain a contradiction does not mean, for Hegel, that it entails both A and $not-A$ for some proposition A. A Hegelian 'contradiction' is better understood as an internal tension. What it means is that the elements of the object grate against each other in some sense appropriate to the kind of object in question. Now, a Lakatosian research programme is a dynamic unit. Its constituent parts interact and modify each other (in particular, the hard core and heuristic combine to act on the 'protective belt'). It may not even be possible to characterise one part of a research programme in isolation from the others. In the jargon, the parts are essentially related. A research programme is in this sense an organic whole. A programme 'contains a contradiction' (in the Hegelian sense) when it becomes unable to protect its hard core without violating the spirit of its positive heuristic. (Larvor, 1998, p. 70).

However confused or historically motivated he might be, Lakatos's methodology has some strong dialectic tones. Does it, however, suffice to make it *dialetheic*?

Priest argues strongly that dialectic theories are dialetheic (Priest, 1989). He goes further and claims that the use of dialectic by first Hegel and later Marx *requires* dialetheism [ibid, my emphasis]. Recently, Priest also suggested a dialetheism based formalisation of dialectic (Priest, 2023).[12] Ficara, in particular, argues that Hegelian dialectic can be *re*-interpreted as dialetheism. The Hegelian debate on the details of such arguments falls outside the scope of this paper. However, from a dialetheist position, Hegelian dialectic *is* dialetheist. And the logic of dialetheism is paraconsistency—a logic where inconsistencies do not entail everything[13] (Priest, 2002).

---

[10] "We would like to show that these weaknesses were caused by his confusion of dialectic with logic. In this way, Lakatos developed an appealing and interesting theory, which, at least at first glance, has the advantages of both—the liveliness of dialectic and the soundness of logic. Unfortunately, this attempt to combine dialectic with logic also has one disadvantage. The focus on logic restricts severely the scope of the changes, to which this method can be applied. That is why Lakatos was forced to neglect in his rational reconstruction many episodes in the history of mathematics, which simply do not fit into his scheme. But on the other hand, dialectic gives his theory the illusion of universality, for which reason, perhaps, he seemed to be unaware of his omissions." (Kvasz, 2002).

[11] "The use of historical narrative as philosophical argument is part of Lakatos' Hegelian inheritance. For Hegel, history is a bit like a huge Platonic dialogue. Just as a dialogue starts with simple ideas and progresses dialectically towards a sophisticated conception of whatever happens to be under discussion, so the history of humanity begins with simple forms of consciousness and develops towards a perfect final state." (Larvor, 1998, p. 65).

[12] This paper appeared in a special issue of the journal "History and Philosophy of Logic" which was dedicated to various formalisations of dialectic.

[13] Direct applications of paraconsistency include ontology, mereology and belief revision (Priest, 2001).

Therefore, PR is dialectic. As a dialectic theory, it is dialetheic. As such, it must admit a logic of dialetheism, which is the paraconsistent logic. This is an interesting direction to take. Because, as Larvor argued above, it is not possible to reach the conclusion that "PR is paraconsistent" by means of Hegel in the way that Lakatos understood him, as Hegel's interpretation of contradictions is not entirely logical. Instead, we reach this conclusion by means of dialectic and dialetheism.

Let us now consider an example from PR to illustrate Lakatos's interpretation of reasoning with inconsistencies.

> ALPHA: I have a counterexample which (...) will be a counterexample to the main conjecture, i.e. this will be a global counterexample as well. (...)
> Imagine a solid bounded by a pair of nested cubes—a pair of cubes, one of which is inside, but does not touch the other. This hollow cube falsifies your first lemma, because on removing a face from the inner cube, the polyhedron will not be stretchable on to a plane. Nor will it help to remove a face from the outer cube instead. Besides, for each cube $V - E + F = 2$, so that for the hollow cube $V - E + F = 4$.
> (...)
> GAMMA: (...) Hands up! You have to surrender. Scrap the false conjecture, forget about it and try a radically new approach.
> (...)
> DELTA: But why accept the counterexample? We proved our conjecture—now it is a theorem. I admit that it clashes with this so-called 'counterexample'. One of them has to give way. But why should the theorem give way, when it has been proved? It is the 'criticism' that should retreat. It is fake criticism. This pair of nested cubes is not a polyhedron at all. It is a *monster*, a pathological case, not a counterexample.
> (Lakatos, 2015, p. 14-5, Lakatos's emphasis)

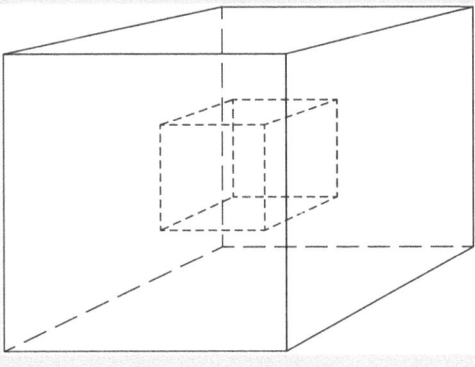

Following this dialogue, PR continues with a discussion of *monster-barring*. The method of monster-barring suggests that the object, which has been put forward as a counter-example, is actually not a counter-example but a *pathological case* (Lakatos, 2015, p. 15).

When the nested-cube contradicts the Conjecture, not everything follows proof-theoretically. Certain strategies are employed in order to reach a temporary state of consistency or even an equilibrium. In the example above, this strategy is monster-barring. Later on, PR suggests another strategy, called *exception-barring*. The meta-theory of PR, therefore, strictly suggests that not-everything goes. Only *some* strategies to resolve the puzzle can be employed.

The contradiction in the example above is worth discussing.

The conjecture suggests that $\forall x. \chi(x) = 2$ where x varies over a domain of geometric objects and $\chi(x)$ represents the Euler characteristics of object x. The existence of the nested-cube, denoted by n, on the other hand, shows that $\chi(n) = 4$. And, as $2 \neq 4$ and n lies within the domain of geometric objects, the following two sentences are inconsistent.

$$\forall x. \chi(x) = 2$$

$$\chi(n) \neq 2$$

It is important to note that Lakatos's end goal is not to leave the state of affairs there, but resolve the contradiction. This, however, still does not refute our claim that the system remains paraconsistent. What follows or does not follow from a contradiction defines the Lakatosian methodology—if everything did follow, there would be no Lakatosian methodology. All would be trivial. But, they are not.

Let me clarify this argument further.[14] Paraconsistent reasoning suggests a rational way of reasoning under inconsistencies. Priest, for example, discusses Bohr's theory of atom, Dirac's δ-function and Newton/Leibniz's infinitesimal calculus as examples of inconsistent scientific theories (Priest, 2007). Yet, one needs to "reason in a non-trivial way from inconsistent information" [ibid]. And this way of thinking can also be the rational way of reasoning. Priest argues that "it seems reasonable to hold that if one theory is sufficiently better than all of its competitors on sufficiently many of the criteria, then, rationally, one should believe this rather than the others." [ibid].

Bueno and da Costa argue along the same lines, and defend the "view that if scientific theories are taken to be quasi-true, and if the underlying logic is paraconsistent, it's perfectly rational for scientists and mathematicians to entertain inconsistent theories without triviality. As a result, as opposed to what is demanded by traditional approaches to rationality, it's not irrational to entertain inconsistent theories" (Bueno & da Costa, 2007).

---

[14] I am thankful to the anonymous referee for asking for further clarifications for this argument.

Therefore, I maintain that it is possible that one can be rational whilst reasoning with inconsistencies. Furthermore, the Lakatosian method of proofs and refutations is an example of this. Opponents of this thesis may maintain that this can also be handled within the realm of classical logic. This, however, is far from true. The reason is that the Lakatosian method relies on inconsistencies. If there were no inconsistencies, there would be no Lakatosian method. If there were no inconsistencies, there would be no "guilty lemmas", no "proofs that do not prove". This defines the Lakatosian methodology. If all followed from a contradiction, it would be impossible to identify what characterises Lakatosian methodology. This is the reason why there would be no Lakatosian methodology in a classical logic.

Finally, one can argue that, syntactically, the Lakatosian contradictions in PR are conditional contractions, therefore can easily be handled within the power of classical logic and belief revision methodologies.[15] This argument, however, is not necessarily true. If a contradiction appears to follow from $\varphi_1 \wedge \cdots \wedge \varphi_n$, and if removing φ_1 from the conjunction eliminates the contradiction, then one can argue that this is good enough to maintain a consistent theory. There are, however, some problems with this approach. First, the theory loses information as φ_1 is now *not* part of the conjunction. There may be some informational cost attached to removing a conjunct. Second, it is not always possible to identify the "trouble-maker conjunct" as it may not be theoretically possible.[16] Three, the elimination based methods to maintain logical consistency argue not only against paraconsistency but also against non-classical conjunctions. One can desire to maintain consistency, but this does not immediately entail that one needs to stick with classical conjunction *and* classical entailment. They are often separate issues. Moreover, one can also maintain a classical theory by introducing different conjunctions (including multiplicative conjunctions, for example) or conditionals (including the magic-wand operator in separation logics, for example). In conclusion, there are various logical *tricks* that one can use to maintain a classical theory. Yet, they do not refute our central arguments as to why PR needs a paraconsistent theory. Because the reason is not the lack of logical and mathematical methods to eliminate inconsistencies. The reason is that PR needs inconsistencies to operate.

In PR, Lakatos offers various methods to work with inconsistencies. And these methods characterise PR. Let us now consider another example which reiterates some of the points I have made earlier.

GAMMA: I have just discovered that my *Counterexample 5*, the cylinder, refutes not only the naive conjecture but also the theorem. Although it satisfies both lemmas, it is not Eulerian.

(continued)

---

[15] I am thankful to the anonymous referee for pointing this out.
[16] In Priest (2007), Priest list some examples where it is not simply possible to follow the elimination strategy in some scientific theories.

> ALPHA: Dear Gamma, do not become a crank. The cylinder was a joke, not a counterexample. No serious mathematician will take the cylinder for a polyhedron.
> GAMMA: Why didn't you protest against my *Counterexample 3*, the urchin? Was that less 'crankish' than my cylinder? *Then* of course you were *criticising* the naive conjecture and welcomed refutations. Now you are *defending* the theorem and abhor refutations! *Then*, when a counterexample emerged, your question was: *what is wrong with the conjecture*? *Now* your question is: *what is wrong with the counterexample*?
> DELTA: Alpha, you have turned into a monster-barrer! Aren't you embarrassed?
> (Lakatos, 2015, p. 45, Lakatos's emphasis)

The above (counter-) example, *cylinder*, once again, is used to establish a methodology to ward off contradictions. Using the terminology above, now we have the following, where c denotes cylinder.

$$\forall x . \chi(x) = 2$$
$$\chi(c) \neq 2$$

The method of monster-barring argues that c does not lie within the scope of the universal quantifier $\forall x$, hence the theorem would not apply to c. Once again, we have a rational and logical strategy to follow when contradictions emerge. Not everything goes.

As Lakatos puts it in PR, "*[t]his revolution in mathematical criticism changed the concept of mathematical truth, changed the standards of mathematical proof, changed the patterns of mathematical growth!*" (Lakatos, 2015, p. 110–111, Lakatos's emphasis). This change, however, may not work the way Lakatos imagined.

Non-classical, in particular paraconsistent, reading of the Lakatosian method is arguably another point of support for Lakatos's view regarding mathematics as quasi-empirical. Lakatos asked a similar question when he was discussing falsification in mathematics: "[W]hat is the nature of potential falsifiers of mathematical theories?" (Lakatos, 1979, p. 35). His immediate answer is amusing but indeed accurate: "The very question would have been an insult in the years of intellectual honeymoon of Russell or Hilbert. After all, *Principia* or the *Grundlagen der Mathematik* were meant to put an end—once and for all—to counterexamples and refutations in mathematics." (Lakatos, 1979, p. 36). Granted, Lakatos's understanding of formal theories and their quasi-empirical counterparts is not paraconsistent. Nevertheless, his reasoning, the way the method of proofs and refutations work, is. Counterexamples and refutations are essential parts of his methodology and the theory of PR relies on them. This makes it paraconsistent—reasoning with inconsistencies, refutations, contradictions and counter-examples in a non-trivial way.

The opponents of this idea may suggest that there are subtle differences between inconsistencies and counter-examples. The former is a death sentence for a theory, whereas the latter allows us to improve the theory, promoting knowledge growth. From a classical point of view, there is some truth in this line of thought. After all, working with inconsistencies in a theory provides us with feedback to *fix* the theory. Yet again, the way the theory can be *fixed* or improved depends on the meta-theory of the methodology one needs to endorse amongst others. One way or another, revisionist theories still work with inconsistencies—until the next counter-examples emerge. Furthermore, the existence of the plethora of revisionist theories for mathematical practice is yet another argument supporting paraconsistency. Once inconsistencies emerge, there may be more than one way of fixing the theory, if that is the interim goal. Not everything goes as revisionists might argue, but only a selected few methodologies.

A similar approach has been taken by paraconsistent mathematicians and philosophers (Mortensen, 1994, 2010; Weber, 2021). Particularly, Weber's recent work develops a basic theory of sets that is "axiomatically in a paraconsistent logic", and extends the discussion to a few other foundational issues in mathematics, including algebra and topology. Therefore, once met with the paradoxes of set theory, one can choose to follow the Euclidean methodology or the Lakatosian one or the paraconsistent one, amongst others.[17] One thing in common with all such methods is, however, that all work with inconsistencies. Not everything follows from a contradiction or a paradox.

In PR, Lakatos offers many methods and techniques to understand contradictions. The way Lakatos attempts to *resolve* contradictions is, however, game theoretical.

5.4 Solution: Paraconsistent Games for the Lakatosian Method

With a slight abuse of terminology, it is possible to define a paraconsistent game: a game where, under contradictions, players are able to make rational moves to increase their pay-offs and form rational strategies. There are various ways a game can be *paraconsistent*. It can be about what players know—epistemic paraconsistent games. It can also be about inconsistent strategies and moves—behavioural paraconsistent games. In this work, when I say paraconsistent games, I will use it to mean either of the aforementioned games.

---

[17] This is a good point for logical pluralism, as discussed by Beall and Restall (2006). Nevertheless, in order not to diverge from my focus, I leave it for a future work to examine the logical pluralism of the Lakatosian thought.

In game theoretical paraconsistency, contradictory strategies (strategies that can lead to both a win and a loss), irrational players (players who do not seek to maximise their utility) and non-classical probabilities are allowed (Bueno-Soler & Carnielli, 2016; Mares, 1997). The aforementioned elements of inconsistencies are also familiar from game semantics which can be a good example of a paraconsistent game. Hintikkan game semantics suggests an interpretation of logical formulas by means of game theoretical choices and operations. As such, loosely put, it *gamifies* logic (Hintikka & Sandu, 1997; Mann et al., 2011).[18] Semantic games, depending on the logics in question, can be paraconsistent: we can have competitive games where both players may admit winning strategies. We can even have a cooperative, multi-player and coalition-based games for various non-classical logics (Başkent, 2016a, 2020; Başkent & Henrique Carrasqueira, 2020). Arguably, paraconsistent games are not a foreign concept in philosophical logic, and in what follows I will show that such elements also exist in PR by focusing on individual tools and techniques employed in PR.

First, *proofs that do not proof* are the proof theoretical equivalent of the game theoretical strategies that *knowingly* produce a loss. Even if they may not bring a win, players may still learn from the plays following such strategies, thus there may be an epistemic gain despite the loss. There might, however, be a cost to this: computational power and resources may be consumed whilst producing a proof that does not prove. Moreover, such proofs may produce a signal: an error in the proof might signal another error in a lemma, for instance. This also makes them a subject of evolutionary game theory.[19] From a paraconsistent perspective, proofs that do not prove is another argument to support paraconsistency. These are the proofs from which some propositions do not follow—even if there is an inconsistency. This is paraconsistency.

On the other hand, considering the constructivist Kolmogorov connection between "truth–proofs–computation–strategies", there is much more to say about proofs that do not prove. Such objects, if they exist, are the strategies that do not bring wins and the programs that do not compute what they are set out to compute. Inconsistent (or dialetheic) truth suggests "proofs that do not prove" exist, which in turn suggest that "programs that do not compute" exist, which finally suggest "winning strategies that do not bring a win" exist as well. The theories of the latter two are not well developed yet. Nevertheless, proofs that do not prove show the existence of such objects by means of the aforementioned Kolmogorov connection.

Second, *re-examining proofs* is strategy pruning. Similar to well-known game theoretical situations, such as the iterated prisoners' dilemma, in strategy pruning,

---

[18] It is important that the research programme of gamifying logic did not start with Hintikka, and can be seen in some works of Hodges (2013).

[19] Games where players knowingly follow a strategy that generate a loss are a common subject of inquiry in evolutionary game theory. Altruism, both in animal and non-animal groups is a well-known example.

the player re-runs the strategy and uses it to improve the same initial strategy by simply removing those alternatives that players know would not work.

Another major example is the "iterated elimination of strictly dominated strategies"—a common solution method in game theory. In this method, the strategies that are eliminated can be the pruned version of the very strategy. In other words, when a given strategy σ is pruned to σ' and played along in the next run of the game, the strictly dominated strategy amongst σ and σ' is eliminated. The strategies can be refined, programs can be made efficient. This, however, contradicts the basic tenets of strategies (Hodges, 2013; van Benthem & Klein, 2022). A strategy is supposed to be pre-defined. Therefore, revising a strategy must already be part of the strategy. This creates self-reference as I argue next.

Third, *revising proofs* is strategy revision, which contradicts the very definition of a strategy. A strategy is defined as "a set of rules that describe exactly how [a] player should choose, depending on how the [other] players have chosen at earlier moves" (Hodges, 2013). That means that strategies are pre-set and pre-defined. They are omniscient and should cover all possible cases and scenarios, including their own revision. Hence, a strategy cannot be revised, it should contain its own revision. This is self-reference (Heifetz, 1996). Harsanyi noted this much earlier in 1967:

> It seems to me that the basic reason why the theory of games with incomplete information has made so little progress so far lies in the fact that these games give rise, or at least appear to give rise, to an infinite regress in reciprocal expectations on the part of the players. In such a game player 1's strategy choice will depend on what he expects (or believes) to be player 2's payoff function U_2, as the latter will be an important determinant of player 2's behavior in the game. But his strategy choice will also depend on what he expects to be player 2's first-order expectation about his own payoff function U_1. Indeed player 1's strategy choice will also depend on what he expects to be player 2's second- order expectation – that is, on what player 1 thinks that player 2 thinks that player 1 thinks about player 2's payoff function U_2... and so on *ad infinitum*.
> (Harsanyi, 1967)

Heifetz argued similarly on the same issue:

> Nevertheless, one may continue to argue that a state of the world should indeed be a circular, self-referential object: A state represents a situation of human uncertainty, in which a player considers what other players may think in other situations, and in particular about what they may think there about the current situation. According to such a view, one would seek a formulation where states of the world are indeed self-referring mathematical entities.
> (Heifetz, 1996)

There is a wide variety of work on self-referential paradoxes in games, including their connections to *non*-self-referential variations (Abramsky & Zvesper, 2015; Brandenburger & Keisler, 2006; Pacuit, 2007; Başkent, 2018). The mathematical details of such ideas, however, fall outside the scope of the current work.

Fourth, *turning contradictions into new examples* and *lemma-incorporation* are methods familiar from belief revision and dynamic logics (Mares, 2002; Priest, 2001). However, for PR it is an embedded and essential part of the heuristic method. It is a meta-model theoretical strategy that changes and updates the model where inconsistencies become consistencies. One can have different metaphysical goals of obtaining a consistent strategy at the end. Nevertheless, as before, the very act of reasoning game theoretically and strategically under the inconsistencies is paraconsistent.

There is yet another angle. The epistemics of players and strategies can be analysed further using epistemic game theory. For instance, what players know about each other and their preferences and how they revise their strategies after a new piece of information is introduced, fall within the scope of epistemic game theory. PR is no exception to this approach.

Epistemic game theoretical elements in PR are plentiful. In the following passage from PR, some epistemic elements and the preferences based on them are revealed step by step.

> GAMMA: (...) *A polyhedron is a solid whose surface consists of polygonal faces*. And my counterexample is a solid bounded by polygonal faces.
> TEACHER: Let us call this definition *Def. 1*.
> DELTA: Your definition is incorrect. A polyhedron must be a surface: it has faces, edges, vertices, it can be deformed, stretched out on a blackboard, and has nothing to do with the concept of 'solid'. *A polyhedron is a surface consisting of a system of polygons*.
> TEACHER: Call this *Def. 2*.
> DELTA: So really you showed us two polyhedra—two surfaces, one completely inside the other. A woman with a child in her womb is not a counterexample to the thesis that human beings have one head.
> (...)
> DELTA: (...) By polyhedron I meant *a system of polygons arranged in such a way that (1) exactly two polygons meet at every edge and (2) it is possible to get from the inside of any polygon to the inside of any other polygon by a route which never crosses any edge at a vertex*. (...)
> TEACHER: *Def. 3*.
> ALPHA: (...) Why don't you just define a polyhedron as a system of polygons for which the equation $V - E + F = 2$ holds? This Perfect Definition ...
> KAPPA: *Def. P*.
> (Lakatos, 2015, p. 15-7, Lakatos's emphasis)

Above, players reveal what they know and understand about polyhedron and what they think and believe about each other's knowledge about polyhedron. It is interactive and it is strategic.[20]

PR presents some more instances of epistemic games where players learn from each other and develop their ideas further. In such instances what they know

---

[20] Lakatos's subsection titles in PR make the classification about the role of strategies very easy to follow.

about their opponents and what they know about what their opponents know about themselves change, improve and evolve. The way they execute their strategies and make moves directly depends on how their knowledge about the aforementioned situations is formed.

Let us now see some examples from PR in order to illustrate the game theoretical elements at work.

First, the method of exception-barring.

> BETA: (...) It now seems to me that no conjecture is generally valid, but only valid in a certain restricted domain that excludes the *exceptions*. I am against dubbing these exceptions 'monsters' or 'pathological cases'. That would amount to the methodological decision not to consider these as interesting *examples* in their own right, worthy of a separate investigation. But I am also against the term *'counterexample'*; it rightly admits them as examples on a par with the supporting examples, but somehow paints them in war-colours, so that, (...), one panics when facing them, and is tempted to abandon beautiful and ingenious proofs altogether. No: they are just *exceptions*.
> (Lakatos, 2015, p. 26, Lakatos's emphasis)

The above quote discusses which moves are or can be admissible for players at certain positions in the game. Following PR, if some "counterexamples" are to be excluded, this simply means that there is no available move in that position in the game that allows the players to admit those objects as counterexamples. They are, then, "exceptions", and excluded.

Moreover, if some moves are not available, we can then discuss *admissible strategies*. A strategy s is admissible if and only if there is a strictly positive probability measure on the strategy profiles for the other players, under which s is optimal (Brandenburger et al., 2008). Admissibility is important as it is "a prima facie reasonable criterion: It captures the idea that a player takes all strategies for the other players into consideration; none is entirely ruled out" (Brandenburger et al., 2008). This concept explains exception-barring well. In order to identify a geometric object as an exception, it must be compared and contrasted against the counterexamples, without immediately being ruled out. Moreover, for some pupils in PR, exception-barring as a strategy must be optimal.

Second, the method of monster-adjustment.

> RHO: I agree that we should reject Delta's monster-barring as a general methodological approach, for it doesn't really take 'monsters' seriously. Beta doesn't take his 'exceptions' seriously either, for he merely lists them and then

(continued)

retreats into a safe domain. Thus both these methods are interested only in a limited, privileged field. *My* method does not practise discrimination. I can show that 'on closer examination the exceptions turn out to be only apparent and the Euler theorem retains its validity even for the alleged exceptions'.
(...)
ALPHA: How can my counterexample 3, the 'urchin', be an ordinary Eulerian polyhedron? It has 12 star-pentagonal faces...
RHO: I don't see any 'star-pentagons'. Don't you see that in actual fact this polyhedron has ordinary triangular faces? There are 60 of them. It also has 90 edges and 32 vertices. Its 'Euler characteristic' is 2. The 12 'star-pentagons', their 30 'edges' and 12 'vertices', yielding the 'characteristic'—6, are only your fancy. Monsters don't exist, only monstrous interpretations. One has to purge one's mind from perverted illusions, one has to learn how to see and how to define correctly what one sees. My method is therapeutic: where you—erroneously—'see' a counterexample, I teach you how to recognise—correctly—an example. I adjust your monstrous vision...

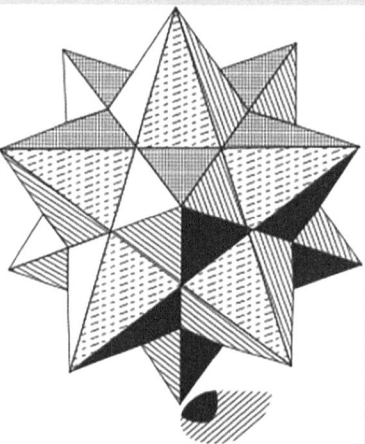

(Lakatos, 2015, p. 33, Lakatos's emphasis)

Monster-adjustment then is a case of evolutionary game theory. Evolutionary games are about change. Species can evolve, norms may change, definitions can be revised. Evolutionarily, a predator can evolve to be a prey in a different eco-system. And, as such, monsters can be adjusted.

Third, the method of monster-barring.

> GAMMA: *Then* of course you were *criticising* the naive conjecture and welcomed refutations. *Now* you are *defending* the theorem and abhor refutations! Then, when a counterexample emerged, your question was: *what is wrong with the conjecture*? *Now* your question is: *what is wrong with the counterexample*?
> (Lakatos, 2015, p. 45, Lakatos's emphasis)

Winning a game, that is proving a proposition φ, does not always mean disproving $\neg\varphi$. Logic must allow this and must have a non-classical negation. Similarly, the *game* of proving a proposition does not have to be zero-sum. Hintikka's game theoretical semantics is a good example of this once used for non-classical logics, as I argued earlier (Başkent, 2016a). For some non-classical logics, for example, winning a game (showing the truth of a proposition φ) does not entail that the opponent loses (thus $\neg\varphi$ does not have to fail).

Similarly, in PR, defending the conjecture and refuting the counterexamples do not necessarily suggest the same strategy. This is another way of seeing non-classical logical elements at work in PR—both semantically and game theoretically.

Fourth, the method of proofs and refutations.

> LAMBDA: All this shows that one cannot put proof and refutations into separate compartments. This is why I would propose to rechristen our '*method of lemma-incorporation*' the '*method of proof and refutations*'. Let me state its main aspects in three heuristic rules:
> *Rule 1. If you have a conjecture, set out to prove it and to refute it. Inspect the proof carefully to prepare a list of non-trivial lemmas (proof-analysis); find counterexamples both to the conjecture (global counterexamples) and to the suspect lemmas (local counterexamples).*
> *Rule 2. If you have a global counterexample discard your conjecture, add to your proof-analysis a suitable lemma that will be refuted by the counterexample, and replace the discarded conjecture by an improved one that incorporates that lemma as a condition. Do not allow a refutation to be dismissed as a monster. Try to make all 'hidden lemmas' explicit.*
> *Rule 3. If you have a local counterexample, check to see whether it is not also a global counterexample. If it is, you can easily apply Rule 2.*
> (Lakatos, 2015, p. 53, Lakatos's emphasis)

This is the construction of a (recursive) strategy that directly relies on opponents' moves and strategies. This is in line with Harsanyi and Heifetz's thesis, presented earlier.

Examples can be multiplied. Let us stop here.

Clearly, Lakatos would like these games to terminate: "Unlimited concept-stretching destroys meaning and truth" (Lakatos, 2015, p. 105). This means, game theoretically, that there is a non-zero cost attached to concept-stretching in the method of proofs and refutations. And, it seems that the *reason* for that cost is metaphysical rather than mathematical as "*[r]ationality, after all, depends on inelastic, exact, concepts!*" (Lakatos, 2015, p. 108, Lakatos's emphasis). Lakatos explains it further in a footnote: "Gamma's demand for a crystal-clear definition of 'counterexample' amounts to a demand for crystal-clear, inelastic concepts in the metalanguage as a condition of rational discussion" [ibid].

Nevertheless, a game theoretical perspective remains to be the best way to analyse Lakatos's heuristic recursion. The aforementioned rigidity in the concepts of metalanguage finds some resonance in non-classical and heterodox approaches to game theory by means of behavioural game theory and applied psychology. The way classical and traditional game theory evolved into behavioural game theory resembles how the method of proofs and refutations *may* evolve, breaking the rigid expectations for the ontology of the metalanguage in question to understand rationality. Game theoretical methodologies to dissect PR further can produce a behavioural account of the method of proofs and refutations. Such behavioural and strategic accounts can explain the goal-oriented, opportunist and rational approaches demonstrated by the players (that are the pupils and the Teacher in PR, who often represent historical mathematicians). This would be another line of inquiry to explore the non-classical elements in PR.

5.5 Conclusion

The theoretical richness of PR allows us to approach it from a variety of perspectives (Başkent, 2009, 2012; Başkent, 2015a). A game theoretical analysis of PR, using contemporary logical advancements, builds on this tradition and sheds further light on PR.

What separates the Lakatosian philosophy of mathematics from the others is how it reacts to inconsistencies and how it channels inconsistencies to mathematical discovery. It is important to underline that this procedure is paraconsistent, and can be extended to the debates within some other schools of thought in philosophy of mathematics. One can argue that what we call mathematical discovery is the result of applying a fit-for-purpose paraconsistent logic to inconsistencies at hand. The examples we discussed illustrate how Lakatos's paraconsistent logic works. It is not difficult to apply the same approach to other methods of mathematical discovery in order to unearth the game theoretical elements in them.

Reasoning with strategic inconsistencies and inconsistent strategies allows us to present a new body of evidence for the dialetheic agenda. In addition to epistemological, truth theoretical and logical arguments for it, dialetheism can also be approached from a game theoretical perspective. This paper aims at presenting a case study to establish an argument *for* dialetheism as well as shedding light on the Lakatosian methodology from a non-classical logical perspective.

This work fits within a broader research programme, called "Paraconsistent Games". Behavioural economics, applied psychology and artificial intelligence provide a plethora of cases where rational people (and machines) may behave, reason or think inconsistently in a coherent way. However, it is important to reinforce this line of thought using ideas from the philosophy of practice-based mathematics, particularly the Lakatosian thought. This allows us to see the remits of the aforementioned research programme as well as the not-well-studied aspects of the Lakatosian methodology. Discussing rationality within this context enriches such debates. We leave it for future work.

There is more to be done. An examination of the direct connection between the method of proofs and refutations and dialetheism within the context of practice-based mathematics is an immediate next step. The way that contradictions play a *constructive* role in the generation and discovery of mathematical knowledge is important for dialetheism and formal theories of rationality. Second, on a broader scale, relating PR to logical pluralism by examining the constructivist elements in it remains a big task. Third, it is important to compare and contrast our approach with various other revisionist philosophies of mathematics, including Hintikka's, as we already touched upon. Fourth, as argued earlier very briefly, game theory is the study of strategies and rational behaviour. A "behavioural" approach to PR would allow us to understand the very mathematical methodologies used to construct various proofs of the Euler's Conjecture. Characterising, for instance, Cauchy's approach to the conjecture and its proof using game theoretical elements, and then contrasting it with the strategy followed by, for instance, Lhuilier or Gergonne would shed some light on PR—both historically and mathematically.

Such future work would enrich the debates in both non-classical logic and the philosophy of mathematics, providing a long-overdue connection.

Acknowledgments This paper is a much expanded version of a talk delivered at the Imre Lakatos Centenary Conference at the London School of Economics in November 2022. I am very thankful to the organisers.

I very much appreciate the patient comments of the anonymous referee(s), which helped improve the paper.

A substantially earlier version of this text was presented at a special session on "*How to Solve It?*: Heuristics and Inquiry Based Learning" at the American Mathematical Society's Fall Western Sectional (remote) meeting in October 2020. I am grateful to the organisers for the opportunity.

After his mere 51 years in this world, I still find inspiration and research ideas in Lakatos's work, particularly in *Proofs and Refutations*, which I think is the best gift given to mathematicians from a philosopher.

References

Abramsky, S., & Zvesper, J. (2015). From Lawvere to Brandenburger-Keisler: Interactive forms of diagonalization and self-reference. *Journal of Computer and System Sciences, 81*(5), 799–812.

Ariely, D. (2008). *Predictably irrational: The hidden forces that shape our decisions*. Harper-Collins.

Ariely, D. (2010). *The upside of irrationality*. Harper.
Başkent, C. (2009). A geometrical - epistemic approach to Lakatosian heuristics. In C. Drossos, P. Peppas, & C. Tsinakis (Eds.), *Proocedings of Seventh Panhellenic Logic Symposium* (pp. 8–14). Patras University Press.
Başkent, C. (2012). A formal approach to Lakatosian heuristics. *Logique et Analyse, 55*(217), 23–46.
Başkent, C. (2015a). Inquiry, refutations and the inconsistent. *Perspectives on interrogative models of inquiry - Developments in inquiry and questions* (pp. 57–71). Springer.
Başkent, C. (Ed.). (2015b). *Perspectives on interrogative models of inquiry: Developments in inquiry and questions*. Springer.
Başkent, C. (2016a). Game theoretical semantics for some non-classical logics. *Journal of Applied Non-Classical Logics, 26*(3), 208–239.
Başkent, C. (2016b). Towards paraconsistent inquiry. *Australasian Journal of Logic, 13*(2), 21–40.
Başkent, C. (2017). A non-classical logical approach to social software. In C. Başkent, L. Moss, & R. Ramanujam (Eds.), *Rohit Parikh on logic, language and society* (pp. 91–109). Springer.
Başkent, C. (2018). A Yabloesque paradox in epistemic game theory. *Synthese, 195*(1), 441–464.
Başkent, C. (2020). A game theoretical semantics for a logic of nonsense. In J. F. Raskin & D. Bresolin (Eds.), *Proceedings of the 11th International Symposium on Games, Automata, Logics, and Formal Verification (GandALF 2020)* (Vol. 326, pp. 66–81). Electronic Proceedings in Theoretical Computer Science.
Başkent, C., & Bağçe, S. (2009). An examination of counterexamples in *proofs and refutations*. *Philosophia Scientiae, 13*(2), 3–20.
Başkent, C., & Henrique Carrasqueira, P. (2020). A game theoretical semantics for a logic of formal inconsistency. *Logic Journal of the IGPL, 28*(5), 936–952.
Beall, J. C., & Restall, G. (2006). *Logical pluralism*. Clarendon Press.
Brandenburger, A., & Keisler, H. J. (2006). An impossibility theorem on beliefs in games. *Studia Logica, 84*, 211–240.
Brandenburger, A., Friedenberg, A., & Keisler, H. J. (2008). Admissibility in games. *Econometrica, 76*(2), 307–352.
Bueno, O., & da Costa, N. C. A. (2007). Quasi-truth, paraconsistency, and the foundations of science. *Synthese, 154*, 383–399.
Bueno-Soler, J., & Carnielli, W. A. (2016). Paraconsistent probabilities: consistency, contradictions and Bayes' theorem. *Entropy, 18*(9), 325.
Carnielli, W. A., & Coniglio, M. E. (2016). *Paraconsistent logic: Consistency, contradiction and negation*. Springer.
Corfield, D. (1997). Assaying Lakatos's history and philosophy of science. *Studies in History and Philosophy of Science, 28*(1), 99–121.
Ficara, E. (2013). Dialectic and dialetheism. *History and Philosophy of Logic, 34*(1), 35–52.
Harsanyi, J. C. (1967). Games with incomplete information played by "Bayesian" players: I. The basic model. *Management Science, 14*(3), 159–182.
Heifetz, A. (1996). Non-well-founded type spaces. *Games and Economic Behavior, 16*, 202–217.
Hintikka, J., & Sandu, G. (1997). Game-theoretical semantics. In J. van Benthem & A. ter Meulen (Eds.), *Handbook of logic and language* (pp. 361–410). Elsevier.
Hodges, W. (2013). Logic and games. In E. N. Zalta (Ed.), *The Stanford encyclopedia of philosophy*.
Kahneman, D. (2011). *Thinking, fast and slow*. Farrar, Straus and Giroux.
Koetsier, T. (1991). *Lakatos' philosophy of mathematics: A historical approach*. North-Holland.
Kvasz, L. (2002). Lakatos' methodology between logic and dialectic. In G. Kampis, L. Kvasz, & M. Stölzner (Eds.), *Appraising Lakatos: Mathematics, methodology and the man*. Kluwer.
Lakatos, I. (1979). *Mathematics, science and epistemology*. Cambridge University Press.
Lakatos, I. (2015). *Proofs and refutations. Cambridge philosophy classics*. Cambridge University Press.
Larvor, B. (1998). *Lakatos: An introduction*. Routledge.

Mann, A. L., Sandu, G., & Sevenster, M. (2011). *Independence-friendly logic*. Cambridge University Press.

Mares, E. D. (1997). Paraconsistent probability theory and paraconsistent Bayesianism. *Logique et Analyse, 40*(160), 375–384.

Mares, E. D. (2002). A paraconsistent theory of belief revision. *Erkenntnis, 56,* 229–246.

Mortensen, C. (1994). *Inconsistent mathematics*. Springer.

Mortensen, C. (2010). *Inconsistent geometry*. College Publications.

Musgrave, A., & Pidgen, C. (2021). Imre Lakatos. In E. N. Zalta (Ed.), *The Stanford encyclopedia of philosophy*.

Pacuit, E. (2007). Understanding the Brandenburger-Keisler belief paradox. *Studia Logica, 86*(3), 435–454.

Priest, G. (1989). Dialectic and dialetheic. *Science & Society, 53*(4), 388–415.

Priest, G. (2001). Paraconsistent belief revision. *Theoria, 67*(3), 214–228.

Priest, G. (2002). Paraconsistent logic. In D. Gabbay & F. Guenthner (Eds.), *Handbook of philosophical logic* (Vol. 6, pp. 287–393). Kluwer.

Priest, G. (2007). Paraconsistency and dialetheism. In D. M. Gabbay & J. Woods (Eds.), *Handbook of history of logic* (1st ed., Vol. 8, pp. 129–204). Elsevier.

Priest, G. (2023). The logical structure of dialectic. *History and Philosophy of Logic, 44*(2), 200–208.

van Benthem, J., & Klein, D. (2022). Logics for analyzing games. In *The Stanford encyclopedia of philosophy*.

Weber, Z. (2021). *Paradoxes and inconsistent mathematics*. Cambridge University Press.

Open Access This chapter is licensed under the terms of the Creative Commons Attribution-NonCommercial-NoDerivatives 4.0 International License (http://creativecommons.org/licenses/by-nc-nd/4.0/), which permits any noncommercial use, sharing, distribution and reproduction in any medium or format, as long as you give appropriate credit to the original author(s) and the source, provide a link to the Creative Commons license and indicate if you modified the licensed material. You do not have permission under this license to share adapted material derived from this chapter or parts of it.

The images or other third party material in this chapter are included in the chapter's Creative Commons license, unless indicated otherwise in a credit line to the material. If material is not included in the chapter's Creative Commons license and your intended use is not permitted by statutory regulation or exceeds the permitted use, you will need to obtain permission directly from the copyright holder.

Chapter 6
Extending Heuristics: Discovery in Logic, Mathematics, and the Sciences

Otávio Bueno

Abstract Central to Imre Lakatos's philosophy is the methodological continuity between mathematics and the empirical sciences, especially the central role played by counterexamples (refutations) in both fields. It is unclear, however, to what extent the original framework conceived in *Proofs and Refutations* can be extended beyond the particular case study that motivated it in the first place. In this paper, I argue that, with suitable adjustments, the framework not only illuminates important aspects of the discovery of new logics (such as intuitionistic, paraconsistent and quantum logics), but it also can have a far more extensive role than Lakatos himself seems to have acknowledged in the empirical sciences as well.

Keywords Heuristics · Mathematical discovery · Scientific discovery · Nonclassical logics · Lakatos

6.1 Introduction

A crucial feature of Imre Lakatos's philosophy of mathematics is its emphasis on the ways in which mathematics and the sciences are not as far apart as the traditional conception of mathematics suggests (Lakatos, 1976, 1978b). According to this conception, mathematics is a body of immutable, cumulative truths, which increases by the deductive addition of certainty upon certainty, including axioms that are taken to be, if not self-evident, at least constitutively true. In contrast, the sciences evolve dynamically largely in response to empirical refutations from bold hypotheses that increase the content of what was previously taken to be the case, as characterized in Popper (1963).

Lakatos's innovation consists in identifying a process through which mathematical concepts and theories evolve, indicating along the way how this process is remarkably similar to what goes on in the empirical sciences. On his view,

---

O. Bueno (✉)
Department of Philosophy, University of Miami, Coral Gables, FL, USA

proofs do not prove but improve naïve conjectures *via* a sophisticated process involving the careful interplay of naïve mathematical conjectures, counterexamples, proof analyses, lemma reformulations, and resulting improved conjectures (Lakatos, 1976). The complex ways in which concepts are refined in response to counterexamples is insightfully illustrated and examined in the context of a particular case study regarding the classification and characterization of polyhedra. Throughout, the crucial notion of heuristics—procedures aiming to the discovery of novel results *via* problem-solving—plays a decisive role.

When Lakatos turned his attention to the philosophy of science, he brought some of these constituents to the interpretation of scientific practice and the (rational reconstruction of the) history of science. As is well known, negative and positive heuristics are two central components of the methodology of scientific research programs (Lakatos, 1978a). Both heuristic procedures are concerned with strategies aimed at the generation of new hypotheses. In the case of negative heuristics, the aim is to protect the hard core of a research program, which is formed by the constitutive hypotheses whose rejection would ultimately dismantle the program itself. In this instance, the heuristics involves the injunction never to direct potential refutations to the program's hard core, but rather identify auxiliary hypotheses that could take the blame. Of course, some of these hypotheses eventually need to be empirically supported—in the sense that they withstand the outcome of fierce empirical testing and are not empirically refuted straightaway—otherwise, in the long run, the program will degenerate. In the case of positive heuristics, the aim is to articulate the program further with the goal of formulating novel predictions—predictions the program was not initially designed to make but is able to yield given its theoretical resources. Once again, it will be crucial that some of these predictions eventually receive some empirical support—that is, they are not refuted after a series of increasingly probing tests—to prevent the research program from moving into a degenerative phase.

Central to both kinds of heuristics is the treatment of refutations, whether potentially directed to the hard core or to the auxiliary hypotheses that form a program's protective belt. It is similarly crucial to mathematical discovery, as described in Lakatos's dialogue on mathematical heuristics (Lakatos, 1976), that refutations—or counterexamples as they are referred to in mathematics—be treated in ways that increase the content of the relevant conjectures. This means that revised and refined mathematical hypotheses about the domain under study rule out more possibilities than the initial conjectures did. This process is vividly and insightfully described in the case study Lakatos (1976) offers. A significant concern is to determine the extent to which the original framework of mathematical discovery articulated in this work can be extended beyond the specific case study that motivated it in the first place, whether within mathematics or beyond.

In this paper, I argue that, with suitable adjustments, the framework not only illuminates important aspects of the discovery of new logics (such as intuitionistic, paraconsistent and quantum logics), but it also has a far more extensive role than Lakatos himself seems to have acknowledged in the empirical sciences as well.

6.2 Heuristics, New Logics, and the Application of Mathematics

From a historical point of view, logic is not a field especially known for constant changes. Although there have been conceptual and inferential developments since Aristotle first conceived and systematized the area (Kneale & Kneale, 1984), arguably the major substantial change had to wait until Gottlob Frege's invented mathematical logic near the end of the nineteenth century, setting the field up on an extremely fruitful formalization path (van Heijenoort, 1967). Motivated by the goal of systematizing inferences in mathematics, the traditional conception of logic characterized the field as a domain of necessary, universal, and formal inferences centered around the concept of validity. After Frege, with the development and dissemination of the model-theoretic conception, validity has been ultimately characterized in terms of (set-theoretic) models. An argument is valid provided that in every model in which the premises are true, so is the conclusion. Models, being typically taken to be abstract objects, are thought to have the required generality both in terms of their range and in terms of what they encode. As a result, at least in principle, they are taken to be just the right devices to ensure the necessity, universality, and formality of the logical principles they are supposed to sanction (for some concerns, see Etchemendy, 1990).

Classical logic, of course, is far more than a collection of logical principles and inference rules. It also comes with a particular conception of its subject matter. Conceived in terms of principles that are universal, necessary, and formal, classical logic has been modeled, at least since its invention by Frege, on the traditional understanding of mathematics—a subject that is taken to be focused on immutable, abstract, necessary objects. By being true in every model, logical principles are taken to hold no matter what—that is, quite independently of any specificity connected with any given domain. It is this domain independence—which is typically formulated in terms of topic neutrality—that confers logical principles and rules both universality and necessity. Universality emerges since the rules and principles in question are taken to apply to any objects in any domain: the truth of the relevant principles remains invariant under any variation of the models. Necessity results from the fact that the principles are supposed to apply in every possible situation; in other words, they are designed to go through no matter what. Formality, in turn, obtains since an inference's validity is secured from the logical form of the premises and conclusion of the argument under consideration. In other words, the form of the argument alone is supposed to guarantee that the truth of the premises leads to the truth of the conclusion (a searching examination of the issue can be found in Dutilh Novaes, 2012).

Despite their impressive overall success, certain principles and rules from classical logic face counterexamples, which clearly question their general tenability. Several non-classical logics have been designed in response to these counterexamples and in recognition of the limitations they display regarding the proper scope of logical principles. Central to this process is the careful reflection on the proposed counterexamples and how they undermine established logical principles. As might be expected, from the perspective of the classical logician, these are not genuine counterexamples but monstrosities that need to be banned. Given classical logic's necessity and universality, no such counterexamples are possible—even in principle.

Those who think that the counterexamples are real argue, instead, that the refutations that challenge the logical principles indicate genuine possibilities. The counterexamples prompt a reflection on the nature of logical constants and suggest refinements on the way the principles that govern their operation are conceived and understood—with the eventual introduction of more perspicuous constants and refined theorems about them. In this way, new logics can be, and have been, introduced.

Consider, for instance, the principle of *Explosion*, according to which any statement in the relevant language follows from a contradiction. It is, of course, a valid inference in classical logic on the (model-theoretic) grounds that since nothing can be a model of contradictory premises, there is no model of such premises in which the premise is true and the conclusion false—no matter what the conclusion is. Proof-theoretically, a simple, well-known derivation suffices to establish the principle:

1. $P \wedge \neg P$ (Assumption)
2. P (1, Conjunction Elim.)
3. $P \vee Q$ (2, Disjunction Intro.)
4. $\neg P$ (1, Conjunction Intro.)
5. Q (3, 4, Disjunctive Syl.)

Clearly, there is no connection between premises and conclusion. In fact, one could validly derive that the Moon is made of blue cheese from any contradiction about numbers, such as the original logicist reformulation of arithmetic that Frege provided.

If relevance considerations are taken into account, however, and some connection between premises and conclusion is required, Explosion becomes far less obvious. It is unclear why a conclusion that bears no connection with the premises should logically follow from them. This is a point that relevant logicians (and, to a certain extent, paraconsistent logicians as well) do insist on. By devising a three-valued semantics (Priest, 2006), the truth-conditions for logical connectives, including negation and disjunction, can be refined, thus making room for possibilities that are precluded by the two-valued semantics associated with classical logic. This enriched context allows for the reexamination of classically valid arguments and the articulation of counterexamples that ultimately question Explosion's validity. On the three-valued semantics for the paraconsistent logic LP (Logic of Paradox), three truth-values are recognized: true (t), false (f), and both true and false $\{t, f\}$, with t being the designated value (Priest, 2006). The truth conditions and the falsity conditions for the connectives are precisely those from classical logic, with the difference that such conditions need to be specified separately, since with three truth-values, the negation of the connectives' truth conditions need not be equivalent to their falsity conditions. Now, suppose that 'P' is a statement that is both true and false, such as the Liar sentence ("This sentence is not true"). In this case, 'P' is, in particular, true, and so is '$\neg P$', since it is similarly both true and false. Suppose that 'Q' is (just) false. As a result, in the derivation of Explosion above, Disjunctive Syllogism fails, given that '$P \vee Q$' is true, on the grounds of the truth of P, and '$\neg P$' is similarly true, but 'Q' is false.

The result is a more refined account of logical constants and logical consequence, leading to the formulation of relevant and paraconsistent logics. In contrast with classical logic, these logics do not identify contradiction with triviality. In this way, they allow for the accommodation of inferences involving inconsistency without logical anarchy (for further details, references, and a survey of different systems of paraconsistent logics, see da Costa et al., 2007).

In other words, a revision in the semantics allows for the reinterpretation of a traditional proof of the validity of a principle of classical logic (Explosion) and the introduction of a more refined understanding of negation (paraconsistent negation) and logical consequence (a paraconsistent consequence relation). This, in turn, makes room for the expression of possibilities that are foreclosed by classical negation—reasoning nontrivially from inconsistencies—which, in light of the validity of Explosion, cannot get off the ground in classical logic. New logical constants and corresponding logics can be formulated as a result.

Note that the articulation of more refined concepts of negation and of logical consequence emerge from considerations of counterexamples. It is the counterexample to the validity of Explosion that opens the way to the possibility of formulating a logic in which inconsistency is not identified with triviality—a characterizing feature of paraconsistent logics in contrast with classical logic. One of the central motivations that, in the 1960s, led Newton da Costa to develop an entire hierarchy of paraconsistent logics (the C logics) resulted precisely from counterexamples of this kind. As he notes, when faced with inconsistencies in our theories:

> Usually, we try to change the inconsistent theories to transform them into consistent ones. It is clear that under this transformation, some characteristic properties of a given inconsistent theory must be preserved; for instance, the common formal systems of set theory preserve certain traits of inconsistent naïve set theory (da Costa, 1974, p. 497).

The recognition of an inconsistency in one's theory is often the expression of an underlying counterexample. Consider set theory—a central motivation for the development of paraconsistent logics in da Costa's hands. The unrestricted comprehension principle that is characteristic of naïve set theory states that:

$$\exists y \forall x \ (x \in y \leftrightarrow Fx).$$

Despite its initial plausibility, the principle is unquestionably inconsistent. If the formula 'F' is interpreted as '$x \notin x$', Russell's paradox is immediately derived. For, the set $R = \{x : x \notin x\}$ is both a member of itself ($R \in R$) and not a member of itself ($R \notin R$). This fact yields a significant counterexample to the suggestive, but ultimately troublesome, principle to the effect that every property determines a set (of the objects that have that property). The need then emerges to find a proper refinement of the naïve comprehension principle so that some of its plausibility is preserved without the unwanted contradictions. It is in this sense that da Costa notes that "common formal systems of set theory preserve certain traits of inconsistent naïve set theory" (da Costa, 1974, p. 497): the comprehension principle is, in part, preserved—with suitable conceptual and inferential refinements.

How can set theorists restrict the naïve comprehension principle along these lines? Various strategies are available to achieve that. Here are three of them (da Costa & Bueno, 2001):

(a) In Zermelo-Fraenkel (ZF) set theory, the comprehension principle is restricted to properties whose bearers are members of sets:

$$\exists y \forall x \ (x \in y \leftrightarrow (Fx \wedge x \in z)).$$

The restriction is quite substantial, and several additional axioms need to be introduced to ensure that the resulting theory has enough resources to characterize mathematics more broadly (see Jech, 2006).

(b) In Kelley-Morse (KM) set theory, the restriction is formulated somewhat differently by invoking properties whose bearers are members of a given set:

$$\exists y \forall x \ (x \in y \leftrightarrow (Fx \wedge \exists z \ (x \in z))).$$

This allows for the introduction of classes, in addition to sets, and expand the ontology of set theory in a significant direction. As opposed to what goes on in Zermelo-Fraenkel set theory, a universal set can be formulated and studied (Kelley, 1955).

(c) In W.V. Quine's New Foundations for Mathematical Logic (NF), the concept of stratification is advanced, and much of naïve comprehension is preserved—at least on the surface (Quine, 1937). The restriction goes as follows:

$$\exists y \forall x \ (x \in y \leftrightarrow Fx), \text{ as long as } `Fx' \text{ is stratified (in addition to the standard conditions on the variables)}.$$

The notion of stratification entails that a formula such as '$x \in x$' is not stratified (and, thus, Russell's paradox is avoided), while '$x \in y$' is stratified. NF is an intriguing set theory which, as opposed to both ZF and KM, is incompatible with the axiom of choice (Quine, 1937; Forster, 1995).

As is well known, these are only a few of the nonequivalent ways in which it is possible to restrict the naïve comprehension principle. All these ways provide refinements of this principle and of the underlying concept of set. But there is one additional refinement that animated da Costa in particular, motivated by a remark of David Hilbert's. In his celebrated 1900 address concerning mathematical problems, Hilbert emphasized the significance of exploring *all logical possibilities* when examining a mathematical problem. He noted:

> The mathematician will have also to take account not only of those theories coming near to reality, but also, as in geometry, of all logically possible theories. He must be always alert to obtain a complete survey of all conclusions derivable from the system of axioms assumed (Hilbert, 1902, p. 454).

One possibility that should be examined in the context of set theory consists in keeping the unrestricted comprehension principle, and its significant plausibility, intact (da Costa, 1974; da Costa & Bueno, 2001). However, to prevent triviality, a change in the underlying logic to a paraconsistent one is in order, so that inconsistent but nontrivial set theories become logically possible (da Costa et al., 2007). As a result, a strategy is advanced in which even a classical mathematician who takes seriously the significance of exploring *all* logical possibilities can adopt a nonclassical point of view. Paraconsistent logics allow for the expression of an expanded scope of what is logically possible, since the possibility of inconsistent but nontrivial theories is foreclosed by classical logic. Given that it is precisely the exploration of such theories that is at issue, changing the underlying logic is called for.

This was indeed a key motivation for da Costa's formulation of paraconsistent logic. As he points out:

> [...] there are certain cases in which we might think of studying directly an inconsistent theory. For example, a set theory containing Russell's class (the class of all classes which are not members of themselves) as an existing set (da Costa, 1974, p. 497).

If one's goal is to study properties of the Russell set, this cannot be done in classical set theories, such as ZF, KM, or NF, since all of them have been designed to prevent the formulation of this set (assuming that these theories are consistent). A framework in which inconsistencies are not rule out by fiat, but allows for their study, needs to be in place.

Along these lines, da Costa also introduced four additional constraints that shaped the formulation of the hierarchy of paraconsistent logics C_n, $1 \leq n \leq \omega$, that he proposed:

> I) In these calculi the principle of contradiction, $\neg (A \wedge \neg A)$, must not be a valid schema; II) From two contradictory formulas, A and $\neg A$, it will not in general be possible to deduce an arbitrary formula B; III) It must be simple to extend C_n, $1 \leq n \leq \omega$, to corresponding predicate calculi (with or without equality) of first order; IV) C_n, $1 \leq n \leq \omega$, must contain the most part of the schemata and rules of C_0 [classical propositional calculus] which do not interfere with the first conditions. (Evidently, the last two conditions are vague.) (da Costa, 1974, p. 498).

The first constraint indicates the significance of restricting the validity of the law of non-contradiction: otherwise, the very formulation of inconsistent sets would be ruled out. The second constraint highlights the importance of restricting Explosion:

otherwise, given the presence of inconsistencies (allowed by the first constraint), logical anarchy would result. If the first two constraints indicate restrictions on the validity of certain principles from classical logic, the following two are meant to increase the inferential resources available, within the boundaries imposed by these restrictions. The third constraint emphasizes the importance of having at least logical inferences at the predicate level to accommodate and sanction inferences in mathematics (higher-order logics are, of course, often needed as well). Finally, the fourth constraint highlights the need to preserve classical logic as much as possible while still allowing for the study of the inconsistent. This is analogous to the way in which da Costa considered naïve comprehension: classical set theories aim to preserve as much as they can of the naïve principle. Similarly, on his view, paraconsistent logics aim to preserve as much of classical logic as is feasible given the goal of studying inconsistencies without logical disorder.

Throughout this process, it becomes clear that the development of set theory and the articulation of a significant class of nonclassical logics move along parallel lines—and both crucially rely on the role of counterexamples to provide a source of conceptual and inferential refinements. The realization that naïve comprehension in set theory is inconsistent prompts the search for suitably refined set-theoretic principles; the realization that not every property determines a set prompts the search for suitably refined logical inferential devices that would allow for the study of inconsistent sets. The outcomes are more refined concepts and inferences in set theory and logic. In this way, the process described by Lakatos (1976, 1978b) in the formulation of more refined concepts in geometry and analysis—where different conceptions of polyhedron or of the continuum are involved—applies to the invention of nonclassical logics and set theories as well.

One could argue that, in the case of Explosion, there is no counterexample. The requirement of relevance is not part of classical logic's validity. The fact that the premise and the conclusion of Explosion have nothing in common does not violate the validity of the rule. Since there are no models of a contradiction in classical logic, there is no model in which a contradiction is true and the conclusion is false. And without a counterexample to begin with, a Lakatosian approach cannot get off the ground. As opposed to what happens in mathematics and the sciences, where counterexamples can be found, this is not the case in logic.

In response, the objection assumes not only classical logic's conception of validity but also its underlying semantics, none of which recognizes the possibility of a statement and its negation being true. However, this is precisely the possibility that needs to be acknowledged, as just discussed, in order to explore all logically possible theories in the context of set theory, following Hilbert's motto. In other words, by the very standards of a classical logician who acknowledges the need for exploring all logical possibilities, one needs to appreciate a possibility in the logical space that classical logic closes by fiat, namely, the possibility of violating the law of non-contradiction. It is not surprising that this is precisely the first constraint that da Costa introduced when he articulated paraconsistent logic. Once this possibility is recognized, a counterexample to Explosion immediately emerges, since the premise (a contradiction) can be true and the conclusion (an untrue sentence in the language) is false. The Lakatosian approach can then get off the ground.

Similar considerations can also be found in the study of the role of mathematics in the formulation of physical theories. In 1928, Paul Dirac found a significant differential equation that now bears his name: the Dirac equation. It offered a relativistic wave equation for the electron, accommodating the spectral features involved in an atom's emission and absorption of radiation. The equation, however, had negative energy solutions, and as is common in similar situations elsewhere in physics, Dirac initial response was to ignore these solutions, since they did not have any physically meaningful interpretation. By revising this initial reaction, Dirac would eventually manage to offer an account of the equation that is compatible with the presence of a new particle, the positron. Interestingly, he only arrived at this account after two failed attempts to make sense of the negative energy solutions: one, in 1928, in which these solutions, as noted, were ignored as being physically irrelevant, and another, in 1930, in which the negative energy solutions were interpreted as "holes" in spacetime. As it turns out, according to this "hole" theory, electrons and protons would have the same mass—something that, clearly, is empirically inadequate.

Pressured by empirical and conceptual counterexamples, in 1931, Dirac managed to advance a new interpretation of the negative energy solutions, which involved positing a new particle that has the same mass as the electron but the opposite charge. This particle's presence was eventually supported, in 1932, by empirical considerations, and it is now called "positron" (for references and details, see Schweber (1994), Bueno (2005) and Bueno and French (2018)).

Significant throughout this episode is the importance of managing counterexamples and the way in which they may lead to conceptual refinement. Ignoring a solution that lacks a proper physical interpretation is a way of addressing a counterexample that arises from the application of mathematics, provided that it is unclear what should be done with the structural surplus generated by the solutions to the relevant differential equation. However, by advancing a suitable physical interpretation, one offers a more refined alternative to the counterexample. Clearly, the proposed interpretation needs to find empirical support. Since interpreting the negative-energy solution as a "hole" in spacetime faces both conceptual and empirical challenges, it is seriously inadequate. But improving this interpretation further and positing a new particle offers a better refinement, and the fact that the novel interpretation has empirical support closes the deal.

Once again, by emphasizing the significance of conceptual refinement *via* the analysis of counterexamples, Lakatos's approach offers key resources to make sense of this episode of the application of mathematics to physics. As this case illustrates, Lakatos's framework in *Proofs and Refutations* is just as relevant to the empirical sciences as it is to mathematics.

6.3 Extending Heuristics

The considerations thus far suggest that a Lakatosian process of conceptual refinement in light of counterexamples can be in place in mathematics, in logic, and in the empirical sciences. This allows for a more refined account of heuristics beyond the negative and positive heuristics advanced in the methodology of scientific research programs (Lakatos, 1978a), which preserves Lakatos's continuing interest in mathematical and scientific discovery.

Negative heuristics, as noted above, is basically a rule to the effect that one should not direct potential counterexamples, or empirical refutations, to the principles, laws, hypotheses, or theories that form the core of a research program. Positive heuristics is a group of rules directing empirical refutations, or counterexamples, to the principles that form the protective belt of a research program so that, ideally, novel predictions are generated through the revisions required from the uncovering of counterexamples.

The crucial contribution of the dynamics of proofs and refutations is the conceptual innovation that emerges from counterexamples. Concepts are "stretched" to accommodate possibilities that were not recognized prior to the counterexamples. Lakatos's suggestive metaphor of concept "stretching" can be understood in modalist terms (Bueno, 2021; Bueno & Shalkowski, 2015): revised concepts—in light of counterexamples—express unacknowledged possibilities, possibilities whose recognition was *precluded* by previously inadequate, limited concepts. This can be illustrated with three examples:

(a) As indicated above, the study of properties of the Russell set (the set of all non-self-membered sets) is impossible in classical set theories, whose underlying logic is classical. But this study is perfectly possible in paraconsistent set theories, and their underlying paraconsistent logics, which distinguish inconsistency and triviality (da Costa et al., 2007).

(b) The study of quantum particles for which identity is not defined, given a particular interpretation of quantum mechanics, is possible only in set theories in which the identity of sets is not characterized in terms of the identity of its members. These (quasi-)set theories restrict the principle of extensionality: only the indistinguishability of the relevant particles is required, not their identity (French & Krause, 2006). The possibility of such non-individuals (objects for which identity does not apply) is precluded by classical logic's principle of identity, according to which every object is identical to itself. Quasi-set theory does not state that certain objects are not self-identical (which would be inconsistent), but rather that identity does not apply to them, which is still prevented by the identity principle. To explore this possibility, a non-reflexive logic is in order.

(c) The study of constructive features of mathematical reasoning is possible only if mathematical objects are not assumed to have certain properties just because of their completeness. In other words, it is not assumed that they either have or do not have the properties in question—a construction exhibiting that the relevant properties obtain is required (Dummett, 2000). Excluded middle undermines this possibility by ensuring that the relevant mathematical objects have the properties independently of any construction that establish that they do. To investigate this possibility, an intuitionistic logic is required.

Here the conceptual innovations emerge from the recognition of counterexamples, respectively, to the principles of explosion, extensionality, and excluded middle. New possibilities are then recognized: (a) the distinction between inconsistency and triviality in paraconsistent logics; (b) the reformulation of extensionality for indistinguishable, but distinct, objects in quasi-set theory and non-reflexive logics; and (c) the study of constructive features of mathematical reasoning in intuitionistic logics.

The recognition of new possibilities in light of counterexamples is also crucial in the sciences. Faced with an empirical refutation of his "hole" theory, according to which there are holes in spacetime, Dirac was led to conceive of a positive electron (the positron)—a theoretical possibility that would have been unconceivable prior to his own interpretation of the Dirac equation.

In some instances, it is precisely the lack of conceptual innovation in light of counterexamples that brings the downfall of a theory. For instance, Bohr's atomic model was known to be inconsistent with electrodynamics: the little planetary model was completely unstable given the latter theory. To avoid refutation, Bohr introduced a clearly *ad hoc* postulate:

> [e]nergy radiation [within the atom] is not emitted (or absorbed) in the continuous way assumed in the ordinary electrodynamics, but only during the passing of the systems between different 'stationary' states (Bohr, 1913, p. 874).

Given the inconsistency, Bohr's model was not ignored only because of its empirical success: it accounted for the spectral emission lines of the hydrogen atom. However, *despite the empirical success*, the model was eventually abandoned with the development of the new quantum theory, which accommodated more broadly and accurately the phenomena that Bohr's model aimed to account for. The fact that Bohr's model lacked conceptual resources to accommodate electrodynamics impoverished significantly the model. It was an important component in its eventual downfall.

6.4 Conclusion

In both empirical and nonempirical domains, the conceptual contribution of heuristics is central. In nonempirical domains, such as in logic and mathematics, we find: (i) the recognition of new possibilities that were previously unacknowledged is made possible by conceptual innovation, and (ii) the resulting opening of new avenues of research are signs of progress, such as in the cases of paraconsistent set theories, quasi-set theories, and constructive mathematics.

In empirical domains, such as in the sciences, the recognition of new possibilities via conceptual innovation opens the way for the generation of novel predictions, some of which will need to be borne out for progress to take place, such as in Dirac's case of the positron.

In the absence of a proper way of accommodating possibilities—when there is some lack of conceptual innovation—an empirical proposal may become unmotivated and be eventually abandoned, despite its empirical success. To a certain extent, this was the fate of Bohr's atomic model.

Whether the domains are empirical or not, conceptual innovation—the key element of the dynamics of proofs and refutations—is, thereby, crucial. In this way, Lakatos' insightful account of mathematical discovery can be extended to a much broader range of cases than perhaps considered feasible, from logics to the empirical sciences.

Acknowledgments My thanks go to Roman Frigg, Marcus Giaquinto, John Worrall, the audience at the Lakatos@100 Conference at the London School of Economics and Political Science on November 3-4, 2022, and an anonymous reviewer for this chapter for extremely helpful and perceptive comments on an earlier version of this work. Their feedback led to significant improvements.

References

Bohr, N. (1913). On the constitution of atoms and molecules. *Philosophical Magazine, 26*, 1–25. 476–502; 857–875.
Bueno, O. (2005). Dirac and the dispensability of mathematics. *Studies in History and Philosophy of Modern Physics, 36*, 465–490.
Bueno, O. (2021). Modality and the plurality of logics. In O. Bueno & S. Shalkowski (Eds.), *Handbook of modality* (pp. 319–327). Routledge.
Bueno, O., & French, S. (2018). *Applying mathematics: Immersion, inference, interpretation*. Oxford University Press.
Bueno, O., & Shalkowski, S. (2015). Modalism and theoretical virtues: Toward an epistemology of modality. *Philosophical Studies, 172*, 671–689.
da Costa, N. C. A. (1974). On the theory of inconsistent formal systems. *Notre Dame Journal of Formal Logic, 15*, 497–510.
da Costa, N. C. A., & Bueno, O. (2001). Paraconsistency: Towards a tentative interpretation. *Theoria, 16*, 119–145.
da Costa, N. C. A., Krause, D., & Bueno, O. (2007). Paraconsistent logics and paraconsistency. In D. Jacquette (Ed.), *Philosophy of logic* (pp. 791–911). North-Holland.
Dummett, M. (2000). *Elements of intuitionism* (2nd ed.). Clarendon Press.
Dutilh Novaes, C. (2012). *Formal languages in logic: A philosophical and cognitive analysis*. Cambridge University Press.
Etchemendy, J. (1990). *The concept of logical consequence*. Harvard University Press.
Forster, T. (1995). *Set theory with a universal set*. Second, revised edition. Oxford University Press.
French, S., & Krause, D. (2006). *Identity in physics: A historical, philosophical, and formal analysis*. Oxford University Press.
Hilbert, D. (1902). Mathematical Problems. *Bulletin of the American Mathematical Society, 8*, 437–479. (Translated from the German by Mary Newson. The original article was published in

Göttinger Nachrichten, 1900, pp. 253–297, and in the *Archiv der Mathematik und Physik* (third series) *1*, 1901, pp. 44–63 and 213–237.).

Jech, T. (2006). *Set theory*. (Third, revised and expanded, edition). Springer.

Kelley, J. (1955). *General topology*. D. van Nostrand.

Kneale, W., & Kneale, M. (1984). *The development of logic*. Clarendon Press.

Lakatos, I. (1976). In J. Worrall & E. Zahar (Eds.), *Proofs and refutations: The logic of mathematical discovery*. Cambridge University Press.

Lakatos, I. (1978a). The methodology of scientific research Programmes. In J. Worrall & G. Currie (Eds.), *Philosophical papers* (Vol. 1). Cambridge University Press.

Lakatos, I. (1978b). Mathematics, science and epistemology. In J. Worrall & G. Currie (Eds.), *Philosophical papers* (Vol. 2). Cambridge University Press.

Popper, K. (1963). *Conjectures and refutations: The growth of scientific knowledge*. Routledge.

Priest, G. (2006). *In contradiction: A study of the Transconsistent*. Second, expanded edition. Oxford University Press.

Quine, W. V. (1937). New foundations for mathematical logic. *American Mathematical Monthly*, *44*, 70–80.

Schweber, S. (1994). *QED and the men who made it*. Princeton University Press.

van Heijenoort, J. (Ed.). (1967). *From Frege to Gödel: A source book in mathematical logic, 1979–1931*. Harvard University Press.

Open Access This chapter is licensed under the terms of the Creative Commons Attribution-NonCommercial-NoDerivatives 4.0 International License (http://creativecommons.org/licenses/by-nc-nd/4.0/), which permits any noncommercial use, sharing, distribution and reproduction in any medium or format, as long as you give appropriate credit to the original author(s) and the source, provide a link to the Creative Commons license and indicate if you modified the licensed material. You do not have permission under this license to share adapted material derived from this chapter or parts of it.

The images or other third party material in this chapter are included in the chapter's Creative Commons license, unless indicated otherwise in a credit line to the material. If material is not included in the chapter's Creative Commons license and your intended use is not permitted by statutory regulation or exceeds the permitted use, you will need to obtain permission directly from the copyright holder.

Chapter 7
The Case of Early Copernicanism: Epistemic Luck *vs*. Predictivist Vindication

Vincenzo Crupi

Abstract From Copernicus himself up to Kepler and Galilei, Copernicans have been "right for the wrong reasons" (Finocchiaro, *Defending Copernicus and Galileo: Critical Reasoning in the Two Affairs*. Springer, Dordrecht, 2010), because *there were* no epistemically compelling reasons objectively favoring the Copernican position at that stage—a good deal of research in the history and philosophy of science has converged on this claim. The situation of early Copernicans would then be regarded as one of "epistemic luck". Lakatos and Zahar (*Why did Copernicus's research programme supersede Ptolemy's?* University of California Press, 1975) have featured in one relatively rare contemporary episode of sustained opposition to the epistemic luck thesis about early Copernicanism. Although known and appreciated in certain philosophical circles, it is fair to say that Lakatos and Zahar's predictivist vindication has remained quite unsuccessful. The goal of this paper is to revive it in updated form. My analysis will support two general claims. First and foremost, previous limitations of a predictivist account of the Copernican controversy can be amended to counter the epistemic luck thesis. And second, consideration of the vindication thesis highlights certain important but neglected elements of the historical narrative.

Keywords Copernican revolution · Epistemic luck · Predictivism · Use-novelty

---

Forthcoming in R. Frigg, J. Alexander, L. Hudetz, M. Rédei, L. Ross, and J. Worrall (eds.), *The Continuing Influence of Imre Lakatos's Philosophy: A Celebration of the Centenary of his Birth*. Springer.

V. Crupi (✉)
Department of Philosophy and Education Sciences, University of Turin, Torino, Italy
e-mail: vincenzo.crupi@unito.it

7.1 Introduction: The Epistemic Luck Thesis

From Copernicus himself up to Kepler and Galilei, Copernicans have been "right for the wrong reasons" (Finocchiaro, 2010), because *there were* no epistemically compelling reasons objectively favoring the Copernican position at that stage—a good deal of research in the history and philosophy of science has converged on this claim. In the jargon of contemporary analytic epistemology, the situation of early Copernicans would then be regarded as one of *epistemic luck*. Roughly, epistemic luck characterizes an agent who happens to have a true belief without adequate justification.[1] The precise scope of the *epistemic luck thesis* about early Copernicanism may vary significantly. For our present purposes, it is safe to focus on a version of the thesis which appears particularly sound and popular. According to such version, Copernicanism has been a matter of epistemic luck *at least* from 1543 (the publication of Copernicus's *De Revolutionibus*) up to, say, 1600, namely a moment in which the Copernican allegiance of both Kepler and Galilei is already documented while their own scientific achievements in astronomy were yet to come. Some authors would be happy to say that Copernicanism eventually got to be vindicated with Newton, as it was subsumed under a more comprehensive theory of unrivalled success (e.g., Salmon, 1990, p. 190). Others might want to insist that heliocentric astronomy remained ultimately unsteady until more "direct" and "physical" evidence of the Earth's motion became available in the eighteenth and nineteenth centuries (see Graney, 2015, ch. 10).

The epistemic luck thesis may perhaps look puzzling to those who are not much versed in the specialized scholarship. After all, its impact on textbooks, encyclopedias, and the general culture has been limited. The celebration of Copernicus, Kepler, and Galilei as founding figures for modern science has survived the Twentieth century largely unscathed and remains prevalent to this day. However, the presence of the epistemic luck thesis among experts actually has a sensible explanation. An advocate of the epistemic luck thesis would typically argue along the following lines, with a remarkable combination of historical and philosophical insights. First, they would note that each major phenomenon about the heavens that was known around 1600 and could be logically derived from the Copernican system could also be derived from an alternative, geostatic system, and viceversa (of course, with relevant background assumptions). They would go on to point out that, therefore, the Copernican view could not be distinguished from competing approaches on the basis of the available empirical evidence. As concerns further potentially discriminating criteria ("simplicity" is of course a recurrent example), they would deny that such criteria could have served as an effective basis to favor Copernicanism as objectively and epistemically superior to its competitors at least

---

[1] On the assumption that Copernicanism is fundamentally correct, the most relevant specification is probably *veritic* (epistemic) luck: "a person S is *veritically lucky* in believing that p in circumstances C iff, given S's evidence for p, it is just a matter of luck that S's belief that p is true in C" (Engel, 2022, p. 36).

at the time when Kepler and Galilei decided to join the Copernican camp (pointing out, for instance, that minor epicycles remained a customary and pervasive device in Copernicus' work, or that other virtues would support geocentrism instead).

The textual evidence about the popularity of the epistemic luck thesis is sparse but consistent, spanning now more than a century. According to Pierre Duhem's thoughtful discussion in *To Save the Phenomena*, a considered attitude of antirealism fostered by the astronomical tradition led competent observers such as Andreas Osiander and Cardinal Bellarmine to duly appreciate that heliocentric and geocentric systems were empirically on a par at the time, and therefore scientifically on a par too. As Duhem famously and firmly concluded, we are "compelled to acknowledge and proclaim that logic sides with Osiander, Bellarmine, and Urban VIII, not with Kepler and Galilei – that the former had understood the exact scope of the experimental method and that, in this respect, Kepler and Galilei were mistaken" (Duhem, 1908, p. 113). Fifty years on, another seminal reference is of course Thomas Kuhn. In a key passage of *The Copernican Revolution*, he notes that "each argument" originally put forward by Copernicus "cites an aspect of the appearances that can be explained by *either* the Ptolemaic *or* the Copernican system". The insistence of Copernicus on the greater "harmony" of heliocentrism, Kuhn points out, could only be appealing to a "limited and perhaps irrational subgroup of mathematical astronomers". Only in hindsight can one appreciate that some of them "fortunately" did follow their "Neoplatonic ear" (Kuhn, 1957, p. 181). And a major theme of Kuhn's view of science is of course that one should strenuously resist turning the benefit of scientific hindsight into a form of hindsight bias in historical matters.

Notably, unlike other implications of Duhem's or Kuhn's work, the epistemic luck thesis about early Copernicanism does not seem to have lost ground over time.[2] As recently as 2011, historian Robert Westman introduced his impressive reconstruction of *The Copernican Question* noting that "Copernicus had opened a question [...] which previously had not been seen to possess far-reaching consequences: how to choose between different models of heavenly motion *supported indifferently* by the same observational evidence" (Westman, 2011, p. 5, emphasis added). Recent extensive work on anti-Copernican astronomy *after* Kepler and Galilei (especially the interesting case of Riccioli, 1651) yielded even stronger claims, if anything. According to Graney, for instance, "in the middle of the seventeenth century [...] science backed geocentrism" (Graney, 2015, pp. 144–145; and also see Marcacci, 2015). As for late Twentieth century philosophy of science, Wesley Salmon provides a striking example: "until Newton's dynamics came upon the scene, it seems to me, Thyco's [geostatic] system was clearly the best available theory" (Salmon, 1990, p. 190). And physicists themselves are apparently no exception: according to Carlo Rovelli, for instance, "Kepler trusted Copernicus'

---

[2] Swerdlow (2004, p. 88) seems to offer a forceful but occasional exception: "There is altogether too much literature today—ultimately, I think, inspired by Duhem and his nonsense about 'saving the phenomena'—that holds that Copernicus had no good reasons to believe his theory to be a true description of the world. He had very good reasons and quite a lot of them."

theory *before* its predictions surpassed Ptolemy's" (Rovelli, 2019, p. 120; also see Timberlake & Wallace, 2019, pp. 144–145).

Imre Lakatos and Elie Zahar have featured in one relatively rare contemporary episode of sustained opposition to the epistemic luck thesis about early Copernicanism. Although known and appreciated in certain philosophical circles, it is fair to say that their attempt has remained quite unsuccessful. In essence, the goal of this paper is to revive it.

7.2 Background and Outline

Lakatos and Zahar (1975) carried out a comprehensive methodological assessment of the Copernican revolution. In their view, Copernicus' programme had a remarkable amount of "*immediate* support" from known phenomena that was not matched by the traditional geostatic approach, even if both parties were able to account somehow for all essential facts established in the late Sixteenth century. This showed that "there were good objective reasons for Kepler and Galilei to adopt the heliostatic assumption" (Lakatos & Zahar, 1975, p. 188). Such claim, I submit, clashes with the epistemic luck thesis. According to Lakatos and Zahar's analysis, it is not the case that Kepler and Galilei (not even Copernicus, in fact) were "right for the wrong reasons", or at least not because good reasons were lacking. To this extent, given the information that was actually available in the relevant historical context, it was *not* just a matter of luck that their theoretical allegiance turned out to be correct. It was instead a matter of plausible epistemic justification through sound scientific methodology. Let us call this the *vindication thesis*.

Lakatos and Zahar's analysis did not remain unchallenged, however. According to Thomason (1992), in particular, their methodological reconstruction can not justify the desired conclusion. Given that no articulated criticism of Thomason (1992) has emerged in more than 30 years, one is tempted to see such apparently old-fashioned controversy as settled, and perhaps too arcane to be reconsidered anyway. Yet this temptation should be resisted, I submit, if one is concerned about the epistemic luck thesis. In fact, Thomason did not challenge Lakatos and Zahar's dismissal of alternative philosophical reconstructions of the Copernican arguments. This, in turn, leaves the early Copernicans' choice with no plausible methodological justification. Moreover, there is at least one major reason why this case deserves a fresh look: in fact, the arguments in the debate may well need to be assessed anew as concerns their philosophical grounds. Let us briefly see why.

Both Lakatos and Zahar (1975) and Thomason (1992) relied on Lakatos's methodology of scientific research programmes as integrated by Zahar's "new conception" of "novel fact" (Lakatos & Zahar, 1975, p. 185). Such conception had been put forward in Zahar (1973) as a solution to what arguably is the first clear statement of the "problem of old evidence", drawing from the now classic historical case of Einstein's general theory of relativity and the Mercury perihelion. But a good deal of additional work has been done meanwhile on this crucial topic (see Barnes,

2022, for a valuable survey). This raises the question whether the integration of an updated analysis of the accommodation *vs.* prediction distinction may help revamp Lakatos and Zahar's original verdict and subvert the premises of the epistemic luck thesis, thus vindicating the vindication thesis.

The rest of this contribution is organized as follows. First, we will have to set the stage for an assessment of Lakatos and Zahar's line of argument in updated form. This will include a revised discussion of the use-novelty of empirical facts in science, which actually amounts to a relatively new tentative demarcation between empirical success and mere accomodation of known phenomena. The next section will lay out such proposal and also provide a characterization of the two contenders, namely, Copernicanism and Sixteenth century geocentrism. The methodological exercise will then ensue, outlining a replay of the original debate between Lakatos and Zahar and Thomason. The results will lead us to explore some implications of the vindication thesis and how it arguably sheds light on certain important methodological remarks to be found in the writings of leading figures of the Copernican controversy. In the last part of our discussion, however, we will see how the case of *ancient* heliocentrism may represent a significant, if indirect, challenge to vindicationism. This issue does not seem to have received due attention in the philosophical literature, but it may turn out to be no less serious and interesting than more familiar topics of contention.

7.3 Logical Predictivism?

Let S be a set of empirical findings established by scientific observation and let T be a theory (virtually *any* theory) postulating principles, structures, and/or processes underlying the "phenomena" encoded in S. As it turns out, it is a crucial fact of the philosophical analysis of science that, as a matter of logic, it will always be possible to derive all elements in S as consequences of a "theoretical cohort" integrating T with a relevant set of auxiliary assumptions. But this means that an alternative theory T^* could also be aligned with S in the same way, namely as embedded in a suitable theoretical cohort.[3] Duhem (1906) is of course a seminal source for this paramount methodological circumstance (see Laudan, 1990, p. 274, for a more recent statement), which also serves as an undisputed starting point for Lakatos and Zahar. As they say, "any two rival research programmes can be made observationally equivalent by producing observationally equivalent falsifiable versions of the two with the help of suitable *ad hoc* auxiliary hypotheses" (Lakatos & Zahar, 1975, p. 180).[4] *Duhemian corollary* will work as a convenient shorthand for this statement.

---

[3] One such expanded set including theoretical principles and various auxiliary assumptions is sometimes just called a "system". "Theoretical cohort" is a nice terminological variant which I draw from Strevens (2020).

[4] Here, by "observationally equivalent" one should read "such that all *known* observable facts are accounted for by each theory as embedded in its own theoretical cohort".

Zahar's "new conception" of "novel fact" was meant to go beyond this kind of "uninteresting" empirical equivalence and to specify how the same evidence may still give more support to one theory against another "depending on whether the evidence was, as it were, 'produced' by the theory or explained in an *ad hoc* way". In what follows, much in line with important work by Worrall (2002, 2006), I will employ a minimal implementation of use-novelty which—unlike Zahar's (1973)— squarely avoids reliance on dubious psychological and historical contingencies such as "the reasoning which [the scientist] used to arrive at a new theory" (Zahar, 1973, p. 219). Consider the following, admittedly basic, characterization of an observable fact F as *strongly* confirming a scientific theory T:

(a) there exist other observable facts, E, such that F follows from T and E; but
(b) F does *not* follow from T alone; and
(c) E and F are logically independent.

Each one of clauses (a)–(c) should be meant to apply on the background of further contextually unchallenged assumptions.[5] On this basis, there are two key scenarios in which a researcher will be able to conclude that T is strongly confirmed by F. One amounts to purely temporal novelty: here, the elements in E happen to be already known at a given moment, F is logically derived and *then* established by observation. (In an experimental setting, for instance, the facts in E will typically reflect certain conditions that have been purposely designed and realized in order to check for the occurrence of F, which is expected under those conditions on the basis of T, and ideally not otherwise.) But a situation in which *both* E and F happen to be known is just as much compatible with the fulfilment of (a)–(c), and it arguably captures the idea of so-called *use-novelty*.[6] In Zahar's original cornerstone case, for instance, observationally established facts about the solar system turn out to be sufficient and non-redundant to derive from Einstein's theory of general relativity the already known and otherwise independent fact of Mercury's precessing perihelion and its observable consequences. As all three clauses above are satisfied

---

[5] The historical evidence in the philosophy of science suggests that a definition of this kind must be liable to charges of triviality. What if T amounts to the combination of $E \supset X$ and $X \supset F$ for arbitrary X and otherwise independent phenomena E and F, for instance? Or what if T combines arbitrary X with the factitious auxiliary $X \supset (E \supset F)$? Here I will not try to develop a formal treatment to neutralize all such frivolous counterexamples (although a subtle potential triviality objection raised by Jason Alexander helped me with the formulation of clause (c)). They will be of no consequence for the subsequent discussion, however. In all cases of interest for us, T will include categorical and unverifiable claims about the world (such as "the Earth revolves around the Sun") that are relevant in the derivation of F from T and E. See Lange's (2004, p. 208) objection to Myrvold (2003) for a related debate.

[6] As far as I can tell, a confirmation theorist who relies on (a)–(c) will elude the troubles raised by Votsis (2014) for "incidental predictivists". Consider the potentially problematic hypothetical case of two scientists A and B such that A derives known fact X from T and known fact Y whereas B derives known fact Y from T and X. If clauses (a)–(c) are satisfied in both cases, my proposal implies that *both X and Y* strongly confirm T. So Votsis's objections do not seem to apply here (Votsis, 2014, pp. 75–76).

in this case, evidence about Mercury's perihelion qualifies as an empirical success of the theory regardless of whether Einstein himself may have hoped or even planned to address that problem better than it was handled by classical Newtonian means (see Earman & Janssen, 1993, for a thorough reconstruction). Another related way to look at clauses (a)–(c) is to see them as implying $T \vDash E \supset F$ but ruling out each of $\vDash E \supset F$, $T \vDash E$, and $T \vDash F$. This may be regarded as a situation in which the *connection itself* between E and F is made sense of by T, not the brute fact of their joint occurrence.[7]

To be sure, this characterization is fully consistent with the Duhemian point that virtually any theory can be tailored and refined to recover known phenomena such as E and F (see Crupi, 2021), and it is also consistent with the idea that verified observable consequences, even if merely accommodated, can still provide *weak* support for a theory. However, the fact that a key piece of theory (e.g., a Lakatosian hard core, or part thereof) enables the derivation of some of the available evidence from other independent parts of it is arguably contingent on what the theory actually says and is taken as a distinctive element of empirical success. An analogy with evidential reasoning in statistical settings may be helpful. Surely a good measure of fit between, say, a linear model and a relevant data set speaks in favor of a linear interpretation of the underlying process at least to some extent. However, the more stringent demand of so-called *cross-validation* is routinely applied to guard against "overfitting", namely to go beyond the limited support that mere accommodation can provide. If a subset of the data constrains a specification of the model parameters which in turn fares well on a separate subset, the support achieved is taken as clearly stronger (see Schurz, 2014, p. 92, for a similar remark).

7.4 A Reconstruction Reconstructed

An updated account of use-novelty is the first step in my project to recast Lakatos and Zahar's (1975) analysis in a new form, and to counter Thomason's (1992) criticism. The second step needed is of course a characterization of the theories to be compared. Here, the heliocentric "rough model" or framework (the Lakatosian core of Copernicanism, as it were) will be meant as implying the following claims[8]:

---

[7] As concerns clauses (a)–(c) themselves, I'm really not claiming much originality. In Niiniluoto's (2016) terminology, for instance, the fulfilment of (a)–(c) implies that T achieves "deductive systematization" or complies with a "linking up" variant of the notion of "unification" with regards to E and F. Similar conditions have been also employed to explicate Whewell's celebrated idea of "consilience": see McGrew (2003) and Myrvold (2003). Also see Alai (2014) for a related discussion and proposal.

[8] My reconstruction here is largely consistent with Lakatos and Zahar's (1975) and similar to Carman's (2018). Point (vii), in particular, is explicitly stated early on in the *Commentariolus* as a basic feature of the heliocentric system. In *De Revolutionibus* (Book I, Chap. X), it is presented as the consequence of more fundamental assumptions that are shared through the astronomical

(i) the Sun is stationary;
(ii) the sphere of the fixed stars, centered (approximately) in the Sun, is at rest;
(iii) the Earth revolves around the Sun;
(iv) the Earth rotates around its own axis;
(v) the Moon orbits the Earth (closely);
(vi) planets other than the Earth also revolve around the Sun;
(vii) planets are ordered from the center outward by (strictly) increasing revolution periods.

As concerns the core commitments of the Ptolemaic approach, here is a fitting list for our purposes:

(i*) the Earth is stationary;
(ii*) the sphere of the fixed stars revolves around the central Earth;
(iii*) the Sun revolves around the Earth;
(iv*) all planets (including the Moon) revolve around the Earth with a combination of (few) circular motions;
(v*) heavenly bodies are ordered from the outer sphere inward by decreasing overall rotating speed.

Of course, (i*)–(v*) are all consequences of the *full* Ptolemaic theory that was taught in the schools in Copernicus's time including the sophisticated machinery of deferents and epicycles as appropriately specified.

A key complaint by Thomason is that Lakatos and Zahar (1975) "ignored Tycho Brahe" (Thomason, 1992, p. 161). The claim is not unfounded at first sight, for in Lakatos and Zahar (1975) Brahe's case is taken on in the criticism of competing methodological reconstructions (especially the falsificationist and simplicist), but otherwise relegated in minor footnote remarks. Indeed, a reference to Tycho Brahe's theory occurs almost invariably whenever the epistemic luck thesis arises. It is therefore crucial to emphasize that in the current context Brahe's model is nothing but a specification of the core claims (i*)–(v*) above and indeed no more than a variant of the traditional, full Ptolemaic system. In fact, for *any* "planet", the *actual trajectory* postulated by Brahe *around the (stationary) Earth* is demonstrably identical to the corresponding Ptolemaic trajectory. The only caveat is that the Sun is not always further away than Mercury and Venus, but rather at the center of their epicycles. This difference is of course interesting but immaterial for all astronomical evidence available between *De Revolutionibus* and Galilei's discovery of Venus's phases (in 1610), and thus immaterial for our purposes too. In this perspective, to describe the Thyconic system as a "mixed" model, "combining" Ptolemy and Copernicus is quite misleading. At least in terms of Lakatos and Zahar's methodological question about "immediate support" favoring Copernicus's

---

tradition (also see Lakatos & Zahar, 1975, p. 185). This elucidation was prompted by a remark from John Worrall.

theory, one could just as well regard Brahe's theory as "Ptolemaic" throughout—or, equivalently, consider Brahe's theory as *the* key geocentric counterpart of Copernicanism at least by the time of Kepler and Galilei.[9] In any event, the post-Lakatosian reconstruction outlined above thoroughly includes the Tychonic system as a specific model entailing the pillars of Ptolemaic geocentrism (i*)–(v*).[10] Let us now check the implications.

Fact 1: *Stations and retrogressions are observed for each of Mercury, Venus, Mars, Jupiter, and Saturn*. Thomason (1992) questions that this major point from Lakatos and Zahar (1975, p. 185) may strongly support the Copernican framework on the grounds that, historically and psychologically, Fact 1 (a "*dominant* problem in Western astronomy", Lakatos & Zahar, 1975, p. 182) was something that Copernicus definitely *did* want to account for when devising his theory. By our criterion of empirical success (as distinct from accommodation), this is irrelevant, however. Logically, as soon as observational evidence E indicates the non-redundant fact of the very existence of a (Copernican) planet (i.e., a major heavenly body other than Moon, Sun, and fixed stars), the Copernican framework (i)-(vii) immediately entails Fact 1 as concerns that object. On the other hand, Fact 1 does *not* follow from core Ptolemaic assumptions (i*)–(v*) as conjoined with E or any other independent observable fact, so in this case clause (a) is violated. Of course, according to the Duhemian corollary, Fact 1 follows from a full Ptolemaic theoretical cohort (Brahe's is one example), but then clause (b) above is violated. As a consequence, Fact 1 does provide strong and immediate support to the Copernican position against the Ptolemaic approach in our revised reconstruction.[11]

---

[9] See Margolis (1991) for a revealing independent argument supporting this move on purely historical grounds.

[10] One objection here might be that the Tychonic model cannot entail (iv*) just because for Tycho the Sun is the center of simple circular epicycles for each planet. I take this to be an inconsequential semantic issue, however. In both (i)–(vii) and (i*)–(v*), I employ "to revolve" to denote a periodic motion around a stationary center. This is quite consistent with the planets "orbiting" the Sun for Tycho, much as the Moon orbits the Earth for Copernicus. Once this innocent stipulation is clarified, I submit that the traditional Tychonic model does verify (i*)–(v*). In any event, regardless of the terminological preferences, the fact remains that one can turn a full classical Ptolemaic model into a full Tychonic model by just moving the Sun to the center of Mercury's and Venus's (embedded) epicycles, and leaving all actual trajectories otherwise unaltered. (I thank José Díez for pressing me on this point.) Note that such variation is arguably no larger departure from Ptolemy's original full theory than the "inverted-direction epicycle" approach, which is thoughtfully discussed by Carman and Díez (2015, pp. 30–31) and explicitly and sensibly taken to belong squarely to the Ptolemaic tradition.

[11] Fact 1 is a qualitative statement. However, in an insightful footnote, Thomason (1992, p. 181, n. 19) makes a striking observation concerning a more quantitative aspect of these phenomena: in the Copernican approach, the appearance of retrograde motion for superior planets such as Saturn can be large enough to be easily detected only in presence of a "considerable gap" with the fixed stars. The fascinating implication is that, conversely, the observable amplitude of the retrogressions of superior planets may be a basis for a Copernican to infer a large distance of the fixed stars. This in turn would potentially make an empirical success of a fact that no vindicationist seems to have

Fact 2: *Mercury and Venus are never seen to go in opposition.* Thomason (1992) does not address this point from Lakatos and Zahar (1975, p. 186), but he could have easily objected that, here again, Fact 2 was an established phenomenon that Copernicus did want to account for when devising his theory. Yet Fact 2 is entailed by the Copernican framework (i)-(vii) along with observational evidence E such as a small observed interval between two successive conjunctions (less than a year) for Mercury and Venus, implying the non-redundant statement that both planets are internal. On the other hand, Fact 2 does *not* follow from core Ptolemaic assumptions (i*)–(v*) as conjoined with E or any other independent observable fact, so in this case clause (a) is violated. Of course, according to the Duhemian corollary, Fact 2 follows from a full Ptolemaic theoretical cohort (Brahe's is one example), but then clause (b) above is violated. As a consequence, Fact 2 does provide strong and immediate support to the Copernican position against the Ptolemaic approach in our revised reconstruction.

Fact 3: *Mercury's retrogressions are seen to be more frequent than Venus's.* Thomason addresses a closely related point from Lakatos and Zahar (1975, p. 186) and questions that it may strongly support the Copernican approach for "it seems plausible to hold that [it] played some role guiding Copernicus to the view that the Sun was in the center of the planets' orbits" (Thomason, 1992, p. 185). Yet Fact 3 is entailed by the Copernican framework (i)–(vii) along with evidence E such as a smaller observed interval between two successive conjunctions for Mercury than for Venus, implying the non-redundant statement that the former must be the innermost internal planet. On the other hand, Fact 3 does *not* follow from core Ptolemaic assumptions (i*)–(v*) as conjoined with E or any other independent observable fact, so in this case clause (a) is violated. Of course, according to the Duhemian corollary, Fact 3 follows from a full Ptolemaic theoretical cohort (Brahe's is one example), but then clause (b) above is violated. As a consequence, Fact 3 does provide strong and immediate support to the Copernican position against the Ptolemaic approach in our revised reconstruction.

Fact 4: *Intervals between successive conjunctions are smaller for Mercury than for Venus.* This point is not addressed by either Lakatos and Zahar (1975) or Thomason (1992), but it is of interest in our perspective. We have seen that observational information about successive conjunctions can complement the Copernican framework (i)–(vii) entailing the ordering of internal planets, by which Fact 3 can then be derived. In addition, this situation is largely symmetric: indeed, Fact 4 is entailed by the Copernican framework (i)–(vii) along with E now meant as known observable facts mentioned above. More precisely, because Mercury and Venus are never seen to go in opposition (Fact 2), the theory entails that they must be internal planets, and because retrogressions are seen to be less frequent for Venus than for Mercury (Fact 3), the latter must be the innermost, with a shorter orbital period and thus more frequent conjunctions. On the other hand, Fact 4 does *not*

---

ever dared to classify as more than a (reasonable) accommodation, namely the failed detection of stellar parallax (see, e.g., Worrall, 2002, p. 198).

follow from core Ptolemaic assumptions (i*)–(v*) as conjoined with either E or any other independent fact, so in this case clause (a) is violated. Of course, according to the Duhemian corollary, Fact 4 follows from a full Ptolemaic theoretical cohort (Brahe's is one example), but then clause (b) above is violated.

Fact 5: *The length of Venus's retrograde arc is seen to be greater than Mercury's.* This is a case that Thomason himself allows as use-novel for Copernicus (from *De Revolutionibus*, Book I, Chapter X) because, although of course known, it does "not seem obviously relevant to the structure of the cosmos" (Thomason, 1992, p. 188), and thus to the guiding explanatory aims of Copernicus' inquiry. In our perspecrive, Fact 5 is entailed by the Copernican framework (i)–(vii) along with observational evidence E such as the interval between two successive conjunctions and relevant angular measurements implying a non-redundant assessment of the magnitude and period of Mercury's and Venus's motion as referred to the Sun. On the other hand, Fact 5 does *not* follow from core Ptolemaic assumptions (i*)–(v*) as conjoined with E or any other independent observable fact, so in this case clause (a) is violated. Of course, according to the Duhemian corollary, Fact 5 follows from a full Ptolemaic theoretical cohort (Brahe's is one example), but then clause (b) above is violated.

Fact 6: *Mars, Jupiter, and Saturn are all seen to always retrogress at opposition.* Fact 6 is considered but dismissed by Thomason (1992, p. 188). Yet Fact 6 is entailed by the Copernican framework (i)–(vii) along with known evidence E such as the observation of a quadrature for each of Mars, Jupiter, and Saturn, implying the non-redundant fact that all three planets are external. On the other hand, Fact 6 does *not* follow from core Ptolemaic assumptions (i*)–(v*) as conjoined with E or any other independent observable fact, so in this case clause (a) is violated. Of course, according to the Duhemian corollary, Fact 6 follows from a full Ptolemaic theoretical cohort (Brahe's is one example), but then clause (b) above is violated.

Fact 7: *Jupiter's retrogressions are seen to be more frequent than Mars's, and Saturn's more frequent than Jupiter's.* This point (from *De Revolutionibus*, Book I, Chapter X) is not addressed by either Lakatos and Zahar (1975) or Thomason (1992). Fact 7 is entailed by the Copernican framework (i)–(vii) along with evidence E such as a larger observed interval between two successive conjunctions for Mars than for Jupiter, and for Jupiter than for Saturn (all of which greater than a year), implying the non-redundant statement that Mars must be the innermost external planet, and Saturn the outermost. On the other hand, Fact 7 does *not* follow from core Ptolemaic assumptions (i*)-(v*) as conjoined with E or any other independent observable fact, so in this case clause (a) is violated. Of course, according to the Duhemian corollary, Fact 7 follows from a full Ptolemaic theoretical cohort (Brahe's is one example), but then clause (b) above is violated.

Fact 8: *Intervals between successive conjunctions are smaller for Saturn than for Jupiter, and smaller for Jupiter than for Mars.* This point is not addressed by either Lakatos and Zahar (1975) or Thomason (1992), but it is of interest in our perspective. We have seen that observational information about successive conjunctions can complement the Copernican framework (i)–(vii) entailing the ordering of external planets, by which Fact 7 can then be derived. In addition, this situation is largely symmetric: indeed, Fact 8 is entailed by the Copernican framework (i)–(vii) along with E now meant as known observable facts mentioned above. More precisely, because Mars, Jupiter, and Saturn are all seen to go in opposition (Fact 6), the theory entails that they must be external planets, and because retrogressions are seen to be less frequent for Mars than for Jupiter, and for Jupiter than for Saturn (Fact 7), the former must be the innermost and the latter the outermost, with decraeasing orbital periods and thus increasingly frequent conjunctions. On the other hand, Fact 8 does *not* follow from core Ptolemaic assumptions (i*)–(v*) as conjoined with either E or any other independent fact, so in this case clause (a) is violated. Of course, according to the Duhemian corollary, Fact 8 follows from a full Ptolemaic theoretical cohort (Brahe's is one example), but then clause (b) above is violated.

Fact 9: *The length of Mars' retrograde arc is seen to be greater than Jupiter's, which is seen to be greater than Saturn's*. Thomason pairs this with Fact 5 as use-novel for Copernicus (Thomason, 1992, p. 188). In our perspective, Fact 9 is entailed by the Copernican framework (i)–(vii) along with observational evidence E such as the interval between two successive conjunctions and relevant angular measurements implying a non-redundant assessment of the magnitude and period of Mars's, Jupiter's, and Saturn's motion as referred to the Sun. On the other hand, Fact 9 does *not* follow from core Ptolemaic assumptions (i*)–(v*) as conjoined with E or any other independent observable fact, so in this case clause (a) is violated. Of course, according to the Duhemian corollary, Fact 9 follows from a full Ptolemaic theoretical cohort (Brahe's is one example), but then clause (b) above is violated.

Although surely incomplete, the reconstruction above concerning facts (1)–(9) is sufficient to license a key conclusion for our purposes: according to our characterization of empirical success (which recovers Zahar's original motivation, as illustrated by the Einstein/Mercury example), and despite the uncontested truth of the Duhemian corollary, the Copernican view was indeed "immediately supported" by various known facts which did not support geocentric competitors in the same way.[12] It should be clear—but it's worth emphasizing—that this conclusion relies on a broadly Lakatosian distinction between core *vs.* full models.[13] Again following Lakatos and Zahar, I'm not committed to deny the ("uninteresting")

---

[12] One may wonder whether my approach leaves *any* room for strong support in favor of the geocentric position. A fascinating example can be drawn from Carman and Díez (2015, pp. 26–28) and concerns a pattern of phases for a superior planet such as Mars. In our terms, from the observationally established fact that Mars is sometimes found at opposition, one can infer by *either* the heliocentric postulates (i)–(viii) *or* the geocentric postulates (i*)–(v*) the observation of a waxing *vs.* waning gibbous disk before and after opposition, respectively. In this sense, my reconstruction converges with Carman and Díez's (2015) point that a geocentric system gets strong empirical success in a case like this (even if the phenomenon happened to be unobserved before modern times).

[13] In a similar fashion, Myrvold's (2003) assessment of the Copernican controversy relied on the contrast of "a bare-bones Ptolemaic hypothesis with a bare-bones Copernican hypothesis" rather than the corresponding "fully specified models of the heavens, with all parameters filled in". A Lakatosian approach, equipped with a core/programme distinction, can provide a motivation for this move.

traditional remark that, unlike core models, full models of either strain (heliocentric or geocentric) with all their parameter values specified end up being empirically indistinguishable around 1600 in a relevant sense. In particular, one can see that, for all of them, clause (b) of my criterion of strong support is invariably violated.

7.5 Methodological Issues in Clavius *vs*. Kepler and Galilei

Strictly speaking, Lakatos and Zahar did not want to commit to any specific account of "why Kepler and Galilei actually became 'Copernicans'" (Lakatos & Zahar, 1975, p. 188, footnote 1). This notwithstanding, it seems clear that at least some methodological issues did play a role in the arguments and choices of major figures in the Copernican controversy. Consider for instance the following important quote from a most distinguished post-Copernican astronomer of geocentric allegiance, German Jesuit Christopher Clavius (1538–1612):

> That Copernicus should have succeeded in saving the phenomena in a different way is not at all surprising. The motions of the eccentrics and epicycles taught him the times, the magnitudes, and the quality of appearances, future as well as past. Since he was exceedingly ingenious, he was able to conjure up a new method, in his opinion more convenient, of saving the appearances. [...] Just as, when we know a correct conclusion, we can construct a chain of syllogisms which derive that conclusion from false premises. [...] All that can be concluded from Copernicus' assumption is that it is not absolutely certain that the eccentrics and epicycles are arranged as Ptolemy thought, since a large number of phenomena can be defended by a different method. (Clavius, 1581, quoted in Duhem, 1908)

Clavius' sources and references may be remote, but his logical and methodological insight is neat. He knows from the astronomical tradition that alternative models can account for the same phenomena. The implications of this fact converge with a sound principle of (Aristotelian) logic, namely that true (observational) statements can logically follow from false (theoretical) claims. And Clavius goes on to note that, for someone who wants to devise a theoretical system or model by which a known conclusion follows, the task is virtually only a matter of ingenuity and dedication. The latter point essentially reflects the Duhemian corollary, and it is not by chance, therefore, that Duhem himself cites and much appreciates this passage in *To save the phenomena*.

There is one subtle but crucial step in Clavius' line of argument that is flawed, however. Granted, that one may come up with *some* novel theory saving all known phenomena is indeed "not at all surprising" for an "exceedingly ingenious" scholar such as Copernicus. This is a completely general existential claim, though. It does not imply in any way that ingenuity and dedication are enough to generate a theory *such as Copernicus'*, namely such that its core claims disclose substantial logical connections between already known facts. While ingenuity and dedication *alone* do plausibly account for the accommodation of known observational facts, they do not account for crucial additional features that a theory may have, such as the successful derivation of use-novel data from further and otherwise independent evidence, as it happens with Facts (1)–(9) above. In modern terms, the idea of the comparative strength of the Copernican approach does not arise from a neglect of the Duhemian corollary (which Clavius essentially, and appropriately, endorses and emphasizes) but from the methodological relevance of the prediction / accommodation distinction (which Clavius apparently disregards).

Let us imagine a late Sixteenth Century astronomer who thinks that Clavius is unduly dismissive of Copernicus' achievement precisely because in her/his view facts such as (1)–(9) do provide strong and selective support to heliocentrism. How would such a scholar phrase her/his position? "Use-novelty" is of course an esoteric term of art of contemporary philosophy of science and "prediction" would be too much of a stretch of ordinary parlance, for virtually all relevant observations describe long known phenomena. In such a predicament, our target methodological point would have to be glimpsed through periphrases, metaphors, and tentative terminology, such as a reference to "explanation": use-novel facts would be regarded as "explained" by the theory in a way that merely accommodated phenomena can not attain. Arguably, Kepler was precisely one such astronomer, and the opening remarks of his first important work, the *Mysterium* (1596, Chapter I), look strikingly like a direct response to Clavius:

> I have never been able to agree with those who rely on the model of accidental proof, which infers a true conclusion from false premises by the logic of the syllogism. Relying [...] on this model they argue that it was possible for the hypotheses of Copernicus to be false and yet for the true phenomena to follow from them as if from authentic principles. (Kepler, 1981, p. 75, translation slightly adapted)

By contrast, the most important point by which his confidence in Copernicus' theory was established, Kepler says here, is a matter of explanation: "for the things at which from others we learn to wonder, only Copernicus magnificently gives the explanation [*rationem reddit*], and removes the cause of wonder, which is not knowing causes". In a laborious but forceful attempt to motivate such claim, Kepler goes on to mention a series of manoeuvres that only a Copernican ("someone who places the Sun at the center") can perform:

> If you tell him to derive from the hypothesis [...] any of the phenomena which are actually observed in the heavens, to argue backwards, to argue forward, *to infer one phenomenon from another* [*unum ex alio colligere*], and to perform *anything that the truth permits* [*quae veritas rerum patitur*], he will have no difficulty with any point. (Kepler, 1981, p. 75, emphasis added)

"To infer one phenomenon from another"—in informal terms, this is pretty much as close as one can get to what we see with Facts (1)–(9) above. And this, according to Kepler, is something that "the truth permits", a sign of the status of the theory itself, not just of its inventor.

Soon after receiving Kepler's *Mysterium*, Galilei himself replied with a famous letter. "I have adopted Copernicus' opinion many years ago", Galilei writes to Kepler, "and from that I've been able to find the causes of many natural effects, which are doubtless inexplicable [*inexplicabiles*] by the conventional hypotheses" (Galilei, 1890–1909, vol. X, p. 68). Much has been written about which "effects"

Galilei may be referring to in this important but elusive passage. A recurrent conjecture (starting from Kepler himself) is that he might already have been thinking about his later (and mistaken) argument that tides prove the Earth's motion (see Voelkel, 2001, pp. 71–72). But even if tides are included in Galilei's "many effects", his claim may well have a wider scope and resonate with Kepler's own remarks on explanatory success: Copernicus' theory had distinctive support from the start on the basis of long known phenomena that geocentric approaches could not explain (only accommodate). One crucial point should be emphasized here. It may appear that these remarks attribute epistemological or methodological significance to metaempirical virtues ("explanatory power", one might say). But this would be a rather misleading impression, in my reading. To the extent that "explanation" really amounts to a paraphrase to capture and convey the idea of support from use-novel evidence as contrasted with plain accommodation, the arguments at issue are fully reducible to a straight and monist empiricist view of scientific confimation. After all, according to the spirit of predictivism, verified use-novel consequences provide nothing but supporting *empirical* evidence, and the distinction between use-novel facts and facts that are temporally novel is meant to be a largely inconsequential contingency in methodological terms. It is just because some known facts are still "predicted" in this broad sense that one is led to regard them as "explained" in a distinctive way.[14]

In a later exchange with Balliani, the missing pieces of Galilei's view seem to emerge much more explicitly: the variety of reasons favoring heliocentrism is unpacked, including both new *and old* items; no less important, Thycho's system is said to face virtually the same hurdles that plague Ptolemy's original approach:

> As for Copernicus' opinion, truly I take it as certain, and not only for the observations of Venus, of sunspots, and of the Medicean moons, but for *his own other reasons*, as well as for many more particular reasons of mine which I regard as conclusive [...] In Thyco's opinion there remain, I find, those utmost difficulties which lead one to depart from Ptolemy, whereas in Copernicus I have nothing at all to raise the slightest qualm. (Galilei, 1890–1909, vol. XII, pp. 34–35, emphasis added.)

It is quite difficult to make sense of this passage (from 1614) unless it implies that facts such as (1)–(9) above (Copernicus' "own other reasons") do carry substantial evidential weight, favoring heliocentrism against geocentrism in either Ptolemy's *or Thyco's* variants.

In principle, even if *there were* good reasons to prefer the heliocentric system to its geocentric competitors at the end of the Sixteenth Century, they might have played no role in the scientific choices of early Copernicans. This does not seem to be the case, however. The textual evidence provides at least some significant hints that Kepler and Galilei, unlike their opponents, did appreciate the relevance of use-novel data as distinctively supporting heliocentrism.

---

[14] A similar reductionist strategy may well be pursued for other alleged "theoretical" virtues. As for "simplicity", for instance, Sober (2015, pp. 12–21) offers much relevant material. Hall (1970), by contrast, seemed to imply that simplicity retains a distinctive epistemic role.

7.6 The Aristarchus Puzzle

Our line of argument so far was meant to divert certain objections to vindicationism about early Copernicanism along the lines of Lakatos and Zahar (1975) and thereby to challenge the popular epistemic luck thesis. Apparently, an updated and sharpened construal of Lakatos and Zahar's basic approach can survive Thomason's (1992) criticism and even enlighten certain plausible underlying motivations for Kepler's and Galilei's scientific engagement in heliocentric astronomy. This section will be devoted to the reconstruction of a rather different source of concern for vindicationists. It may be called the *Aristarchus puzzle*, and it arises from the following argument.

1. Facts such as (1)-(9) above indicate that Copernicus' heliocentrism was better than post-Ptolemaic geocentrism.
2. *Ceteris paribus*, working scientists have accepted a theory X as better than a competing theory Y if and only if X was better than Y given the available evidence.
3. Facts (1)–(9) above were available evidence in Aristarchus' time (third century BC).
4. Pre-Ptolemaic geocentrism (from Eudoxus) is no better than post-Ptolemaic geocentrism given the available evidence in Aristarchus' time.
5. The *ceteris paribus* clause in 2. is fulfilled in Aristarchus' case.
6. Copernicus' heliocentrism was essentially the same theory as Aristarchus'.
7. Aristarchus' theory was scientifically unpopular in his time.
8. Copernican heliocentrism is better than post-Ptolemaic geocentrism even only by the available evidence in Aristarchus' time [from 1 and 3].
9. Aristarchus' heliocentrism is better than post-Ptolemaic geocentrism even only by the available evidence in Aristarchus' time [from 6 and 8].
10. Aristarchus' heliocentrism was better than pre-Ptolemaic geocentrism given the available evidence in Aristarchus' time [from 4, 9, and the transitivity of "being a better theory than"].
11. Working scientists have accepted Aristarchus' theory as better than pre-Ptolemaic geocentrism if and only if Aristarchus' theory was better than pre-Ptolemaic geocentrism given the available evidence in Aristarchus' time [from 2 and 5].
12. Aristarchus' heliocentrism was *not* better than pre-Ptolemaic geocentrism given the available evidence in Aristarchus' time [from 7 and 11].
13. Contradiction [from 10 and 12]

Although informal, I will take the above argument as essentially valid. Indeed, trying to retain all the premises 1 to 7 and reject the conclusions seems an unlikely and contrived manoeuvre. Accordingly, some of the premises must be given up. The problem then is that the argument can be legitimately regarded as a *reductio* of vindicationism, and especially of a Lakatosian strain. To clarify this, let us consider the premises more closely.

Premise 1 is a straightforward implication of both Lakatos and Zahar's (1975) and our methodological analysis. The overarching principle stated in premise 2 is taken almost literally from Worrall (1976, pp. 164–165), where it features as "both a clarification of, and an improvement on, the account already given by Lakatos" of "how methodologies can be tested using history of science" (Worrall, 1976, p. 168). Premise 3 is an historical truism: as early as the fourth century BC, Eudoxus' model must have integrated basic observations such as (1)–(9). Premise 4 seems a methodological assessment which none of the parties would challenge (see Lakatos & Zahar, 1975, p. 180, and Thomason, 1992, p. 191, for instance), while rejecting premise 5 would commit one to invoke a rather massive interference of extra-scientific factors in the development of early Hellenistic culture, for which no historical evidence seems in sight. This survey leaves the (post)Lakatosian vindicationist in a quite uncomfortable position, for Lakatos and Zahar also apparently accept premise 6 (Lakatos & Zahar, 1975, p. 188), while premise 7 is the traditional consensus view in the history of ancient science.

The Aristarchus puzzle is a neglected but serious difficulty for vindicationists. We know that the Greeks developed astronomy in a mathematical and empirically testable form. According to vindicationists, Copernicus and early Copernicans had strong empirical arguments for heliocentrism. But if heliocentrism was already devised by Aristarchus in presence of at least some of the same crucial empirical evidence, then why it was not accepted in ancient times?[15]

According to a Lakatosian view as developed by Worrall (1976), methodology is regarded as testable against the history of science. Ideally, in a successful methodological programme difficulties are addressed in a way that gets independent support by historical research. Arguably, the most appealing solution of the Aristarchus puzzle for a Lakatosian methodologist would be to find good reasons to reject premise 7. Consider Lakatos and Zahar's (1975, p. 181) insistence against Kuhn's remark that "there were no good reasons for taking Aristarchus seriously" (Kuhn, 1962/70, p. 75) and their explicit statement that "Copernicus' (*and indeed, Aristarchus'*) rough model had excessive predictive power over its Ptolemaic rival" (Lakatos & Zahar, 1975, p. 188, emphasis added). In light of the Aristarchus puzzle, these claims surely are a source of discomfort for a (post-)Lakatosian vindicationist. However, if one is willing to entertain the hypothetical scenario that heliocentrism

---

[15] Beyond both Aristarchus and Ptolemy, historical research over the last decades has addressed developments leading to Copernicus across the high and late medieval period, with a special emphasis on the role of Arabic astronomy (see Swerlow & Neugebauer, 1984, for a key discussion). Surely Ptolemy was known and much criticised by Arabic intellectuals not only for philosophical reasons, but also on scientific grounds, some of which feature prominently in Copernicus' work too. Some Arabic astronomic treatises include non-Ptolemaic technical solutions which also appear in Copernicus with no known documentation in other earlier sources, either ancient or Christian, and even the Earth's motion is occasionally discussed as a scientific possibility (Ragep, 2007). As the evidence for actual transmission to Copernicus is so far inconclusive, however, the implications of these facts remain a matter of controversy (see Blåsjö, 2018). Moreover, whereas advanced non-geocentric astronomical approaches in antiquity are ostensibly acknowledged by Copernicus, it is not clear to what extent they were appreciated in the Arabic tradition.

was well received after all (or even actively developed) after Aristarchus, this would offer a valid way out of the difficulty. Such scenario is just too good to be true, though.

Or is it?

7.7 Forgotten Heliocentrism?

There is at least one fascinating and independent strain of recent historical research indicating a clear solution of the Aristarchus puzzle for vindicationists. In a series of contributions from the 1990s on, physicist and historian Lucio Russo has advocated a highly innovative view of the development of science in the Hellenistic period in which astronomy plays an important role (Russo, 1994, 2004). This approach implies, among other things, that premise 7 of the Aristarchus puzzle above needs to be radically revised: indeed, it is submitted, third century BC heliocentrism was not discarded at all, but rather seriously considered by contemporaries of Aristarchus such as Archimedes (287–212 BC) and Erathostenes (276–194 BC), and actively endorsed and developed by major later figures including Seleucus (born 190 BC) and above all Hipparchus (190–120 BC). But how can this amazing claim be supported?

The awful methodological hurdle here is the painful lack of a firm textual basis for a scholar of our age. With the exception of two minor writings by Aristarchus and Hipparchus themselves, respectively, *no* scientific work of astronomical content has reached us from the timeframe between these two prominent figures. Under this grim predicament, the task of collecting evidence to assess Russo's hypothesis is bound to follow a rather thin indirect route, namely, to rely on passages of scientific relevance in non-scientific pre-Ptolemaic sources on the plausible assumption that they may bear traces of proper scientific work which is nowadays inaccessible. For our purposes, a valuable illustration comes from a passage in Seneca's *Naturales Quaestiones* (mid-first century AC), which is thoroughly analysed by Russo (1994, pp. 221–223). Here Seneca — a distinguished Roman intellectual with no direct involvement in scientific research — is reporting the position of "some people" as concerns "the five planets" by which one can understand "why they move backward" (*quare agantur retro*):

> [They] would say to us: You are wrong if you judge that any star either stops or alters its orbit. It is not possible for celestial bodies to stand still or turn away. They all move forward. [...] What is the reason, then, that some celestial bodies appear [*videantur*] to move backward? The encounter with the Sun imposes upon them the appearance of slowness, as well as the nature of their paths and their orbits which are so placed that at a fixed period they deceive [*fallant*] observers. In the same way ships seem to be standing still even though they are moving under full sail.

That retrogressions occur at inferior conjunctions (e.g., for Venus) or oppositions (e.g., for Mars), namely at a certain kind of "encounter with the Sun" is of course a common notion across geocetric *and* heliocentric accounts of the phenomena. Only a heliocentric theorist, however, would go on making a sharp distinction between real motion along the orbits and purely apparent motion "backward", insisting that retrogressions must be explained as the effect of the illusory perception of a moving observer as one who sees a sailing ship as apparently still. A bold but disregarded theoretical proposal would have hardly found its way from an isolated early Hellenistic scholar on the Eastern side of the Hegean Sea to a Hispanic playwright working in Imperial Rome three centuries later unless some influential figure in between had accepted and reported it as a successful account of the relevant phenomena. Overall, Russo's comprehensive reconstruction spans a couple of centuries, including texts from Vitruvius, Lucretius, Cicero, Manilius, Pliny, and Plutarch, beyond Seneca. Each piece of evidence is only circumstantial by itself, but a systematic pattern is clearly discernible: while active astronomical research remained ostensibly silent through all this period (Russo, 1994, p. 211), earlier Hellenistic doctrines appear to have occasionally surfaced as pieces of shallow erudition in loosely related contexts. In both Pliny and Cicero, for instance, one finds the casual but explicit statement that "planet" is a misleading label, for so-called "wandering stars" are in fact not wandering in any literal sense (Russo, 1994, p. 225).

Currently, Russo's position remains way out of mainstream perspectives in the history of science.[16] Moreover, a historical resolution of the Aristarchus puzzle is bound to trigger a "revenge" problem, namely to raise a counterpart "Ptolemy puzzle": assuming that Aristarchus' insight was indeed successfully developed in Hellenistic science, then why did *Ptolemy* not recover heliocentrism in the second century AC? (Statement 9 would then become the pillar of the new *reductio* argument.) The externalist strategy is comparatively stronger in this case, though. It seems established that Ptolemy had to refound the astronomical tradition in Alexandria without the benefit of a continuous intellectual lineage from his Hellenistic predecessors and indeed after a rather dramatic and durable halt. Russo also makes a case that Ptolemy's sources were themselves incomplete, not including the latest and more advanced fruits of Hellenistic science (Russo, 1994, pp. 210–213; also see Amabile, 2020). Indeed, he finds it "not too surprising if traces of some of Hipparchus' ideas might be found more easily in the Rome of the first century BC [. . .] than in the Alexandria of the second century AC" (Russo, 1994, p. 232).

Logically, rejecting premise 7 above surely is one way to meet the challenge of the Aristarchus puzzle. For a predictivist like Lakatos and Zahar (1975), given all other plausible premises 1–6, Aristarchus' theory should not have been unpopular in his time. At least if Russo is right, this is just what observant historical inquiry reveals: "Aristarchus' heliocentric model had been given up not in the period

---

[16] Scholz's (2023) fascinating discussion is consistent with some of the unconventional claims by Russo. In Kragh (2007, pp. 24–28), however, one finds the traditional view that "Aristarchus' heliocentric system was not considered a serious rival to the geocentric models and soon went into oblivion", including a reference to the much discussed accusation of impiety from the Stoic philosopher Cleanthes (which is questioned in Russo & Medaglia, 1996, as the misguided effect of a philological accident).

between Aristarchus and Hipparchus, but [...] during the long interruption of the scientific activity that occurred between Hipparchus and Ptolemy" (Russo, 1994, p. 238).[17]

7.8 Concluding Remarks

In an effort to bring together the threads of our discussion, one is tempted to consider that critics of Lakatos and Zahar's (1975) predictivist vindication of early Copernicanism may themselves be right for the wrong reasons. In fact, while the limitations of the original methodological reconstruction can arguably be amended to counter the epistemic luck thesis, the challenge of the Aristarchus puzzle may still undermine a post-Lakatosian account in a way that seems to have largely eluded the scope of the most lively philosophical debates. To this extent, Thomason's (1992, p. 198) remark that "the Copernican Revolution [...] should remain the touchstone for evaluating any philosophy of science claiming to be historically relevant" may well be integrated by an additional suggestion, namely that the interaction between philosophy and history of science should perhaps be more thoroughly explored even beyond the limits of the Modern Age.

Acknowledgments The initial project of this work started in November 2021 from a Facebook discussion with Enzo Fano e Flavia Marcacci, whom I want to thank. It took some effort to turn it into a real paper, but I'm happy to report that something good came out of the time I tend to waste on social media. I'm also grateful for the thoughtful feedback from attendees at the Lakatos Centenary conference at the LSE in November 2022 and at the 2023 Conference of the Italian Society for Logic and the Philosophy of Science in Urbino, to Gustavo Cevolani who kindly read and commented an advanced draft, and to José Díez for an extended and sustained exchange of ideas. The research has been also funded by the European Union, Next Generation EU (PRIN 2022, PNRR—M4—C2—INV 1.1, project code 2022ARRY9N, CUP D53D23009560006), project title: *Reasoning with hypotheses: Integrating logical, probabilistic, and experimental perspectives.*

References

Alai, M. (2014). Novel predictions and the no-miracle argument. *Erkenntnis, 79*, 297–326.
Amabile, A. (2020). Le fonti ellenistiche dell'*Almagesto* di Tolomeo (The Hellenistic sources of Ptolemy's *Almagest*). *Quaderni di Storia della Fisica, 23*, 43–77.
Barnes, E. C. (2022). Prediction versus accommodation. In E. N. Zalta & U. Nodelman (Eds.), *The Stanford encyclopedia of philosophy (Winter 2022 edition)*. Stanford University. URL: https://plato.stanford.edu/archives/win2022/entries/prediction-accommodation.

---

[17] While rejecting premise 7 happens to be my favourite post-Lakatosian rescue strategy from the Aristarchus' puzzle, it is not the only logical possibility that can be pursued on historical grounds. A very different (and indeed quite incompatible) alternative might be to give up premise 6. According to Carman (2018, p. 1), for instance, "given that ancient astronomers were perfectly capable of understanding the great advantages of heliocentrism over geocentrism [...], it seems difficult to explain why heliocentrism did not triumph over geocentrism or even compete significantly with it before Copernicus". Carman's remarkable solution is that "the first Copernican was Copernicus".

Blåsjö, V. (2018). A rebuttal of recent arguments for Maragha influence on Copernicus. *Studia Historiae Scientiarum, 17*, 479–497.

Carman, C. C. (2018). The first Copernican was Copernicus: The difference between pre-Copernican and Copernican heliocentrism. *Archive for the History of Exact Sciences, 72*, 1–20.

Carman, C. C., & Díez, J. (2015). Did Ptolemy make novel predictions? Launching Ptolemaic astronomy into the scientific realism debate. *Studies in the History and Philosophy of Science A, 52*, 20–34.

Clavius, C. (1581). *In Sphaeram Ioannis de Sacro Bosco commentarius: nunc iterum ab ipso auctore recognitus, et multis ac variis locis locupletatus*. Ex Officina Dominici Basae.

Crupi, V. (2021). Confirmation. In E. N. Zalta (Ed.), *The Stanford encyclopedia of philosophy (Spring 2021 edition)*. Stanford University. https://plato.stanford.edu/archives/spr2021/entries/confirmation

Duhem, P. (1906). *The aim and structure of physical theory*. Princeton University Press. 1991.

Duhem, P. (1908). *To save the phenomena*. University of Chicago Press. 1969.

Earman, J., & Janssen, M. (1993). Einstein's explanation of the motion of Mercury's perihelion. In J. Earman, M. Janssen, & J. Norton (Eds.), *The attraction of gravitation: New studies in the history of general relativity* (pp. 129–172). Birkhäuser.

Engel, M. (2022). Evidence, epistemic luck, reliability, and knowledge. *Acta Analytica, 37*, 33–56.

Finocchiaro, M. A. (2010). *Defending Copernicus and Galilei: Critical reasoning in the two affairs*. Springer.

Galilei, G. (1890–1909). *Le opere di Galileo Galilei: Edizione nazionale*. Barbera.

Graney, C. M. (2015). *Setting aside all authority*. University of Notre Dame Press.

Hall, R. (1970). Kuhn and the Copernican revolution. *British Journal for the Philosophy of Science, 21*, 196–197.

Kepler, J. (1981). *(1596). Mysterium Cosmographicum (The secret of the universe)*. Abaris.

Kragh, H. S. (2007). *Conceptions of cosmos*. Oxford University Press.

Kuhn, T. (1957). *The Copernican revolution*. Harvard University Press.

Kuhn, T. (1962/70). *The structure of scientific revolutions*. University of Chicago Press.

Lakatos, I., & Zahar, E. (1975). Why did Copernicus's research programme supersede Ptolemy's? In R. Westman (Ed.), *The Copernican achievement* (pp. 354–383). University of California Press. Reprinted in I. Lakatos, *Philosophical Papers* I: *The Methodology of Scientific Research Programmes* (pp. 168-192). Cambridge University Press, 1978.

Lange, M. (2004). Bayesianism and unification. *Philosophy of Science, 71*, 205–215.

Laudan, L. (1990). Demystifying underdetermination. In C. W. Savage (Ed.), *Minnesota studies in the philosophy of science* (pp. 267–297). University of Minnesota Press.

Marcacci, F. (2015). The world-system of Giovanni Battista Riccioli and the phases of Venus and mercury. *Advances in Historical Studies, 4*, 106–117.

Margolis, H. (1991). Tycho's system and Galileo's *Dialogue*. *Studies in History and Philosophy of Science, 22*, 259–275.

McGrew, T. (2003). Confirmation, heuristics, and explanatory reasoning. *British Journal for the Philosophy of Science, 54*, 553–567.

Myrvold, W. C. (2003). A Bayesian account of the virtue of unification. *Philosophy of Science, 70*, 399–423.

Niiniluoto, I. (2016). Unification and confirmation. *Theoria (Spain), 31*, 107–123.

Ragep, F. J. (2007). Copernicus and his Islamic predecessors: Some historical remarks. *History of Science, 45*, 65–81.

Riccioli, G. B. (1651). *Almagestum Novum*. Bologna.

Rovelli, C. (2019). The dangers of non-empirical confirmation. In R. Dardashti, R. Dawid, & K. Thébault (Eds.), *Why trust a theory?* (pp. 120–124). Cambridge University Press.

Russo, L. (1994). The astronomy of Hipparchus and his time: A study based on pre-Ptolemaic sources. *Vistas in Astronomy, 38*, 207–248.

Russo, L. (2004). *The forgotten revolution*. Springer.

Russo, L., & Medaglia, S. (1996). Sulla presunta accusa di empietà ad Aristarco di Samo. *Quaderni Urbinati di Cultura Classica, 53*, 113–121.

Salmon, W. C. (1990). Rationality and objectivity in science, or Thomas Kuhn meets Thomas Bayes. In C. W. Savage (Ed.), *Minnesota studies in the philosophy of science* (pp. 175–204). University of Minnesota Press.
Scholz, E. (2023). From heliocentrism to epicycles: A commentary on pre-Ptolemaic astronomy. In T. L. Knudsen & J. Carter (Eds.), *Episodes from the history of mathematics: Essays in honor of Jesper Lützen*. De Gruyter.
Schurz, G. (2014). Bayesian pseudo-confirmation, use-novelty, and genuine confirmation. *Studies in History and Philosophy of Science, 45*, 87–96.
Sober, E. (2015). *Okcham's razor: A user's manual*. Cambridge University Press.
Strevens, M. (2020). *The knowledge machine: How irrationality created modern science*. Northon & Company.
Swerdlow, N. M. (2004). An essay on Thomas Kuhn's first scientific revolution. *The Copernican Revolution. Proceedings of the American Philosophical Society, 148*, 64–120.
Swerlow, N. M., & Neugebauer, O. (1984). *Mathematical astronomy in Copernicus's De Revolutionibus*. Springer.
Thomason, N. (1992). Could Lakatos, even with Zahar's criterion for 'novel fact', evaluate the Copernican research programme? *British Journal for the Philosophy of Science, 43*, 161–200.
Timberlake, T., & Wallace, P. (2019). *Finding our place in the solar system: The scientific story of the Copernican revolution*. Cambridge University Press.
Voelkel, J. R. (2001). *The composition of Kepler's Astronomia Nova*. Princeton University Press.
Votsis, I. (2014). Objectivity in confirmation: Post hoc monsters and novel predictions. *Studies in History and Philosophy of Science, 45*, 70–78.
Westman, R. S. (2011). *The Copernican question: Prognostication, skepticism, and celestial order*. University of California Press.
Worrall, J. (1976). Thomas young and the 'refutation' of Newtonian optics: A case-study in the interaction of philosophy of science and history of science. In C. Howson (Ed.), *Method and appraisal in the physical sciences* (pp. 107–179). Cambridge University Press.
Worrall, J. (2002). New evidence for old. In J. Wolenski & K. Kijania-Placek (Eds.), *In the scope of logic, methodology, and philosophy of science* (pp. 191–209). Kluwer.
Worrall, J. (2006). Theory-confirmation and history. In C. Cheyne & J. Worrall (Eds.), *Rationality and reality: Conversations with Alan Musgrave* (pp. 31–62). Kluwer.
Zahar, E. (1973). Why did Einstein's programme supersede Lorentz's? *British Journal for the Philosophy of Science, 24*, 95–123. and 223-262. Reprinted in C. Howson (ed.), *Method and Appraisal in the Physical Sciences* (pp. 211–275). Cambridge University Press.

Open Access This chapter is licensed under the terms of the Creative Commons Attribution-NonCommercial-NoDerivatives 4.0 International License (http://creativecommons.org/licenses/by-nc-nd/4.0/), which permits any noncommercial use, sharing, distribution and reproduction in any medium or format, as long as you give appropriate credit to the original author(s) and the source, provide a link to the Creative Commons license and indicate if you modified the licensed material. You do not have permission under this license to share adapted material derived from this chapter or parts of it.

The images or other third party material in this chapter are included in the chapter's Creative Commons license, unless indicated otherwise in a credit line to the material. If material is not included in the chapter's Creative Commons license and your intended use is not permitted by statutory regulation or exceeds the permitted use, you will need to obtain permission directly from the copyright holder.

Chapter 8
The Bayesian Research Programme in the Methodology of Science, or Lakatos Meets Bayes

Stephan Hartmann

Abstract Lakatos argued that Carnap's research programme in inductive logic was degenerative because it underwent a degenerative problem-shift by dealing with ever more specific internal problems and thereby moving further and further away from its original goals. Here I show that this criticism (which may apply to Carnap) cannot be levelled at the contemporary successor to Carnap's programme, Bayesianism. To this end, I discuss various challenges and show how they can be addressed within the Bayesian research programme. In each of these cases, I argue, one can speak of a 'progressive problem shift'. I therefore conclude that the Bayesian research programme in the methodology of science is progressive. Nevertheless, it is essential to continue to explore alternatives to it and to develop criteria for comparing competing research programmes.

Keywords Lakatos · Bayesianism · No alternatives argument · Learning indicative conditionals

8.1 Introduction

In 1965, Imre Lakatos organised a famous International Colloquium in the Philosophy of Science in London. Of the four conference proceedings, the Lakatos–Musgrave volume *Criticism and the Growth of Knowledge* (1970) is perhaps the best known. This volume contains Lakatos's long essay 'Falsification and the Methodology of Scientific Research Programmes,' in which he develops and defends his response to Kuhn's challenge to the rationality of science. The theory that Lakatos develops in this essay is well known and is still taught today in introductory courses in the philosophy of science. It shifts the focus from the assessment of individual scientific theories to the assessment of whole research programmes. Research programmes are sequences of scientific theories, they have

S. Hartmann (✉)
Munich Center for Mathematical Philosophy, LMU Munich, Munich, Germany
e-mail: S.Hartmann@lmu.de

a positive and a negative heuristic, and they have a hard core (which should not be touched) and a protective belt (which can be modified without abandoning the whole research programme). Lakatos illustrates his ideas with many examples from the history of science. In doing so, he provides rational reconstructions of important episodes and thus pursues (what he calls) an 'internal history of science.'[1]

Two years earlier, in 1968, another volume containing the proceedings of that colloquium had been published. This volume, *The Problem of Inductive Logic*, contains Lakatos's essay 'Changes in the Problem of Inductive Logic,' in which Lakatos attempts to show that Carnap's philosophical-mathematical research programme in inductive logic is not progressive but degenerative. Here is what he writes at the beginning of the paper:

> A successful research programme bustles with activity. There are always dozens of puzzles to be solved and technical questions to be answered; even if *some* of these – inevitably – are the programme's own creation. But this self-propelling force of the programme may carry away the research workers and cause them to forget about the problem background. They tend not to ask any more to what degree they have solved the original problem, to what degree they gave up basic positions in order to cope with the internal technical difficulties. Although they may travel away from the original problem with enormous speed, they do not notice it. Problem-shifts of this kind may invest research programmes with a remarkable tenacity in digesting and surviving almost any criticism.
>
> Now problem-shifts are regular bedfellows of problem-solving and especially of research programmes. One frequently solves very different problems from those which one has set out to solve. One may solve a more interesting problem than the original one. In such cases we may talk about a 'progressive problem-shift'. But one may solve some problems less interesting than the original one; indeed, in extreme cases, one may end up with solving (or trying to solve) no other problems but those which one has oneself created while trying to solve the original problem. In such cases we may talk about a *'degenerating problem-shift'*.
>
> I think that it can do only good if one occasionally stops problem-solving, and tries to recapitulate the problem background and assess the problem-shift. (Lakatos, 1968, 316–317)

Lakatos then continues by applying these general considerations to Carnap's inductive logic:

> In the case of Carnap's vast research programme one may wonder what led him to tone down his original bold idea of an a priori, analytic inductive logic to his present caution about the epistemological nature of his theory; why and how he reduced the original problem of rational degree of belief in hypotheses (principally scientific theories) first to the problem of rational degree of belief in particular sentences, and finally to the problem of the probabilistic consistency ('coherence') of systems of beliefs. (Lakatos, 1968, 317)

Lakatos then shows that Carnap's research programme is degenerative. Not so much because its predictions turned out to be false or because it did not predict new facts: a philosophical-mathematical research programme such as this cannot do that. It could, however, help to address new problems in the philosophy of

---

[1] Nanay (2010) discusses Lakatos's idiosyncratic use of the terms 'internal' and 'external history of science' and how they relate to each other. See also Schindler (in this volume).

science, but did not succeed in this respect. Also, and perhaps more importantly, Carnap's research programme failed according to Lakatos because it underwent a degenerative problem-shift by dealing with ever more specific internal problems and thereby moving further and further away from its original goals.

Lakatos might well be right in his assessment of Carnap's research programme. But what about its contemporary successor, Bayesianism? Can Lakatos's criticism also be levelled against it? To begin with, it is clear that Lakatos was not a Bayesian. However, at least two of his students—Colin Howson and Peter Urbach—became leading Bayesian philosophers of science, and another—John Worrall—sympathises at least somewhat with Bayesianism, as suggested in Worrall (2000), despite any objections he may have. It is doubtful, however, that the views of his students would have changed Lakatos's opinion in this regard. Nonetheless, I will argue below that Bayesianism, when properly understood, is a fine example of a Lakatosian research programme in the methodology of science. This research programme is progressive and can meet many challenges in an elegant way. It also has the capacity to address new and interesting problems in the methodology of science and it helps us to get answers to the big questions about the rationality and objectivity of science. Accordingly, I believe that Lakatos is wrong, at least with respect to contemporary Bayesianism, when he writes that

> [p]robabilism has never generated a programme of historiographical reconstruction; it has never emerged from grappling – unsuccessfully – with the very problems it created. As an epistemological programme it has been degenerating for a long time; as a historiographical programme it never even started. (Lakatos, 1976, 20)

Contemporary Bayesianism is a progressive research programme, but not so much in the historiography of science. There were of course attempts to provide historical reconstructions, but I doubt that Lakatos would have been impressed by the Bayesian solution to the Duhem Problem proposed by John Dorling (1979) and popularised by Howson and Urbach (2006) and others. (Deborah Mayo somewhat pejoratively called it 'Dorling's Homework Problem' in her Lakatos Award-winning 1996 book *Error and the Growth of Experimental Knowledge*.) Rather, I will argue that Bayesianism is a progressive research programme in the methodology of science and that it is not only useful to analyse and reconstruct scientific reasoning, but that it also helps us to assess actual scientific reasoning at the frontier of science.

The rest of this paper is organised as follows: Sect. 8.2 provides a brief introduction to standard Bayesianism and a list of three challenges it currently faces. Sections 8.3 and 8.4 discuss two of these challenges in more detail and show how they can be addressed within the Bayesian research programme. Section 8.5 discusses some further challenges and suggests what a Bayesian solution might look like in each case. In each of these cases, I will argue, one can speak of a 'progressive problem-shift'. Section 8.6 therefore concludes that the Bayesian research programme in the methodology of science is progressive. Nevertheless, it is imperative to continue exploring alternatives to it and to develop criteria for comparing competing research programmes.

8.2 Standard Bayesianism

Bayesianism is a philosophical theory about the statics and dynamics of (partial) beliefs. Its starting point is the psychological truism that we believe different (contingent) propositions more or less strongly, that is, we assign different degrees of belief (or credences) to them. To make the concept 'degree of belief' more precise, we need (1) a calculus for combining different degrees of belief, (2) an algorithm for updating degrees of belief, and (3) a (normative) justification for (1) and (2). Bayesianism offers just this, providing a framework that can be applied to a variety of problems in philosophy, including epistemology and the philosophy of science.

Let's see how the justification of the static and the dynamic part of Bayesianism works. We begin with the static part. Here Bayesianism identifies degrees of belief with (subjective) probabilities, i.e., the (rational) degrees of belief of an agent at a certain time have to satisfy the axioms of probability theory (see also Weisberg, 2011). But what justifies this identification? Bayesians present two types of arguments:

1. Pragmatic arguments ('Dutch book arguments'): these arguments show that an agent with incoherent degrees of belief (i.e., degrees of belief that do not respect the axioms of probability theory) will lose money in a corresponding betting scenario (see Pettigrew, 2020 for details).
2. Epistemic arguments ('Epistemic Utility Theory'): these arguments show that identifying degrees of belief with probabilities makes sure that the *inaccuracy* of an agent's degrees of belief is minimised (see Pettigrew, 2016 for a defense of this approach).

Let us now move on to the dynamic part of Bayesianism. Here we consider an agent who entertains the propositions $A_1, \ldots, A_n$. To proceed formally, one introduces an algebra $\mathcal{A}$ which comprises the propositional variables $A_1, \ldots, A_n$ with the values $A_1, \neg A_1$, etc.[2] over which a prior probability distribution P is defined. The agent then learns a piece of evidence, say, $E = A_1$. That is, the agent learns that proposition A_1 is true. This prompts her to switch from the prior probability distribution P to the posterior probability distribution P' which satisfies $P'(E) = 1$. To make sure that her new degrees of belief are coherent (i.e., consistent with the probability calculus), she applies Bayes' Rule (or the *Principle of Conditionalisation*) to obtain, e.g., the new probability of a proposition A_i:

$$P'(A_i) := P(A_i \mid E) = \frac{P(E \mid A_i) P(A_i)}{P(E)}.$$

---

[2] We use the convention of displaying propositional variables in italics and their values in roman script.

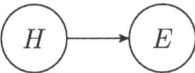

Fig. 8.1 A Bayesian network representing the probabilistic relations between the hypothesis variable H and the evidence variable E

There are pragmatic and epistemic arguments that justify the use of conditionalisation. These arguments are, however, more controversial than in the static case (Pettigrew, 2020).

If the evidence is not fully certain and a further condition (the *Rigidity Condition*) holds, then conditionalisation generalises to Jeffrey conditionalisation (Jeffrey, 2004). In that case, $P'(A_i)$ is determined as follows:

$$P'(A_i) := P(A_i \mid E) \cdot P'(E) + P(A_i \mid \neg E) \cdot P'(\neg E).$$

The most important (though by far not the only) application of Bayesianism in philosophy is confirmation theory, which is concerned with the explication of the notion of 'confirmation': what does it mean that a piece of evidence E confirms a hypothesis H? In a typical scenario, the evidence E is direct, i.e., it is a deductive or inductive consequence of the tested hypothesis H. In this case, the probabilistic relationship between the corresponding propositional variables H and E can be represented by the Bayesian network in Fig. 8.1 (with a probability distribution P defined over it).

Using Bayes' Rule, we can then calculate $P(H)$ and $P'(H) := P(H \mid E)$ (if the evidence becomes certain). We then say that E confirms H if $P'(H) > P(H)$, E disconfirms H if $P'(H) < P(H)$, and E is irrelevant for H if $P'(H) = P(H)$. Confirmation, thus, means probability-raising.

Using the machinery of Bayesian networks, also more complicated testing scenarios (involving, e.g., various auxiliary hypotheses or partially reliable information sources) can be investigated (see, e.g., Bovens & Hartmann, 2003; Osimani & Landes, 2023). Hájek and Hartmann (2010) discuss further epistemological applications of Bayesianism.

Despite these successes, Bayesianism faces a number of foundational problems (see, e.g., Glymour, 1980; Norton, 2011, 2021). In my view, many of these problems are just *modelling challenges* (such as the problem of old evidence, which I will discuss below), while others (such as the possible failure to represent ignorance) may point to a better theory beyond Bayesianism.

I'm not too worried about these difficulties. Bayesianism should be treated just like any other scientific theory (and nothing more), and since all scientific theories are facing problems and challenges, it's hard to expect things to be better in philosophy. At the same time, I think that the successful application of a (scientific or philosophical) theory to many cases speaks largely in favour of the theory in question. In short, when evaluating a philosophical theory, we should also consider its pragmatic utility.

Having said this, I will now identify three further challenges to the Bayesian research programme in the methodology of science and suggest how to address them by extending or modifying the Bayesian research programme.

1. Indirect Evidence

   Standard Bayesianism, as presented so far, assumes that the evidence is direct in the sense that it is a deductive or inductive consequence of the scientific theory under consideration. However, this may not always be the case. Some evidence may be indirect in a sense I will soon make precise. Interesting examples of indirect evidence come from fundamental physics, which is a field where direct empirical evidence is scarce or even non-existent. Here are two examples:

   (a) The no alternatives argument: Does the observation that scientists have not yet found an alternative to string theory (despite a lot of effort and brain power) confirm the theory? Some authors, such as Dawid (2013), think so.
   (b) Analogue simulation: Is it possible to confirm a claim about an empirically unaccessible phenomenon (such as black hole Hawking radiation) by experimenting on a different physical system (e.g., a Bose–Einstein condensate)? Some authors, such as Dardashti et al. (2017), think so.

   Occasionally, indirect evidence has also been called 'non-empirical evidence' (Dawid, 2013). This term is somewhat misleading as in both cases an empirical observation is cited. In the case of the no alternatives argument, it is an observation about the respective scientific community (which has not yet found an alternative theory), and in the case of analogue simulation, it is an observation about another physical system. Hence, 'indirect evidence' seems to be the better term.

   In both cases, it would be very helpful to have other means than providing direct empirical evidence to test the respective theories. But is this really possible? Wouldn't it be too good to be true? Clearly, philosophical theories such as hypothetico-deductivism or Popper's falsificationism dismiss this alleged evidence from the outset. However, this inference may be too quick: While it may well turn out that the alleged examples of indirect evidence are not confirmatory, indirect evidence should not be disregarded because one's favourite theory of confirmation (or corroboration) only allows for deductive evidence. Such theories are not useful for understanding the methodological development of contemporary science. Bayesianism, on the other hand, allows us to analyse confirmation scenarios involving indirect evidence. We will indeed see that indirect confirmation is in principle possible provided that certain conditions hold.

2. New Types of Evidence

   Standard Bayesianism assumes that the evidence is propositional. This is easy to see from an inspection of Bayes' Rule where one has to condition on a proposition representing the evidence. In a learning situation, the probability of this proposition shifts 'by hand' to 1 (in the case of conditionalisation) and the probabilities of all other propositions are in turn updated to make sure that the new probability distribution P' is coherent.

However, there may also be evidence that does *not* lead to a probability shift of any of the propositions in the algebra: some evidence may be non-propositional. Here are two examples:

(a) Structural evidence: The agent may learn, e.g., that the underlying causal network of a set of propositions is such and such. This will lead to an update of the probability distribution. But how could it be modelled? One could add meta-propositions to the algebra which make statements about the causal structure, but this does not seem to be very practicable.

(b) Indicative conditionals: The agent may learn an *indicative conditional* of the form 'If A, then C.' Here the only way to proceed seems to be to condition on the corresponding material conditional, as it can be represented as a Boolean combination of the antecedent and the consequent proposition (and therefore is a proposition itself). But the material conditional is controversial. It is fraught with many problems (but see Williamson (2020) for a recent defense) and, most importantly, it is not at all clear that indicative conditionals are propositions at all (see Douven (2015) for a survey).

We will show below that Bayesianism has the resources to model such learning experiences.

3. Genuinely New Evidence

Standard Bayesianism assumes that the learned proposition is already on the agent's 'radar.' It is expected and is given a prior probability. However, this may not always be the case: some evidence may be genuinely new. Let me explain. In many cases, it is not plausible that agents have prior beliefs about each and every piece of evidence they may learn in the future. However, this is expected from a Bayesian agent. One can only update on a proposition which is already in one's algebra and which has a prior probability attached to it. The following examples from various fields of inquiry raise doubts that this is always possible.

(a) Testimony: Someone told me that there is an excellent new ice cream parlour in my neighbourhood. I update the probability that I get some tasty ice cream today.

(b) Argumentation: We are debating a policy issue and you make a new argument (based, e.g., on a recent scientific finding) which I didn't anticipate at all.

(c) Scientific theory change: An old theory runs into problems and a new theory is proposed. This new theory was unexpected and no one assigned a prior to it. One way that has been proposed to deal with this problem is to argue that the new theory is part of the 'catch-all' of the old theory, i.e., is included in the negation of the old theory (Salmon, 1990). In this case, however, nothing is known about the new theory, and in particular no prior probability is assigned to this new theory (since there may be many other theories in the 'catch-all' set). Accordingly, this proposal is unsatisfactory.

These examples show that the standard Bayesian assumption that the algebra of propositions remains fixed is often a strong idealisation. Logical approaches, such as the AGM model of belief revision (Alchourrón et al., 1985; Hansson, 2022), on the other hand, are not confronted with this problem and can, at least in principle, deal with such cases. This problem for Bayesianism is well known and there is a literature in economics ('awareness') and philosophy (e.g., Bradley, 2017) that deals with it (see also Williamson (2003) and de Canson (2024) for a recent discussion). However, this literature still awaits its application to problems from philosophy of science (such as the problem of theory change).

In the next two sections, I will address the first two challenges in turn. I will give a detailed answer to the third problem on another occasion.

8.3 Challenge 1: Indirect Evidence

The theory of Bayesian networks (Pearl, 1988) is well suited to model confirmation scenarios where there is no direct link between the hypothesis variable H and the evidence variable E. For example, the correlation between H and E may be mediated by a 'common cause' variable X, as illustrated in Fig. 8.2.[3]

To apply this idea to a concrete example, one has to find a variable X which (1) plays an *active role* in the reasoning of the agent and which (2) plausibly screens off H from E. Such variables can indeed be found for the analysis of the no alternatives argument and for the problem of analysing reasoning with analogue simulations. Here is how it works for the no alternatives argument (NAA), which I first present in somewhat more detail than above.

Scientists often argue as follows (P_1 and P_2 are the premises and C is the conclusion of the argument):

P_1: Hypothesis H satisfies several desirable conditions (e.g., it incorporates various scientific principles, it coheres with other established theories, etc.).

P_2: Despite a lot of effort, the scientific community has not yet found an alternative to H.

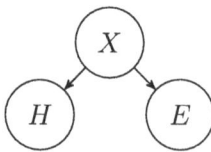

Fig. 8.2 A 'common cause' Bayesian network representing the probabilistic relations between the propositional variables H, E and X

---

[3] For a short introduction to the theory of Bayesian networks, see Hartmann (2021).

Fig. 8.3 The Bayesian network representing the NAA

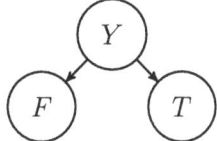

C: H is confirmed.

This argument raises at least two questions: (1) How good are NAAs? And (2) under what conditions, if any, do they work? To address these questions, I propose a Bayesian network model involving the following three variables:

1. The variable T has two values, viz., T: the hypothesis H is true, and ¬T: the hypothesis H is not true.
2. The variable F also has two values, viz., F: the scientific community has not yet found an alternative to H that accounts for the data $\mathcal{D}$ (if there are any) and satisfies the desired constraints $\mathcal{C}$, and ¬F: the scientific community has found an alternative to H that accounts for $\mathcal{D}$ and satisfies $\mathcal{C}$.
3. The variable Y has N values, viz., Y_i: there are exactly i hypotheses which explain $\mathcal{D}$ and fulfil $\mathcal{C}$. (H is one of them.)

Next, we assume that the conditional independencies represented in the Bayesian network in Fig. 8.3 hold. More specifically, we assume that Y screens off T from F, i.e., once the value of Y is known, T and F are independent. I take this to be a plausible assumption.

With this, the following theorem holds. (For details and the proof, see Dawid et al., 2015.)

Theorem 8.1 *We set $P(Y_i) =: y_i$, $P(F \mid Y_i) =: f_i$ and $P(T \mid Y_i) =: t_i$. If (a) f_i and t_i are monotonically decreasing in i, (b) $y_i < 1$ for all i and (c) there is at least one pair (i, j) with $j > i$ such that $y_i, y_j > 0$, $f_i > f_j$ and $t_i > t_j$, then $P(T \mid F) > P(T)$.*

It is interesting to note that the NAA works under rather weak and largely plausible assumptions. Every prior probability distribution that satisfies the conditions stated in the theorem will result in the confirmation of T once F is observed. But what about the assumptions? Are they really plausible? Assumption (a) is plausible if we think of the confirmation situation in terms of a sampling scenario. Assumption (b) is perhaps the weakest. It says that the agent is uncertain about the number of alternatives to the theory under consideration. But doesn't the *underdetermination thesis* teach us that there are always infinitely many alternatives to a given theory that imply the given data (if there are any)? In this case, one should set $P(Y_\infty) = 1$ so that the NAA would work. Clearly, a defender of the NAA has to respond to this worry (see Dawid (2013) for a response). Finally, assumption (c) is related to assumptions (a) and (b), to which it does not add much which could be controversial.

The case of analogue simulation can be analysed in a similar way (Dardashti et al., 2019). In general, the analysis of scenarios involving indirect evidence requires the specification of (1) at least one other 'active' variable (besides H and E) and of (2) a causal structure which represents the conditional probabilistic independencies that hold amongst the variables. I conclude that Bayesianism (unlike deductive theories of confirmation or corroboration) has the resources to model and investigate scenarios involving indirect evidence. The changes or additions that need to be made are rather insignificant and at best concern the 'protective belt' of the Bayesian research programme.

8.4 Challenge 2: New Types of Evidence

Let us now explore how the learning of an indicative conditional can be modelled in Bayesianism. To start with, consider the following example (the 'Ski Trip Example' from Douven and Dietz (2011)):

> Harry sees his friend Sue buying a skiing outfit. This surprises him a bit, because he did not know of any plans of hers to go on a skiing trip. He knows that she recently had an important exam and thinks it unlikely that she passed. Then he meets Tom, his best friend and also a friend of Sue, who is just on his way to Sue to hear whether she passed the exam, and who tells him, 'If Sue passed the exam, then her father will take her on a skiing vacation.' Recalling his earlier observation, Harry now comes to find it more likely that Sue passed the exam.

To model Harry's learning experience, we first note that there are three propositional variables (B, E and S) with the following (positive) values involved here: (1) E: 'Sue passes the exam,' (2) S: 'Sue is invited on a ski trip', and (3) B: 'Sue buys a skiing outfit.' We assume that Harry has a prior probability distribution over these three propositional variables and then learns two items of information: (I_1) B and (I_2) 'If E, then S.' Conditionalising on B and the *material conditional* $E \supset S \equiv \neg E \vee S$, one can show that the probability of E increases under plausible conditions, which is what we—and Harry—expect (Eva et al., 2020). This becomes especially clear if one makes the additional assumption that E is probabilistically independent of B given S, or in more formal terms: $E \perp\!\!\!\perp B \mid S$. This suggests the 'chain structure' depicted in Fig. 8.4.

So far, so good. However, representing the indicative conditional A → C by the material conditional $A \supset C \equiv \neg A \vee C$ has two problems. Firstly, it cannot handle *non-extreme* conditionals, i.e., when there are exceptions and when the conditional is not learnt with certainty. Interestingly, also Jeffrey conditionalising

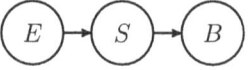

Fig. 8.4 The Bayesian network for the Ski Trip Example

on the material conditional leads to counter-intuitive consequences in these cases. Secondly, it cannot deal with conditionals which are uttered by an only partially reliable information source. This, however, is typically the case and in line with the general idea behind Bayesianism that certainties are hard to find (see Jeffrey, 1983).

The second problem is still unsolved (see Collins et al. (2020) for some preliminary ideas). To address the first problem, the *distance-based approach* to Bayesianism can be adapted (Diaconis & Zabell, 1982). The idea behind this approach is that it is rational to change one's degrees of belief only minimally once one learns new information. Call this the *Principle of Conservativity*, which is also used in other accounts of belief revision (such as the AGM model). More specifically, we consider an agent whose degrees of belief are represented by a prior probability distribution P. The agent then learns some new information that poses *probabilistic constraints* on the posterior probability distribution Q. For examples, if the agent learns that the evidence E obtains, then the corresponding constraint is $Q(E) = 1$.

To work out this proposal, we need to choose a measure for the 'distance' between two probability distributions. For this, the following class of measures turns out to be especially useful:

Definition 8.1 (*f*-**Divergence (Csiszár, 1967)**) Let $S_1, \ldots, S_n$ be the possible values of a random variable S over which probability distributions P and Q are defined and let f be a convex function with $f(1) = 0$, $p_i := P(S_i)$ and $q_i := Q(S_i)$. Then

$$D_f(Q \parallel P) := \sum_{i=1}^{n} p_i \cdot f(q_i/p_i).$$

Many well-known probabilistic divergences are f-divergences. For example, the Kullback–Leibler divergence (KL) obtains for $f(t) = t \log t$. The inverse KL-divergence, the χ^2-divergence and the Hellinger distance follow accordingly. Note that f-divergences are not necessarily symmetrical and that they may violate the triangle inequality. They are therefore not distance functions. And yet, f-divergences are particularly suitable for the present purpose because it can be shown that they yield (Jeffrey) conditionalisation if the agent learns a piece of propositional evidence (Diaconis & Zabell, 1982; Eva et al., 2020).

Theorem 8.2 *An agent considers the propositional variables H and E and has a probability distribution P defined over them. She then learns that $Q(E) =: e' < 1$. Minimising an f-divergence between Q and P taking this constraint into account yields $Q(H) = P(H \mid E) \cdot e' + P(H \mid \neg E) \cdot (1-e')$. This is Jeffrey conditionalisation.*

This is an important result, showing that all f-divergences imply Jeffrey conditionalisation (which I regard as a plausible learning rule) when the agent learns a piece of propositional evidence. (Interestingly, the *Rigidity Condition* is automatically satisfied in this case and does not need to be imposed as an additional constraint.) At the same time, it turns out that all f-divergences are indistinguishable in the case of learning propositional evidence. It is therefore not necessary to decide in favour of a particular f-divergence.

If one learns the (strict) indicative conditional 'If A, then C' from a perfectly reliable source, then the probabilistic constraint on Q is simply $Q(C\,|\,A) = 1$. Nothing more is required. In particular, nothing needs to be said about the propositional status of an indicative conditional. One only needs to specify which probabilistic constraint applies to Q when learning an indicative conditional. Minimising an f-divergence between Q and P taking this constraint into account then yields the same new probability distribution for all f-divergences. The situation is therefore similar to learning a piece of propositional evidence. Interestingly, the new probability distribution is identical with the one which one obtains by conditioning on the corresponding material conditional: $Q = P'$. This is easy to see by noting that $Q(C\,|\,A) = 1$ if and only if $Q(A \supset C) = 1$ (provided that $Q(A) > 0$). Hence, the distance-based approach to Bayesianism justifies the use of the material conditional if the learnt indicative conditional is strict and if the information source is perfectly reliable.

Let us now consider non-strict indicative conditionals (from a perfectly reliable information source), which are, as I stated already, much more natural from a Bayesian point of view. In this case the constraint is $Q(C\,|\,A) < 1$ and one finds that different f-divergences yield different new probability distributions. We therefore have to 'put our money' on one specific f-divergence if we want to model these cases. But on which? To proceed, we have the following three options: First, one can accept the additional epistemic norm *Minimising Inaccuracy* (as in Epistemic Utility Theory) along with the Principle of Conservativity. Then it can be shown that the inverse KL-divergence is the unique probabilistic divergence (Eva et al., 2020). Second, one can try to identify other diachronic norms which (hopefully) restrict the class of admissible divergences. Third, one can explore *empirically* which f-divergence is best. However, the answer to this question may vary with the respective context. In any case, it is still too early to decide which of these options is the right one. And so it is currently best to continue investigating all three options.

As should be clear by now, I do not think that conditionalisation ('Bayes Rule') or Jeffrey conditionalisation are in the Lakatosian hard core of the Bayesian research programme. The Principle of Conditionalisation often leads to the right results (in particular when the evidence learned is propositional), but it should not be considered one of the central elements of Bayesianism—at least if we want the scope of Bayesianism to extend beyond the learning of propositional evidence. There are many other types of evidence an agent may learn, and the corresponding updating can often not be modelled as an instance of conditionalisation, as we have seen for non-strict conditionals. Modeling the learning of structural evidence also requires that one use a different updating rule, which makes sure that the new probability distribution satisfies various probabilistic conditional independencies. Besides, even if we learn a proposition, there may be other relevant propositional variables involved in the reasoning situation whose probability assignment we might want to consider fixed across the update. Such additional constraints cannot be taken into account when using conditionalisation.

The distance-based approach, on the other hand, justifies (Jeffrey) conditionalisation (if it can be applied) and is more general and accordingly worthy of further investigation. I therefore suggest that the *Principle of Conservativity* for updating is in the Lakatosian hard core of the Bayesian research programme in the methodology of science. It needs to be spelled out in detail, in a given context, by choosing a specific probabilistic divergence. Which one of these divergences is best will probably depend on the context.

8.5 Further Challenges

Bayesianism faces a number of further challenges. Here are some of them.

1. The Problem of Old Evidence
   If the agent assigns a prior probability of 1 to the evidence, i.e., if $P(E) = 1$, then E cannot be learnt (because the probability of E does not change) and it therefore makes no sense to apply an updating rule. Consequently, so-called old evidence (i.e., evidence to which the agent assigns already a prior probability of 1) cannot confirm a hypothesis. This contradicts the practice of science, as Glymour (1980) has pointed out. In response, Bayesians have suggested two ways in which the respective hypothesis can be given a probability increase in scenarios with old evidence:

   (a) Work with a counterfactual probability function that assigns a prior probability of less than 1 to E (e.g., Howson, 1991).
   (b) Argue that the agent learns something else than the old evidence E. For example, Garber (1983) suggested that the agent learns that E is a logical consequence of H, and argued that one should therefore condition on the new proposition X: H → E.

   Glymour (1980) has already anticipated, insightfully discussed, and largely rejected both ways to address the anomaly. The main problem with the first way out is that the proposal seems rather ad hoc and leaves open many questions (such as: how far should we go back in time?). The problems with the second way out are questions regarding the possibility of logical learning and shortcomings of the specific models that have been suggested (see Sprenger and Hartmann (2019) for a discussion). I favour a solution which replaces 'H logically implies E' by X: 'H adequately explains E,' and by introducing another proposition Y: 'The best competitor of H adequately explains E.' One can then formulate a number of plausible conditions under which X confirms H (see Hartmann and Fitelson (2015) and Eva and Hartmann (2020) for details).

   Lakatos might have judged that the problem of old evidence is a problem 'one has oneself created while trying to solve the original problem' (to repeat a quote from the beginning of this paper). However, it should be noted that the problem of old evidence is an important one to solve, and the way in which it can actually be solved not only represents an internal progress, but also helps us to better understand how scientists (should) reason.

2. Scientific Theory Change

Earman (1992) and Salmon (1990) (see also Worrall, 2000) have discussed Bayesian accounts of Kuhn's influential theory of theory change. They were not entirely successful. This is not least due to the fact that they have not considered all aspects of Kuhn's theory (Farmakis, 2008), for example, has noted that they have left out the incommensurability issue. But perhaps a full Bayesian account of Kuhn's theory is not necessary. Kuhn may well be right that there is no 'algorithm' that helps us decide once and for all when a particular theory should be abandoned. Feyerabend also made this point in his response to Lakatos when he wrote, 'if you are permitted to wait, why not wait a little longer?' (1970, 215). And even Lakatos argued that there is no 'instant rationality' and that we can provide a rational and objectivist account of theory change only in retrospect, when the internal history is available in the form of a rational reconstruction. Nevertheless, I would like to argue that Bayesianism can help us in everyday scientific reasoning and argumentation, e.g., when we reason about whether we should abandon a theory or research programme and look for an alternative instead.

Bayesianism lends itself here because theorising takes place in the realm of uncertainty, and scientists, like all of us in everyday life, have to make decisions all the time. These decisions should be rational, and Bayesian decision theory provides a useful and justifiable framework for achieving this while still allowing for subjective judgements by scientists. For example, an individual scientist may be faced with the decision of whether to maintain and continue researching the current theory. Perhaps this will lead to a major discovery? And perhaps an observed anomaly can be explained after all. (Remember that Lakatos taught us that every research programme evolves in an 'ocean of anomalies.') One does not know with certainty in advance. A reconstruction of the decision situation that makes explicit the different propositions that the scientist considers and how they are related, together with the corresponding (subjective) probability distribution, can help the agent to make better-reasoned decisions. For example, in the case mentioned above, consider how likely the agent thinks it is that a model can be found within the given research programme (or paradigm) that explains the evidence. Perhaps the agent initially assigns a fairly high probability to this proposition, which she then updates in the light of her (possibly unsuccessful) attempts to find such a model. At some point she will give up, and if many other scientists do the same, the theory (or research programme) will eventually be replaced by another. This thought process can be modelled, wherein Bayesianism proves useful without promising more than it can deliver, which is what one should expect from a progressive research programme.

3. Collective Reasoning and Argumentation

Standard Bayesianism is a philosophical theory in which a single agent is at the centre. This agent maintains a set of propositions that she believes more or less strongly and updates in the light of new evidence according to a particular rule.

As we have seen, this simple approach can be used to analyse a wide range of issues in the philosophy of science.[4] However, it turns out that science happens in a social context, which should be taken into account if Bayesianism is to critically accompany current science. For example, scientists try to convince each other and then update their individual probability distributions by taking into account the information coming from other scientists. Or a committee chair (debating environmental policy measures, for example) may consult scientific experts to make the best decision on the issue based on the experts' probabilistic judgements. There may also be situations where we want to assign a probability distribution to a group, e.g., a scientific community. Something like this could be helpful, for example, if we want to further reconstruct Kuhn's philosophy of science in Bayesian terms. It will be interesting to address these questions and many others in future work. There is no reason why the Bayesian research programme in the philosophy of science should not be further developed in this direction, especially since much work has already been done on which one can build. This again underlines the main point I want to make in this paper, namely that the Bayesian research programme in the philosophy of science is progressive.

8.6 Conclusion

Bayesianism is a progressive scientific research programme in the methodology of science. It is closely related to other Bayesian research programmes, as Bayesianism is not only flourishing in philosophy, but also in cognitive science ('the new paradigm'), neuroscience ('the Bayesian brain,' 'the free energy principle') and artificial intelligence. Lakatos's philosophy of science is useful in reconstructing these Bayesian research programmes. However, I have argued that it is more plausible to place the Principle of Conservativity at the hard core of the Bayesianism research programme in the methodology of science, rather than conditionalisation ('Bayes' rule') or Jeffrey conditionalisation. I have argued that this principle (if, as suggested, it is specified using f-divergences) justifies (Jeffrey) conditionals and allows updating on the basis of other types of evidence (such as indicative conditionals). The relevant research programme is progressive in that it successfully addresses various anomalies (such as the problem of old evidence) and is able to solve new problems. Many other problems are still open and await a Bayesian treatment.

---

[4] In discussing the NAA, we were dealing with an issue that the scientific community is concerned about. However, we did not model the probability functions of the individual scientists, but considered an external agent who assigns a probability function to the scientific community and updates it accordingly.

Despite these successes of the Bayesian research programme in the methodology of science, it is important to also investigate alternative approaches, such as imprecise probabilities (e.g., Augustin et al., 2014) or ranking theory (Spohn, 2012), and to develop criteria for how to evaluate and compare the results. For a similar plea in relation to Bayesian cognitive science, see Colombo et al. (2021).

Acknowledgments I would like to thank Christopher von Bülow and an anonymous reviewer for several suggestions for improvement. Thanks also go to my co-authors Benjamin Eva, Richard Dawid, Soroush Rafiee Rad and Jan Sprenger, with whom some of the results discussed here were found.

I dedicate this essay to the memory of Colin Howson—a much-missed fellow Bayesian and friend (and not least a source of entertaining stories about Lakatos).

References

Alchourrón, C. E., Gärdenfors, P., & Makinson, D. (1985). On the logic of theory change: partial meet contraction and revision functions. *Journal of Symbolic Logic, 50*, 510–530.
Augustin, T., Coolen, F. P. A., de Cooman, G., & Troffaes, M. C. M. (Eds.). (2014). *Introduction to imprecise probabilities*. Wiley.
Bovens, L., & Hartmann, S. (2003). *Bayesian epistemology*. Clarendon Press.
Bradley, R. (2017). *Decision theory with a human face*. Cambridge University Press.
Collins, P. J., Krzyżanowska, K., Hartmann, S., Wheeler, G., Hahn, U. (2020). Conditionals and testimony. *Cognitive Psychology, 122*, 101529.
Colombo, M., Elkin, L., & Hartmann, S. (2021). Being realist about Bayes, and the predictive processing theory of mind. *The British Journal for the Philosophy of Science, 72*(1), 185–220.
Csiszár, I. (1967). Information-type measures of difference of probability distributions and indirect observation. *Studia Scientiarum Mathematicarum Hungarica, 2*, 229–318.
Dardashti, R., Hartmann, S., Thébault, K., & Winsberg, E. (2019). Hawking radiation and analogue experiments: A Bayesian analysis. *Studies in History and Philosophy of Science Part B: Studies in History and Philosophy of Modern Physics, 67*, 1–11.
Dardashti, R., Thébault, K. P. Y., & Winsberg, E. (2017). Confirmation via analogue simulation: What dumb holes could tell us about gravity. *The British Journal for the Philosophy of Science, 68*(1), 55–89.
Dawid, R. (2013). *String theory and the scientific method*. Cambridge University Press.
Dawid, R., Hartmann, S., Sprenger, J. (2015). The No alternatives argument. *The British Journal for the Philosophy of Science, 66*(1), 213–234.
de Canson, C. (2024). The nature of awareness growth. *Philosophical Review, 133*(1), 1–32.
Diaconis, P., & Zabell, S. L. (1982). Updating subjective probability. *Journal of the American Statistical Association, 77*(380), 822–830.
Dorling, J. (1979). Bayesian personalism, the methodology of scientific research programmes, and Duhem's problem. *Studies in History and Philosophy of Science Part A, 10*(3), 177–187.
Douven, I. (2015). *The epistemology of indicative conditionals*. Cambridge University Press.
Douven, I., & Dietz, R. (2011). A puzzle about Stalnaker's hypothesis. *Topoi, 30*, 31–37.
Earman, J. (1992). *Bayes or bust? A critical examination of Bayesian confirmation theory*. MIT Press.
Eva, B., & Hartmann, S. (2020). On the origins of old evidence. *Australasian Journal of Philosophy, 98*(3), 481–494.
Eva, B., Stephan, H., & Rad, S. R. (2020). Learning from conditionals. *Mind, 129*(514), 461–508.
Farmakis, L. (2008). Did Tom Kuhn actually meet Tom Bayes? *Erkenntnis, 68*(1), 41–53.

Feyerabend, P. (1970). Consolations for the specialist. In Lakatos and Musgrave (1970) (pp. 197–230).
Garber, D. (1983). Old evidence and logical omniscience in Bayesian confirmation theory. In J. Earman (Eds.), *Testing scientific theories*, Volume 10 of *Minnesota studies in the philosophy of science* (pp. 99–131). University of Minnesota Press.
Glymour, C. (1980). Why I am not a Bayesian. In *Theory and evidence* (pp. 63–93). Princeton University Press.
Hájek, A., & Hartmann, S. (2010). Bayesian epistemology. In J. Dancy, E. Sosa, & M. Steup (Eds.), *A companion to epistemology* (2nd ed., pp. 93–106). Blackwell.
Hansson, S. O. (2022). Logic of belief revision. In E. N. Zalta (Ed.), *The Stanford encyclopedia of philosophy* (Spring 2022 edition).
Hartmann, S. (2021). Bayes nets and rationality. In M. Knauff & W. Spohn (Eds.), *The handbook of rationality* (pp. 253–264). MIT Press.
Hartmann, S, & Fitelson, B. (2015). A new Garber-style solution to the problem of old evidence. *Philosophy of Science*, *82*(4), 712–717.
Howson, C. (Ed.). (1976). *Method and appraisal in the physical sciences: The critical background to modern science, 1800–1905*. Cambridge University Press.
Howson, C. (1991). The 'old evidence' problem. *The British Journal for the Philosophy of Science*, *42*(4), 547–555.
Howson, C., & Urbach, P. (2006). *Scientific reasoning: The Bayesian approach* (3rd ed.). Open Court.
Jeffrey, R. (1983). Bayesianism with a human face. In J. Earman (Ed.), *Testing scientific theories*, Volume 10 of *Minnesota studies in the philosophy of science* (pp. 133–156). University of Minnesota Press.
Jeffrey, R. (2004). *Subjective probability: The real thing*. Cambridge University Press.
Lakatos, I. (1968). Changes in the problem of inductive logic. In I. Lakatos (Ed.), *The problem of inductive logic* (pp. 315–416). North Holland.
Lakatos, I. (1976). History of science and its rational reconstructions. In Howson (1976) (pp. 1–39).
Lakatos, I., & Musgrave, A. (Eds.). (1970). *Criticism and the growth of knowledge: proceedings of the international colloquium in the philosophy of science*. Cambridge University Press.
Mayo, D. G. (1996). *Error and the growth of experimental knowledge*. University of Chicago Press.
Nanay, B. (2010). Rational reconstruction reconsidered. *Monist*, *93*(4), 598–617.
Norton, J. D. (2011). Challenges to Bayesian confirmation theory. In P. S. Bandyopadhyay & M. R. Forster (Ed.), *Philosophy of statistics*, Volume 7 of *Handbook of the philosophy of science* (pp. 391–440). Elsevier.
Norton, J. D. (2021). *The material theory of induction*. BSPS open series. University of Calgary Press.
Osimani, B., & Landes, J. (2023). Varieties of error and varieties of evidence in scientific inference. *The British Journal for the Philosophy of Science*, *74*(1), 117–170.
Pearl, J. (1988). *Probabilistic reasoning in intelligent systems: Networks of plausible inference*. Morgan Kaufmann.
Pettigrew, R. (2016). *Accuracy and the laws of credence*. Oxford University Press.
Pettigrew, R. (2020). *Dutch book arguments*. Cambridge University Press.
Salmon, W. C. (1990). Rationality and objectivity in science, *or* Tom Kuhn meets Tom Bayes. In C. W. Savage (Ed.), *Scientific theories*, Volume 14 of *Minnesota studies in the philosophy of science* (pp. 175–204). University of Minnesota Press.
Spohn, W. (2012). *The laws of belief: Ranking theory and its philosophical applications*. Oxford University Press.
Sprenger, J., & Hartmann, S. (2019). *Bayesian philosophy of science*. Oxford University Press.
Weisberg, J. (2011). Varieties of Bayesianism. In D. M. Gabbay, S. Hartmann, & J. Woods (Eds.), *Inductive logic*, Volume 10 of *Handbook of the history of logic* (pp. 477–551). Elsevier.

Williamson, J. (2003). Bayesianism and language change. *Journal of Logic, Language and Information*, *12*(1), 53–97.
Williamson, T. (2020). *Suppose and tell: The semantics and heuristics of conditionals*. Oxford University Press.
Worrall, J. (2000). Kuhn, Bayes and 'Theory-Choice': How revolutionary is Kuhn's account of theoretical change? In R. Nola & H. Sankey (Eds.), *After Popper, Kuhn and Feyerabend: Recent issues in theories of scientific method*, Volume 15 of *Australasian studies in history and philosophy of science*. Springer.

Open Access This chapter is licensed under the terms of the Creative Commons Attribution-NonCommercial-NoDerivatives 4.0 International License (http://creativecommons.org/licenses/by-nc-nd/4.0/), which permits any noncommercial use, sharing, distribution and reproduction in any medium or format, as long as you give appropriate credit to the original author(s) and the source, provide a link to the Creative Commons license and indicate if you modified the licensed material. You do not have permission under this license to share adapted material derived from this chapter or parts of it.

The images or other third party material in this chapter are included in the chapter's Creative Commons license, unless indicated otherwise in a credit line to the material. If material is not included in the chapter's Creative Commons license and your intended use is not permitted by statutory regulation or exceeds the permitted use, you will need to obtain permission directly from the copyright holder.

Chapter 9
Lakatos's Naturalism(s): Distinguishing Between Rational Reconstructions and Normative Explanations

Thodoris Dimitrakos

Abstract In the present paper, I argue that, contrary to the standard critique, the Lakatosian conception of rational reconstructions is far from aiming at rigging the historical record and at turning history of science into parody. On the contrary, it can be a fruitful notion which creates an intermediate logical space between eliminative (or scientific) naturalism and aprioristic philosophy with regard to the philosophical comprehension of scientific change. However, I also argue that in order to be so it needs to be explicated and revised in the light of the examination of Lakatos's attitude against naturalism. In particular, my argumentation is developed in three steps. First, I claim that the main reason for rejecting the criticism against the Lakatosian notion of 'rational reconstruction' is that forging the historical record would ruin the role of rational reconstructions as an arbiter for the assessment of rival philosophical theories of rationality. Second, I argue that we have to distinguish between the use of the term in the historiography of science and the use in the philosophy of science. I also suggest that we have to distinguish between the notion of 'rational reconstruction' and the notion of 'normative explanation,' and that we should employ the latter when it comes to historiography. Finally, I argue that Lakatos's appeal to an alleged (Popperian or Fregean) third world of standards of rationality leads his view into incoherence. I propose a way to avoid incoherence by (a) abolishing the third-word metaphor and (b) replacing the inter-methodology, quasi-empirical process of assessment between rival theories of scientific rationality with an intra-methodology conception of the evaluation process.

Keywords Imre Lakatos · Rational reconstruction · Normative explanation · Philosophy of history · Naturalism

---

T. Dimitrakos (✉)
University of Patras, Patras, Greece
e-mail: thdimitrakos@upatras.gr

9.1 Introduction

Lakatos's philosophical account of scientific change is characterized by the demarcationist ((Lakatos, 1978a, pp. 108–110) defense of scientific rationality and the historicist premise that sophisticated philosophy of science should be history-informed. The key notion for both features is that of 'rational reconstruction,' which entails the well-known internal/external ('i/e' hereafter) distinction in the history of science. In his famous, even if not original,[1] paraphrase of Kant's dictum, (Lakatos, 1978a, p. 102) stated that '[p]hilosophy of science without history of science is empty; history of science without philosophy of science is blind'. The passage indicates a relation of mutual dependence. Philosophy of science depends on the history of science because the latter can be the ground of a quasi-empirical process of assessing rival philosophical theories of scientific rationality (or rival methodologies in Lakatos's terminology). In short, Lakatos claimed that he discovered 'a specific method for using history of science as an arbiter of some authority when it comes to debates in philosophy of science' (Lakatos, 1978a, p. 168). On the other hand, history of science depends on philosophy of science because the latter supplies the former with general conceptions about scientific rationality. '[. . .] [A]ll histories of science are inevitably methodology-laden and [we] cannot avoid "rational reconstructions"' (Lakatos, 1978b, p. 110).

The double use of the term 'rational reconstruction,' along with a characteristically uncharitable interpretation of the Lakatosian view, led to the repudiation (and even mockery) of it ((Kuhn, 1971, p. 143; Holton, 1978, p. 106; McMullin, 1970, p. 32). The critics agreed that the rational reconstructions offered a mere caricature of the actual history of science and therefore the notion could not be taken seriously. I suggest that the fact that Lakatos conflated[2] the use of the term as a historiographical category and as part of a philosophical method for appraising different philosophical theories of scientific rationality contributed crucially to this misinterpretation.

In the present paper, I argue that, contrary to the critics, the Lakatosian notion of rational reconstruction is far from aiming at rigging the historical record and at turning the history of science into 'philosophy fabricating examples' (Kuhn, 1971, p. 143). On the contrary, it can be a fruitful notion which creates an intermediate logical space between eliminative (or scientific) naturalism and aprioristic philosophy ('Euclidean methodologies') with regard to the philosophical comprehension of scientific change. This intermediate logical space consists in a moderate (or liberal[3]) naturalist perspective which avoids the relativism entailed by the eliminative version

---

[1] See (Hanson, 1963).

[2] See (Lakatos, 1978a, p. 138).

[3] 'Liberal naturalism makes room in its vision of the world for nonscientific realities (that are not posits of successful scientific explanations) and nonscientific knowledge or understanding'. On the contrary, 'A scientistic naturalism is a reductive or eliminative scientific naturalism, which treats the full extent of nature as exhausted by, say, the scientific image of the world' (de Caro & Macarthur, 2022, p. 2).

of naturalism and the monolithic conception of scientific rationality stemming from aprioristic (Euclidean) methodologies.[4]

In particular, my argumentation will be developed in three steps. First, I claim that the main reason for rejecting the uncharitable interpretation of the Lakatosian perspective is that forging the historical record would ruin the role of rational reconstructions as an arbiter for the assessment of rival methodologies. Second, I distinguish between the notion of rational reconstruction which should be part of a philosophical method for appraising different philosophical theories of scientific rationality, and the notion of normative explanation which should be taken as a historiographical category. Given this conceptual distinction, I argue, Lakatos's view is sustainable and useful against the contemporary eliminativist trends in historiography of science like the Strong Programme in the Sociology of Scientific Knowledge. Finally, I argue that Lakatos's adoption of the Popperian third world (Lakatos, 1978a, pp. 169,191,209; 1978b, pp. 108–119), along with the Lakatosian way of evaluating rival methodologies, leads his view into incoherence. In particular, it makes his account vulnerable to the so-called 'dilemma of case studies' (Pitt, 2001). However, I suggest that the incoherence can be overcome by adopting the framework of liberal naturalism and by substituting Lakatos's inter-methodology with an intra-methodology testing procedure.

It should be clear that, apart from offering a charitable interpretation of Lakatos's account, I also provide a revision of it. The revisionist successor to the Lakatosian view that I propose is adjusted to the major shifts in the philosophical agenda that followed Lakatos's era. It is, first and foremost, adjusted to the abandonment of the demarcation problem[5] and the relevant task of developing grand theories of scientific rationality. In this sense, it might seem that the purposes of my revisionist account are different or even asymmetrical to the purposes of the initial Lakatosian view. However, as I will attempt to show at the very end of the paper, both my and Lakatos's purposes, despite their differences, are concerned with the interplay between the normative analysis provided by philosophy and the empirical data provided by history.

9.2 Why the Lakatosian Conception of History Cannot Be a Parody

Lakatos's demarcationist perspective comes with a meta-criterion, i.e., a criterion for evaluating different demarcation criteria proposed by different philosophical

---

[4] According to Lakatos (1978b, p. 112), those are two sides of the same coin since relativism or scepticism 'thrives on the defeats of earlier [monolithic] versions of the demarcationist programme'.

[5] See (Laudan, 1983).

accounts.[6] His innovative idea is that conceptual analysis alone is not sufficient for assessing the different philosophical theories of scientific rationality and that history of science can provide the ground for such assessment. Given a philosophical theory of scientific rationality, one can retrospect the historical course of science in order to examine which episodes coincide with the given theory and which do not. The sum total of the former episodes consists in the 'internal' history while the sum total of the latter episodes consists in the 'external' history. Lakatos (1978a, p. 102) explains that the distinction is not merely between the intellectual and sociological history of science.[7] The i/e distinction is that between a normative and an empirical history of science and thus it implies an accordance (or not) with a given methodology.[8] It follows that the distinction is framework specific. It essentially depends on the methodology at hand. Different methodologies draw the line between internal and external history in completely different ways (Lakatos, 1978a, pp. 118–122 and 190). However, the line itself is ineliminable since there are always external influences in the behavior of even the greatest scientists, and therefore there an empirical historical approach to explaining cases of irrationality will always be needed (Lakatos, 1978a, pp. 105, 114). Broadly speaking,[9] methodologies which are capable of presenting a larger set of historical episodes as rational should be evaluated as better in comparison to methodologies which leave too many episodes to the external history. '[T]he hallmark of a relatively weak internal history (in terms of which most actual history is either inexplicable or anomalous) is that it leaves too much to be explained by external history. When a better rationality theory is produced, internal history may expand and reclaim ground from external history' (Lakatos, 1978a, p. 134). The testing process takes the form of successive case studies which, in general, consist in the following procedure: '(1) one gives a rational reconstruction; (2) one tries to compare this rational reconstruction with actual history and to criticize both one's rational reconstruction for lack of historicity and the actual history for lack of rationality' (Lakatos, 1978a, p. 53).

---

[6] Lakatos was particularly influenced by Popper, who used to consider the demarcation problem, i.e., 'the problem of drawing a line (as well as this can be done) between the statements, or systems of statements, of the empirical sciences, and all other statements–whether they are of a religious or of a metaphysical character, or simply pseudo-scientific' ((Popper, 1962, p. 38), as the main problem of philosophy of science. See (Lakatos, 1978a, p. 168).

[7] As Kuhn stresses, the terms 'are used by Lakatos in novel and unexpected ways'. 'The 'internal' in Lakatos's sense [...] is closely equivalent to 'rational' in the ordinary sense'. While, '[i]n standard usage among historians, internal history is the sort that focuses primarily or exclusively on the professional activities of the members of a particular scientific community (Kuhn, 1970, pp. 137, 141 and 140 respectively).

[8] Methodology is Lakatos's term for the philosophical theories of scientific rationality. According to this terminology, methodology lacks its traditional Cartesian dimension. For it does not signify a pre-established set of mechanical or algorithmic rules in order to solve a problem. The term 'normative methodology' means simply 'directions for the appraisal of solutions already there' (Lakatos, 1978a, p. 140).

[9] I will come back to the details of Lakatos's method for appraising different methodologies in the last part of this paper.

This view, combined with the suggested dependence of historiography by the normative methodologies and accompanied by Lakatos's few verbal exaggerations,[10] gave rise to the well-known accusation that his conception of history is a parody rather than actual history. Thomas Kuhn (1971, p. 143) wrote that '[w]hat Lakatos conceives as history is not history at all but philosophy fabricating examples'. Ernan McMullin (1970, p. 32) similarly stressed: 'The notion of a "rational reconstruction" [...] precludes the idea that these examples are to serve as historical illustration in the ordinary sense. Rather, they are imaginary or quasi-imaginary examples, recounting what ought to have happened in the course of development of physical hypotheses [...]'. Both Kuhn and McMullin charge Lakatos's assessment process with circularity. Briefly put, their argument can be reconstructed as follows: The Lakatosian assessment process entails the comparison between the rational reconstruction according to a normative methodology and the actual history. However, given Lakatos's conviction about the methodology-ladenness of the history of science, the 'actual history' is nothing more than a fabricated narrative in light of a given methodology. Therefore, the comparison can never lead to a revision of the methodology in question. The assessment process is rigged because it is circular. Furthermore, this way of practicing history of sciences violates the 'internal criteria of the historian's craft' (Kuhn, 1971, p. 142).

I claim that this is a quite uncharitable interpretation for three main reasons. First, as far as verbal exaggerations are concerned, Lakatos made explicit amendments. In defending himself against the accusation of warping the historical data, he wrote:

> 'This [kind of] charge stems probably from a rather unsuccessful joke of mine. Some years ago I wrote that one way to indicate discrepancies between history and its rational reconstruction is to relate the internal history in the text, and indicate in the footnotes how actual history "misbehaved" in the light of its rational reconstruction. Of course, such parodies may be written, and may even be instructive; but I never said that this is the way in which history actually ought to be written and, indeed, I never wrote history in this way except for one occasion.' (Lakatos, 1978a, p. 192)

Second, it should be clear that Lakatos's main concern was not historiographical but philosophical. Rational reconstructions are first and foremost a philosophical tool for assessing different philosophical theories of scientific rationality or, which is the same in this context, different demarcation criteria. The following passage is clear:

---

[10] Maybe the most famous instantiation of this kind of exaggeration is the footnote quote: 'One way to indicate discrepancies between history and its rational reconstruction is to relate the internal history in the text, and indicate in the footnotes how actual history "misbehaved" in the light of its rational reconstruction' (Lakatos, 1978a, 120). But see also the following: '[...] in constructing internal history the historian will be highly selective: he will omit everything that is irrational in the light of his rationality theory. But this normative selection still does not add up to a fully fledged rational reconstruction [...] Internal history is not just a *selection* of methodologically interpreted facts: it may be, on occasions, their *radically improved version*' (Lakatos, 1978a, 119, emphasis in the original).

'In this paper I have proposed a "historical" method for the evaluation of rival methodologies. The arguments were *primarily* addressed to the philosopher of science and aimed at showing how he can – and should – learn from the history of science.' (Lakatos, 1978a, p. 138, emphasis added)

If this is the case, suggesting that the historical data should be forged seems insanely inconsistent. The criticism implies that Lakatos holds, let me say by paraphrasing Putnam's well-known metaphor, a dough-view of the historical record. According to this metaphor, '[t]he things independent of all conceptual choices are the dough; our conceptual contribution is the shape of the cookie cutter' (Putnam, 1987, p. 32). Respectively, Kuhn and McMullin imply that, according to Lakatos, the historical record is the dough and the philosophical theories of rationality the cookie cutter; hence, the former can neither provide cognitive resistance nor lead to the revision of the latter. But then the notion of discrepancy is simply impossible, and so is evaluating different methodologies on the ground of their capacity to incorporate into their own context-specific internal history a major part of the historical record. Given the dough-view of history, the latter can neither undermine nor corroborate any particular view of scientific rationality. In the light of this incompatibility, we should conclude that a minimally charitable interpretation of the Lakatosian conception of history would never take it as aiming to manipulate the historical data.[11] In short, if rational reconstruction is a tool for assessing competing philosophical accounts of science, then it cannot consist in the warping of the historical record.[12]

Third, and in support of the abovementioned conclusion, Lakatos did provide a revision of his own methodology in light of the historical record concerning the Copernican revolution. According to the initial reconstruction of the episode, Copernicus's research program was presented as fully progressive only after 1610, i.e., after Galileo's discovery of the phases of Venus. Lakatos (1978a, p. 184) was dissatisfied by this kind of reconstruction for two main reasons: First, because it coincided with the corresponding reconstruction based on Popperian falsificationism. Second, because it presented the endorsement of Copernicus's heliocentrism as the rational choice only after 1610. Those unwanted implications forced Lakatos to revise his own methodology by incorporating the reconceptualization of the notion

---

[11] It is quite impressive that Kuhn (1970, p. 142) noted very clearly the incompatibility between a dough-view of the historical record and the role of history as an arbiter. "Only if these and other internal criteria of the historian's craft are used, can the results of historical research react back on and change the philosophical position with which the historian began". However, he chose to uncharitably attribute this inconsistency to Lakatos.

[12] (Kuukkanen, 2017, pp. 92–95) and Nanay (2010, pp. 602–606) also conclude that Lakatosian rational reconstructions do not entail forging the historical record. Their similar point is that the appeal to external history indicates that Lakatos never intended to forge the historical record. While I do not disagree, I have elsewhere argued (Dimitrakos, 2020b, p. 4) that this is not a sufficient ground for rejecting the accusation that Lakatos's history is 'philosophy fabricating examples'.

of 'novel fact' by Elias Zahar. The details[13] of the revision are not important for my argumentation here. Neither is whether or not the revision was successful. What is important is Lakatos's choice to revise his methodology in light of well-known and, of course, unfabricated historical data.

9.3 The Discomfort Against Scientific Naturalism$_1$: Rational Reconstruction and Normative Explanations

Setting aside the uncharitable predisposition of his critics, we can focus on the actual problems of Lakatos's account which also contributed to the distorted interpretation of his view. I suggest that a fundamental problem is Lakatos's conflation of the historiographical level and the philosophical level of appraising different theories of scientific rationality. As we saw, Lakatos clearly stated that his arguments were primarily addressed to the philosophers of science. However, in the next sentence he added: 'But the same arguments also imply that the historian of science must, in turn, pay serious attention to the philosophy of science and decide upon which methodology he will base his internal history' (Lakatos, 1978a, p. 138). This addition seems to imply that the rational reconstructions involved in the appraisal of different methodologies are also part of the actual historiography of science. That rational reconstruction, aside from being an indispensable part of the assessment method for the philosophical accounts of scientific rationality, is also a historiographical category. I think that this implication is both mistaken and partly responsible for the uncharitable charge of circularity. I also think that the source of the conflation is Lakatos's discomfort with scientific naturalism.

The idea of a quasi-empirical process of assessing rival normative methodologies on the ground of actual scientific practice in the past is unambiguously naturalistic (Kuukkanen, 2017, p. 92), for it makes empirical investigation, and in particular historical study, part of the procedure of determining what science essentially is. The naturalistic attitude is contradistinguished with the 'aprioristic' or 'Euclidean' epistemologies.

> 'Classical epistemology has for two thousand years modelled its ideal of a theory, whether scientific or mathematical, on its conception of Euclidean geometry. The ideal theory is a deductive system with an indubitable truth-injection at the top (a finite conjunction of axioms) - so that truth, flowing down from the top through the safe truth-preserving channels of valid inferences, inundates the whole system' (Lakatos, 1978b, p. 28).

In this 'apriorist tradition' are included 'Leibniz, Bolzano and Frege [...], and, in [the 20th] century, Russell and Popper' (Lakatos, 1978b, p. 256). It goes without saying that logical empiricism as a general program, independently of the differences among its representatives, is part of this tradition. The main problem with aprioristic epistemologies is that they are irrefutable.

---

[13] See (Dimitrakos, 2020b, pp. 15–17).

'A Euclidean never has to admit defeat: his programme is irrefutable. One can never refute the pure existential statement that there exists a set of trivial first principles from which all truth follows. Thus science may be haunted for ever by the Euclidean programme as a regulative principle, 'influential metaphysics'. A Euclidean can always deny that the Euclidean programme as a whole has broken down when a particular candidate for a Euclidean theory is tottering. In fact rigorous Euclideans themselves constantly reveal that the 'Euclidean' theories of their predecessors were not *really* Euclidean, that the intuition which established the truth of the axioms was inadmissible, misleading, that it was a will-o'-the-wisp, not the truly genuine guiding Light of Reason.' (Lakatos, 1978b, p. 6, emphasis in original)

Lakatos's reaction to aprioristic methodologies is part of a general reaction against the armchair philosophy of science of the post-logical empiricism era, which was marked by the so-called historical turn (Bird, 2008) in the philosophy of science. As the name indicates, history of science was the main tool of this reaction. On the very first lines of *The Structure of Scientific Revolutions*, Kuhn wrote: 'History, if viewed as a repository for more than anecdote or chronology, could produce a decisive transformation in the image of science by which we are now possessed.' (Kuhn, 1996, p. 1). In short, history of science was the main tool for the naturalization of philosophy of science during the 1960s. One main aspect of the naturalization in question took the form of '*confronting* general philosophical frameworks with historical data' (Schickore, 2011, p. 456, emphasis in original). Lakatos's method for assessing rival methodologies is the first explicitly articulated and philosophically elaborated expression of this tendency. It was the first example of the 'confrontation model' for the collaboration between history and philosophy of science. In this sense, it was also an explicit attempt to naturalize the philosophical investigation of demarcating scientific rationality.

However, at the same time, Lakatos was reacting to expressions of full-blown or scientific naturalism, which strip philosophy of science off its normative content. Scientific naturalism in epistemology, philosophy of science, and science studies in general was triggered both by the historical turn and Quine's (1969) work. A paradigmatic case of expression of scientific naturalism in science studies is the *Strong Programme* in the sociology of scientific knowledge (Barnes et al., 1996; Bloor, 1991). Despite internal differences which have to do with favoring different empirical disciplines (sociology, cognitive science, anthropology, etc.) for the study of scientific development, the constitutive principle of radical epistemological naturalism is echoed in Ronald Giere's (2011, p. 61) words:

> philosophers should be in the business of constructing a theoretical account of how science works. Philosophical claims about science would then have the status of empirical theories. In short, the philosophy of science should be naturalized.

Lakatos is clearly opposed to this view, for he thinks that it necessarily ends up at relativism. For instance, against young Kuhn's conclusion[14] that the explanation of scientific change 'must, in the final analysis, be psychological or sociological'

---

[14] In his mature work Kuhn revised essentially this view. See (Dimitrakos, 2018).

(Kuhn, 1970, pp. 20–21), Lakatos commented: '*In Kuhn's view, there can be no logic, but only psychology of discovery* [and according to this view], *scientific revolution is irrational, a matter for mob psychology*' (Lakatos, 1978b, p. 90, emphasis in the original). According to Lakatos, employing exclusively the conceptual tools of the empirical sciences (external history in his terms) in order to make scientific development intelligible makes science look no better than religion or ideology. Empirical investigation can inform us about the belief modification of a scientific community but it cannot inform us whether this belief modification is justified or not. There is no distinction between being taken-to-be-justified (by a community) and being actually justified. But the lack of distinguishability is exactly the 'relativist crux' ((Psillos & Shaw, 2020, p. 408). In other terms, scientific naturalism is eliminative with regard to the normative content and therefore within its context there can be no genuine normative evaluations between different systems of beliefs.

Lakatos understands that rejecting scientific naturalism in philosophy of science also entails rejecting scientific naturalism in historiography of science. If science *per se* is an enterprise that obeys normative standards, then past science should also be made intelligible by employing normative analysis and not exclusively empirical investigation. Lakatos labels the expression of scientific naturalism in historiography of science as 'historiographical positivism' (1978a, p. 135 fn 4) and he rejects it on the ground of what I will call the 'argument from identification': The historians cannot identify what is science and what is not, let alone make historical sense of it, without a set of normative standards of scientific rationality at hand.

> [...] [W]hatever specific form they take, psychologism and sociologism [i.e., scientific naturalism] both seem to me to be open to the following fundamental objection. Everyone [...] is bound to use normative third-world[15] criteria, whether explicit or hidden, in establishing criteria for a scientific community. Merton, for example, no doubt decided what theories to select as scientific before he characterized the institutionalizations of science. He must have already decided that Darwinian biology was scientific, while Catholic theology was not, before he specified his four norms. [Similar considerations] apply to Polanyi and Kuhn. But why do Merton, Polanyi, Kuhn and Toulmin all exclude Catholic theology and astrology from science? [...]
> But if one must have some idea of what constitutes science before one knows which communities ought to count as scientific, then one must first decide what constitutes scientific progress. From the solution of this normative problem one can then proceed to the empirical problem of what socio-psychological conditions are necessary (or most favourable) for producing scientific progress. (Lakatos, 1978b, p. 114)

Based on the argument from identification Lakatos concludes that 'all histories of science are inevitably methodology-laden and that one cannot avoid "rational reconstructions"' (Lakatos, 1978b, p. 110). It is precisely in this conclusion that there is a conflation between the philosophical and the historiographical level. Lakatos suggests that both assessments of rival philosophical theories of scientific rationality and historiography make use of rational reconstructions. The following passage is even more indicative of the conflation:

---

[15] I will come back to the notion of the third world in §4.

'[I]t is not the case that I propose a rational reconstruction of history of science as opposed to describing and explaining it. Rather I maintain that all historians of science who hold that the progress of science is progress in objective knowledge, use, willy-nilly, some rational reconstruction.' (Lakatos, 1978a, p. 192)

If this is the case, the threat of circularity emerges. As we saw, the proposed procedure for writing a historical case study includes two main steps: (1) one gives a rational reconstruction; (2) one tries to compare this rational reconstruction with actual history. But if actual history cannot avoid rational reconstructions then how can this comparison be fruitful?

I suggest that the threat can be avoided by providing a distinction between the notion of rational reconstruction and the notion of normative explanation.

A normative explanation[16] can make an epistemic fact, and therefore a historical episode about science, intelligible by showing how this episode conforms to a norm (or a set of norms). Normative explanations are contrasted to empirical-scientific explanations, that is, explanations which follow the explanatory patterns of the empirical sciences (e.g., psychology, sociology, etc). Normative explanations are, by definition, based on some type of epistemic norm and hence they are mediated by our general philosophical views of scientific rationality. Epistemic norms are fallible ways of grasping truth.[17] If an epistemic norm is valid and its application is correct, then usually this application leads to the apprehension of truth. Given an episode of belief modification (B1 → B2), the explanation for the transition from B1 to B2 can be either empirical-scientific or normative. The empirical-scientific explanation would give an account of the transition based on the causal order provided by one or more empirical sciences (sociology, psychology, etc). In this case we are 'blind' with regard to whether or not the transition is actually justified, i.e., rational. The only information we can get by a set of empirical-scientific explanations is about what the scientific community in this particular episode considered as justified, i.e., what was taken as rational. Normative explanation, in contradistinction, would give an account based on one or more epistemic norms. For instance, B2 was corroborated by new observations (accuracy) or B2 was more compatible with the other theories of the field than B1 (coherence). Normative explanations reveal the relation between the scientists as rational subjects and the part of the world which is their scientific object in a normative way. Hence, they provide at the same time a justification of this

---

[16] The term 'explanation' is used in its broadest sense. It means making something thoroughly intelligible.

[17] Truth is also used in the broadest possible sense. It doesn't preclude a traditional correspondence-theory of truth nor a coherentist conception of it.

relation. It follows that they can sustain the distinction between being-taken-to-be-justified (by a community) and being actually justified. The domain[18] of normative explanations is essential in order to grasp epistemic facts as genuinely normative and therefore allow the possibility of normative evaluation of different systems of past beliefs.

Rational reconstruction serves for the purposes of assessment between rival demarcation criteria and is, so to speak, monolithic, as Lakatos suggests. This means that rational reconstructions depend exclusively on the methodology that they are intended to test. Normative explanation, on the other hand, is a historiographical category that serves for making the actual history of science intelligible and it does not have to be monolithic. The historians are not necessarily committed to a single theory of scientific rationality as Lakatos seems to suggest. They are only committed to the goal of describing what was actually the case with regard to past scientific episodes. The result of the historian's work is not a rational reconstruction. Historians may use both empirical-scientific and normative explanations. The latter can be based on epistemic norms which may be implied by more than one philosophical theory of scientific rationality. They will do whatever the internal criteria of their craft command in order to describe the actual historical course of science. Their goal is essentially different from the philosopher who seeks to evaluate rival theories of scientific rationality and so their cognitive means. Historians' work does not require the monolithicity[19] that a rational reconstruction requires. On the contrary, pluralism of the kind as employed by normative explanations may be presupposed in order to make some episodes intelligible. If this pluralism is incoherent in light of a given methodology, then the problem is on the side of methodology. The methodology needs revision. This is what makes the result of historian's work actual history and for that reason it can be unproblematically compared with a rational reconstruction.

Apart from monolithicity, I suggest that another specific difference between rational reconstructions and normative explanations can be found in the self-conscious character of the application of normative standards by the actual scientists. According to Lakatos (1978b, p. 110), the philosophers (demarcationists) who defend the possibility of demarcating science from other cognitive activities 'reconstruct universal criteria which great scientists have applied *sub- or semi-*

---

[18] Note that normative explanations, just like the empirical-scientific ones are fallible, that is they can be revised in the course of the historiographical research. It is possible for each and every normative explanation to be reduced to a set of empirical-scientific explanations. There is no a priori demarcation criterion that indicates which one is genuine and therefore irreducible to a set of empirical-scientific explanations, and which one can be reduced. This is an essential feature of their fallibility. However, the point is not any particular normative explanation but their domain as a whole. The fallibility of each and every normative explanation does not entail the eliminability of their domain as a whole. See (Dimitrakos, 2024, 2020a).

[19] Monolithicity here does not mean, I want to stress, based on a single epistemic norm. A rational reconstruction can employ several epistemic norms depending on the complexity of the philosophical theory of scientific rationality that are based on. Rational reconstructions are monolithic because they necessarily have to be based on one theory of rationality, i.e., the theory they are meant to test.

consciously in appraising particular theories or research programmes' (emphasis added). This means that rational reconstructions can ignore what the individual scientists of a particular episode think about what they were actually doing. A normative methodology can be corroborated by an episode if the belief modification in this episode coincides with the normative standards proposed by the methodology, regardless if the actual scientists self-consciously applied the standards in question or not. Normative explanations, on the contrary, cannot be applied in the absence of a conscious application of the corresponding norms by the actual scientists. If they are to explain how the belief modification actually took place, there must be evidence that actual scientists consciously applied the epistemic norms which correspond to the normative explanations employed. In order to make this distinction more clear let me use Aristotle's help. In Nicomachean Ethics (1144b25), he argues that the expression of rationality is not found in actions that are performed solely "in accordance with good reason" (κατὰ τὸν ὀρθὸν λόγον) but in actions that are performed out of a regard for right reason (μετὰ τοῦ ὀρθοῦ λόγου). Aristotle's theory of practical reason is not the point here. Only the logical distinction that he provided. I claim that rational reconstructions require only an accordance with good reason, viz., with the normative standards of a methodology at hand, while normative explanations require the existence of belief modifications out of a regard for right reason. The regard for right reason is the conscious application of one or more epistemic norms.[20]

Having established the distinction between rational reconstructions and normative explanations we can re-examine and revisit Lakatos's account. His unwillingness to adopt a version of scientific or eliminative naturalism made him overreact with respect to historiographic practice. In order to avoid 'historiographical positivism' he ended up conflating the historiographical and the philosophical level of assessment between rival demarcation criteria, implying that in both levels rational reconstructions are indispensable. However, this reaction is unnecessary. The distinction between (monolithic) rational reconstructions and (pluralistic) normative explanations can provide suitable conceptual space within which the Lakatosian desideratum of avoiding eliminative naturalism does not come with the threat of circularity and the accusation of violating the internal criteria of the historians' craft. The historiographical implications of Lakatos's philosophical perspective should be minimal. His account entails just the premise of the ineliminability of normative explanations. This premise springs from a general rejection of the view that episodes of scientific knowledge are mere natural phenomena and have to be made intelligible like all the other happenings in nature, i.e., by appealing exclusively to empirical-scientific explanations. Of course, this kind of rejection needs philosophical justification. This justification should take into consideration the wider debate about the status of normativity and the varieties of naturalism

---

[20] The cases of conscious application of one or more epistemic norms may vary. For instance, they can include cases of strict rule-following but also cases that a scientist may be convinced that a model is simpler than another, like Copernicus.

in contemporary analytic philosophy. But the ground for rejecting eliminative naturalism is not the point here. The point is that the ineliminability of normative explanations as a historiographical category does not entail that historical work provides rational reconstructions. Thus, avoiding 'historiographical positivism' entails neither the threat of circularity in the assessment of rival methodologies nor the violation of the internal criteria of the historian's craft. By distinguishing between rational reconstructions and normative explanations we can satisfy Lakatos's main desideratum of preserving the normative character of science and of avoiding, at the same time, the threat of circularity.

To sum up, the accusation of vicious circularity against the Lakatosian assessment process for rival methodologies can be avoided if we distinguish between the notion of 'rational reconstruction,' which is a philosophical tool for assessing different theories of scientific rationality, and the notion of 'normative explanation,' which is a historiographic category. Given this distinction, the assessment process involves comparing a rational reconstruction –in light of a specific theory of scientific rationality– with the actual historical narrative, which consists of normative along with empirical-scientific explanations. It is not the case that we must compare a rational reconstruction with an already rational-reconstructed actual history, which would render the process viciously circular. In short, distinguishing between rational reconstruction and normative explanation can prevent the dough-view of history, which inevitably leads to circularity.

9.4 The Discomfort Against Scientific Naturalism: The Third World and the Dilemma of Case Studies

Another byproduct of Lakatos's overreaction to eliminative naturalism is the adoption of the notion of a third world.

> Kuhn certainly showed that the psychology of science can reveal important and, indeed, sad truths. But the psychology of science is not autonomous; for the - rationally reconstructed - growth of science takes place essentially in the world of ideas, in Plato's and Popper's 'third world', in the world of articulated knowledge which is independent of knowing subjects. (Lakatos, 1978a, p. 92)

Adopting the third world is Popper's direct influence[21] and, as the passage above indicates, it is also an immediate response to psychologism or, more generally, to scientific naturalism. The introduction of the term aims at securing the autonomy

---

[21] 'To explain this expression I will point out that, without taking the words 'world' or 'universe' too seriously, we may distinguish the following three worlds or universes: first, the world of physical objects or of physical states; secondly, the world of states of consciousness, or of mental states, or perhaps of behavioural dispositions to act; and thirdly, the world of *objective contents of thought*, especially of scientific and poetic thoughts and of works of art' (Popper, 1972, p. 106, emphasis in original).

and objectivity of scientific knowledge. It provides a clear-cut distinction from the second world, the world of consciousness, and the first world, the physical world, which are cognitively grasped by physics, psychology, sociology, or other scientific disciplines. The imagery of three worlds has an immediate implication:

> A theory may even be of supreme scientific value even if no one understands it, let alone believes it. Thus the cognitive value of a theory has nothing to do with its psychological influence on people's minds. (Lakatos, 1978b, p. 109)

Once again, the discomfort with scientific naturalism causes an overreaction. The philosophical need for securing the autonomy of epistemic normativity—in the sense of proclaiming its irreducibility to the explanatory patterns of empirical science—takes the form of postulating an autonomous realm absolutely separable from the empirically accessible realms of physical objects and human mental states.

However, there are two major problems here. First, the imagery of different worlds has the same problems as every traditional rationalist approach which equates the methodological distinctions between different modes of intelligibility with ontological divisions. It must answer how those different worlds interact and also if they do interact why they are different worlds. Second, if a theory may be of any scientific value even if no one understands or believes it, then it seems completely unnecessary to appeal to the actual history of science in order to refine the philosophical accounts of scientific rationality. The study of actual history can provide knowledge of the second world and therefore it is not conceivable how this knowledge can inform our perspective about scientific rationality which belongs to the third world. Lakatos wants to avoid the scientific or eliminative version of naturalism but in adopting the notion of the third world he abandons naturalism altogether.

Even if we ignore the problems with the imagery of three worlds as another unneeded verbal exaggeration, the Lakatosian procedure of assessing rival methodologies is not free of problems. The procedure stipulates that every philosophical theory of scientific rationality should turn into a historiographical program which provides rational reconstruction of the history of science. While it should be clear by the previous section that I think the term 'historiographical' is not appropriate for this use, I will stick to the Lakatosian terminology for sake of brevity. The rival historiographical programs are evaluated according to Lakatos own methodology (Methodology of Scientific Research Programs).[22] This 'methodology of scientific research programs of second order' entails that

> While maintaining that a theory of rationality has to try to organize basic value judgments in universal, coherent frameworks, we do not have to reject such a framework immediately

---

[22] Actually, Lakatos's assessment procedure takes place in two stages. The stage that I describe is only the second. The first stage (Lakatos, 1978a, pp. 123–131) takes as meta-criterion, for the evaluation of rival methodology-based historiographies, the Popperian falsificationism. I have argued (Dimitrakos, 2020b, pp. 19–20) for the problems of the first stage. Here I am going to confine myself to the second stage which is the most crucial since, in this stage, Lakatos employs as meta-criteria his own criteria of demarcating scientific rationality.

merely because of some anomalies or other inconsistencies. We should, of course, insist that a good rationality theory must anticipate further basic value judgments unexpected in the light of its predecessors or that it must even lead to the revision of previously held basic value judgments. We then reject a rationality theory only for a better one, for one which, in this 'quasi-empirical' sense, represents a *progressive* shift in the sequence of research programmes of rational reconstructions. Thus this new - more lenient - meta-criterion enables us to compare rival logics of discovery and discern growth in "meta-scientific" - methodological—knowledge.' (Lakatos, 1978a, p. 131, emphasis in the original)

In other words, Lakatos attempts to apply exactly the same criteria for evaluating rival theories in the empirical sciences in the process of evaluating rival philosophical theories of scientific rationality. In adopting the 'three worlds' view he seems to abandon naturalism altogether. In applying the same criteria of evaluation, both to empirical and philosophical theories, he seems to embrace a quite radical version of naturalism. Too little naturalism in theory, too much naturalism in practice!

Setting aside this discord, the naturalizing practice alone faces a major problem shared by most versions of the confrontation model for the relation between history and philosophy of science. The problem has been described as the dilemma of case studies (Pitt, 2001) and it has been repeated by many HPS scholars over the years. Jutta Schickore (2011, p. 469) writes: 'The trouble is that if the scope of "genuine philosophical" theories of science is not clear, it is impossible to decide whether particular pieces of information should be taken as counter-evidence for philosophical theories or whether they are simply irrelevant'. In terms adjusted to Lakatos's account, the problem, briefly put, is that given two theories of scientific rationality, A and B, and a rational reconstruction R of a historical episode E which fits B but not A, one can take R either as a corroboration of B or as an instantiation of irrationality by the standards of A, and therefore irrelevant for the assessment. But then it is disputable (at the very least) whether E counts as evidence for B's progressiveness or as evidence for the inherent irrationality of E. In other words, we must make a prior decision about whether the episode is indeed rational (independently of A's and B's criteria of rationality) if this episode is going to count in favor of B and against A. Without a prior decision[23] of this kind the reconstructed historical episode cannot play the role of the arbiter.

I suggest that neither the 'three worlds' metaphor nor the dilemma of case studies are insurmountable defects of the Lakatosian perspective or at least of a revisionist successor of it. The Lakatosian notion of rational reconstruction can be revised in a way that avoids incoherence. The revision presupposes (a) the abolition of the 'three worlds' view for the standards of scientific rationality and (b) the replacement of the inter-methodology, quasi-empirical process of assessment with an intra-methodology conception of the evaluation process. In this way, we can retain the relation of mutual dependence between history and philosophy of science without ending up at either incoherence or relativism.

---

[23] This kind of prior decision could be made on the basis of Laudan's (1977) "pre-analytic intuitions about scientific rationality'. But this would be unacceptable in Lakatos's view. See (Dimitrakos, 2020b, pp. 18–19).

Regarding the three worlds, as I already said, it is just an unnecessary over-reaction to the threat of scientific naturalism. The three-worlds metaphor can maintain the autonomy of scientific rationality only at the cost of the anti-naturalistic consequence that 'a theory may be of supreme scientific value even if no one understands it, let alone believes it'. This consequence makes the Lakatosian project incoherent. If belief modifications can be considered rational even when no one acknowledges it as such, then the naturalistic aim of refining theories of rationality through the study of actual scientific practices becomes at least questionable. However, securing the autonomy of epistemic normativity does not imply an appeal to some kind of spooky or occult realm. There is an intermediate logical space between supernatural aprioristic normativism and scientific naturalism. Liberal naturalism occupies this space as a perspective which aspires to locate normativity in nature. The central position of liberal naturalism is that embracing ontological naturalism, i.e., the view that there are no supernatural or spooky entities, realms, or cognitive capacities does not necessarily entail methodological naturalism, i.e., the view that nature is equated with the subject matter of the empirical sciences. Thus, the mode of intelligibility which is proper to reason and employs essentially the normative concepts cannot be reconstructed 'out of conceptual materials that already belong in a [empirical]-scientific depiction of nature' ((McDowell, 1996, 73). The liberal naturalist strategies to make room for normativity in nature may vary. In my view, the best way to defend the genuineness of epistemic normativity within a naturalistic framework is adopting scientific realism. Scientific realism guarantees the ineliminability of the normative explanations without violating the fundamental premises of naturalism. However, the particular strategy for sketching a liberal naturalist context for epistemic normativity is not the issue her. The issue is that we can remove the supernaturalist imagery of the three worlds from Lakatos's account without jeopardizing the autonomy of scientific rationality which is his fundamental concern.

The problem that stems from the dilemma of case studies, on the other hand, requires a more substantial revision of the Lakatosian confrontation model. I claim that the trouble with the impossibility of deciding whether each part of the historical record should count as counter-evidence for philosophical theories or simply as irrelevant pieces of information can be overcome if we cease to comprehend the assessment process as an intra-methodology process. We should cease, in other words, emulating the evaluation of philosophical theories of scientific rationality with the comparison of rival scientific theories. We should take confrontations between philosophical theories with the historical data as an intra-methodology process. Each methodology should be understood as a set of constitutive standards of scientific rationality. Each historical episode, on the other hand, is a complex phenomenon which includes a set of essential features. Confronting a methodology with a historical episode presupposes that at least part of the essential features of the episode corresponds to at least a part of the constitutive features of the methodology. Otherwise, as Lakatos stresses, the episode does not count as a(n) (scientific) episode. However, in this confrontation procedure between an abstract theory of rationality and a concrete historical episode, it can hardly be conceivable that the episode fits perfectly to the theory. It is more than plausible to assume that there will always be some kind of partial discrepancy. This kind of discrepancy, especially when repeated in many episodes, cannot but push the methodology at hand to revision and finally to refinement.

At this point, one may reasonably object that even in the case of an intra-methodology confrontation process, the same problem may occur. The adherents of a methodology may stubbornly refuse to take lessons from the features of the episode that do not fit to their constitutive standards of rationality. They may treat them as the irrational aspects of the episode and therefore refuse to revise their theory. This possibility cannot be excluded, of course. However, I claim that the image of confrontation that I sketch has less pitfalls, for it does not require that we have to pick one theory of scientific rationality and reject the others. This image, I think, fits better with the actual history of philosophical theories. For the history of the philosophical theories of rationality, unlike the history of scientific theories, does not look like a graveyard of refuted and definitively rejected theories. It looks more like a battlefield, according to Kant's famous metaphor, of constantly reconceptualized rival perspectives. If this is the case, it is more plausible to assume that the constant refinement of each and every theory of rationality through their confrontation with the historical record will lead to convergence among them since they will constantly incorporate the same lessons from the history of science, rather than to assume that there will be one prevailing theory which will displace the others, as the scientific research programmes do. It is also implausible to assume that in the course of the convergence in question, the philosophical attitude which refuses to take into consideration the aspects of the historical record which do not fit the familiar theory of rationality will play any significant role.

Another objection might be that the revisions I propose significantly alter the original purposes of Lakatos's account. Lakatos aimed to establish a criterion for evaluating competing theories of scientific rationality. In contrast, my intra-methodology confrontation process moves away from this goal. In this sense, it does not offer improved philosophical means for Lakatos's original philosophical aim but changes this aim itself. Thus, one may wonder why we should be concerned with Lakatos's account at all. It is true that my revision reflects the significant shifts in the agenda of philosophers of science since Lakatos's era, particularly the abandonment of the demarcation problem, and consequently, the abandonment of grand theories of scientific rationality. Despite these shifts, my perspective shares a common purpose with Lakatos's original approach: It shows how normative philosophical theories can be informed by empirical disciplines such as the history of science. The philosophical necessity to elucidate the interaction between normative analysis in philosophy and the descriptive narrative in history extends beyond the task of forming comprehensive theories of rationality.[24] It is also related to particular problems of the philosophy of science. For instance, the exchange between the philosophical analysis and the data obtained by studying the actual scientific

---

[24] I owe this clarification to the comments of an anonymous referee.

practice has been crucial in refining philosophical theories of scientific explanation. The transition from a monolithic nomological conception to a more pluralistic understanding of scientific explanation cannot be fully grasped without considering this exchange. In this sense, Lakatos's endeavor to provide a criterion for evaluating rival theories of rationality and his confrontation model depicting the interaction between history and the philosophy of science can be seen as part of a broader effort to elucidate the interplay between normative analysis and empirical data derived from studying actual scientific practice. From this point of view, we can both acknowledge the originality and the relevance to contemporary philosophical problems of this endeavor, despite the major shifts in the philosophical agenda.

9.5 Conclusions

In what preceded I argued that Lakatos's conception of history cannot be a caricature, first and foremost because the Lakatosian philosophy of science is essentially history-informed. If we reject this uncharitable interpretation we can focus on the actual problems of the Lakatosian account. These problems stem mainly from Lakatos's oscillation between scientific naturalism and aprioristic normativism. His aspiration to form a genuinely history-informed philosophy of science pushed him to the side of scientific naturalism and led him to emulate the assessment of rival philosophical theories of scientific rationality with the process of testing competitive empirical scientific theories. His willingness, on the other hand, to secure the autonomy of scientific rationality pushed him to the side of a sort of supernaturalist aprioristic normativism. This oscillation causes the problems of circularity in the assessment of rival methodologies, of violating the autonomy of the history of science, and of the so-called dilemma of case studies.

I suggested that a remedy to these problems can be provided by (1) distinguishing between rational reconstruction as a philosophical tool for evaluating rival theories of scientific rationality and normative explanation as a historiographical category, (2) rejecting the 'three worlds' imagery and adopting the framework of liberal naturalism, and (3) turning Lakatos's inter-methodology evaluation process into an intra-methodology confrontation procedure. Focusing on the real problems of Lakatos's account rather than the prejudiced charge that it turns history of science into parody may reveal that it was a pioneering view on the relation between history and philosophy of science and that a charitable interpretation and commentation can play an important role in the relevant contemporary debates.

Acknowledgements The research project is implemented in the framework of H.F.R.I call "Basic research Financing (Horizontal support of all Sciences)" under the National Recovery and Resilience Plan "Greece 2.0" funded by the European Union—NextGenerationEU (H.F.R.I. Project Number: 16362).

References

Barnes, B., Bloor, D., & Henry, J. (1996). *Scientific knowledge: A sociological analysis*. University of Chicago Press.

Bird, A. (2008). The historical turn in the philosophy of science. In S. Psillos & M. Curd (Eds.), *Routledge companion to the philosophy of science*. Routledge.

Bloor, D. (1991). *Knowledge and social imagery*. Chicago University Press. Retrieved from https://press.uchicago.edu/ucp/books/book/chicago/K/bo3684600.html

De Caro, M., & Macarthur, D. (2022). Introduction. In *The Routledge handbook of liberal naturalism* (pp. 1–4). Routledge. https://doi.org/10.4324/9781351209472-1

Dimitrakos, T. (2018). Scientific mind and objective world: Thomas Kuhn between naturalism and Apriorism. *Erkenntnis, 85*(1), 225–254. https://doi.org/10.1007/s10670-018-0025-5

Dimitrakos, T. (2020a). Integrating first and second nature: Rethinking John McDowell's liberal naturalism. *Philosophical Inquiries, 8*(1), 37–68. https://doi.org/10.4454/philinq.v8i1.216

Dimitrakos, T. (2020b). Reconstructing rational reconstructions: On Lakatos's account on the relation between history and philosophy of science. *European Journal for Philosophy of Science, 10*(3), 29. https://doi.org/10.1007/s13194-020-00293-x

Dimitrakos, T. (2024). Historicizing second nature: The consequences for the is/ought gap. In M. Trajkovski & E. McWilliams (Eds.), *Normativity and normativity of art*. University of Belgrade.

Giere, R. N. (2011). History and philosophy of science: Thirty-five years later. In *Integrating history and philosophy of science* (pp. 59–65). Springer. https://doi.org/10.1007/978-94-007-1745-9_5

Hanson, N. R. (1963). *The concept of the positron: A philosophical analysis*. Cambridge University Press.

Holton, G. (1978). *The scientific imagination: Case studies*. Cambridge University Press. https://doi.org/10.1017/S0033291700022078

Kuhn, T. S. (1970). Logic of discovery or psychology of research? In *Criticism and the growth of knowledge* (pp. 1–24). Cambridge University Press. https://doi.org/10.1017/CBO9781139171434.003

Kuhn, T. S. (1971). Notes on Lakatos. In *PSA 1970* (pp. 137–146). Springer. https://doi.org/10.1007/978-94-010-3142-4_8

Kuhn, T. S. (1996). *The structure of scientific revolutions* (2nd ed.). University of Chicago Press. [1970]. Retrieved from https://search.library.wisc.edu/catalog/999466601902121

Kuukkanen, J.-M. (2017). Lakatosian rational reconstruction updated. *International Studies in the Philosophy of Science, 31*(1), 83–102. https://doi.org/10.1080/02698595.2017.1370930

Lakatos, I. (1978a). In J. Worrall & G. Currie (Eds.), *Mathematics, science and epistemology*. Cambridge University Press. https://doi.org/10.1017/CBO9780511624926

Lakatos, I. (1978b). In J. Worrall & G. Currie (Eds.), *The methodology of scientific research Programmes*. Cambridge University Press. https://doi.org/10.1017/cbo9780511621123

Laudan, L. (1977). *Progress and its problems: Toward a theory of scientific growth* (Vol. 87). University of California Press.

Laudan, L. (1983). The demise of the demarcation problem. In *Physics, philosophy and psychoanalysis* (pp. 111–127). Springer. https://doi.org/10.1007/978-94-009-7055-7_6

McDowell, J. (1996). *Mind and world: With a new introduction*. Harvard University Press.

McMullin, E. (1970). The history and philosophy of science: A taxonomy. In *Historical and philosophical perspectives of science* (Vol. 5, pp. 12–67). University of Minnesota Press. Retrieved from https://conservancy.umn.edu/handle/11299/184664

Nanay, B. (2010). Rational reconstruction reconsidered. *The Monist, 93*(4), 598–617. https://doi.org/10.5840/monist201093434

Pitt, J. C. (2001). The dilemma of case studies: Toward a Heraclitian philosophy of science. *Perspectives on Science, 9*(4), 373–382. https://doi.org/10.1162/106361401760375785

Popper, K. R. (1962). *Conjectures and refutations: The growth of scientific knowledge* (Vol. 15). Routledge.

Popper, K. R. (1972). *Objective knowledge: An evolutionary approach* (Vol. 84). Oxford University Press.

Psillos, S., & Shaw, J. (2020). Relativism and scientific realism. In M. Kusch (Ed.), *The Routledge handbook of philosophy of relativism* (pp. 407–415). Routledge. https://doi.org/10.4324/9781351052306-44/RELATIVISM-SCIENTIFIC-REALISM-STATHIS-PSILLOS-JAMIE-SHAW

Putnam, H. (1987). *The many faces of realism*. Open Court.

Quine, W. V. O. (1969). Epistemology naturalized. In *Ontological relativity and other essays* (pp. 69–90). Columbia University Press.

Schickore, J. (2011). More thoughts on HPS: Another 20 years later. *Perspectives on Science, 19*(4), 453–481. https://doi.org/10.1162/posc_a_00049

Open Access This chapter is licensed under the terms of the Creative Commons Attribution-NonCommercial-NoDerivatives 4.0 International License (http://creativecommons.org/licenses/by-nc-nd/4.0/), which permits any noncommercial use, sharing, distribution and reproduction in any medium or format, as long as you give appropriate credit to the original author(s) and the source, provide a link to the Creative Commons license and indicate if you modified the licensed material. You do not have permission under this license to share adapted material derived from this chapter or parts of it.

The images or other third party material in this chapter are included in the chapter's Creative Commons license, unless indicated otherwise in a credit line to the material. If material is not included in the chapter's Creative Commons license and your intended use is not permitted by statutory regulation or exceeds the permitted use, you will need to obtain permission directly from the copyright holder.

Chapter 10
Heuristic, Physics Avoidance and the Growth of Knowledge

Jack Ritchie

Abstract The notion of positive heuristic is crucial to Lakatos's account of the growth of scientific knowledge, but his presentation of the idea is limited to a few suggestive remarks. I try to develop a more detailed account of positive heuristic, what I call model making and improving, through the work of Mark Wilson. I argue this more sophisticated account of heuristic matches well with the few things Lakatos does say on the subject and can help us see important new connections between Lakatos's work in the philosophy of science and the philosophy of mathematics, and Wilson's work in the philosophy of language.

Keywords Positive heuristic · Physics avoidance · Multi-scalar modelling · Refutation · Concept-stretching

10.1 The Methodology of Scientific Research Programmes

A natural way to read Lakatos's Methodology of Scientific Research Programmes (MSRP), one you can find in textbooks[1] and some of his own work,[2] is to think of it as an attempt to synthesise, and thereby improve on the ideas of Popper and Kuhn. According to Lakatos, Popper's method of conjectures and refutations suffers from two serious problems. First, it doesn't fit the history of science, "[scientists] do not abandon a theory merely because the facts contradict it" (Lakatos, 1978:4). Second, it founders on the Duhem-Quine problem. When scientists encounter a falsification,

---

[1] See for example Chalmers (1999: ch. 9), Godfrey-Smith (2003: ch.7), Musgrave and Pidgen (2023).

[2] See for example Lakatos (1978:1–7). See also Lakatos (1978: 90–93).

J. Ritchie (✉)
University of Cape Town, Cape Town, South Africa
e-mail: jack.ritchie@uct.ac.za

Popper offers no guidance[3] as to which of the many theoretical claims needed to generate the prediction should be abandoned. Kuhn's *Structure of Scientific Revolutions*, although it better accords with the history of science, creates a crisis of rationality thinks Lakatos. Kuhn claims rival paradigms are incommensurable, that there is no paradigm-neutral standard by which they can be compared. This, says Lakatos (1978, p. 91), reduces revolutionary scientific change to mere "mob psychology".[4]

MSRP offers a rational reconstruction of the growth of scientific knowledge meant to overcome these deficiencies. Key to this reconstruction is to replace Popper's focus on theories with what Lakatos calls a research programme. Each programme has two elements, a hard core, that includes the central theories that define the programme, and a protective belt, additional assumptions which are needed to generate predictions, and which protect the hard core from falsification. Associated with each part of the research programme are two forms of guidance for the working scientist, what Lakatos calls the negative and positive heuristic. The negative heuristic is straightforward. It tells the scientist not to abandon the claims that make up the hard core of the programme. The positive heuristic "consists of a partially articulated set of suggestions or hints on how to change, develop the 'refutable variants' of the research-programme, how to modify, sophisticate, the 'refutable' protective belt." (Lakatos, 1978, p. 50) When things go well, the research programme, powered by the positive heuristic, will grow or *progress*. It will produce many novel predictive successes and few ad hoc modifications. This provides us with *one part* of Lakatos's account of the rational growth of knowledge. It explains what we might call *intra-programme growth*, the growth within a research programme. But there is also *inter-program growth*. Typically, there will be more than one research programme being pursued in a science. These programs compete. Over time, if science develops rationally, progressive programmes come to replace what Lakatos calls degenerating programmes. A degenerating research programme is swamped by refutations and its heuristic is failing to produce novel predictive successes.

Intra-program growth provides an answer to Popperian difficulties with the Duhem-Quine problem. When we shift our focus from individual theories to research programmes, a structure can be articulated in which guidance can be offered about what to do in light of a falsification. Inter-programme growth offers a rationalist response to Kuhn's picture of incommensurable paradigms.

This quick sketch captures, I think, what many people take to be the essentials of Lakatos's view. MSRP so understood is a sophisticated kind of hypothetico-deductivism. Certain claims taken together (the hard core and elements of the protective belt) entail predictions. These predictions are occasionally refuted. The heuristics and competition between research programmes provide a structure for the

---

[3] Lakatos calls any such decision employing Popper's method "arbitrary". (Lakatos, 1978: 99).

[4] See also Lakatos (1978:4).

scientific community to respond to apparent falsifications and explore in a rational manner the space of alternative theories and ideas.

Clearly on this understanding of MSRP, the positive heuristic is crucial. It is the engine which drives the research programme forward—responding to falsifications and generating novel predictive successes, which in turn make a programme progressive. But when we look at Lakatos's writings, descriptions of the positive heuristic are disappointingly vague. Other than the general outline provided above, we have just a few suggestive remarks like the following:

> One may formulate the 'positive heuristic' of a research programme as a 'metaphysical' principle. For instance one may formulate Newton's programme like this: 'the planets are essentially gravitating spinning-tops of roughly spherical shape.' (Lakatos, 1978, p. 51)

And when we compare what Lakatos says about heuristic here to his writings in the philosophy of mathematics, it is not just disappointing how vaguely the concept is articulated but surprising too. Heuristic as described in *Proofs and Refutations* is a rich and detailed notion. On encountering a counterexample to a proof, mathematicians are offered specific advice: divide your conjecture into subconjectures and lemmas; look to see if the counterexample is global or local; improve the proof by various procedures like lemma-incorporation and concept-stretching; and avoid tactics like monster-barring which hinder the growth of knowledge. In the course of the dialogue, Lakatos provides wonderfully rich illustrations of the use of these strategies in the development of a series of proofs of Euler's conjecture.[5] Nothing of similar detail can be found in his account of heuristic in MSRP.

My aim here is to give a fuller and better account of positive heuristic by making use of certain ideas of Mark Wilson. The ideal account of heuristic would provide a universal recipe for growing knowledge; and that indeed seems to be the way Lakatos understands heuristic in his work on mathematics. In MSRP, Lakatos implies (although it is not clear given the general vagueness of his discussion)[6] that the positive heuristic varies from research programme to research programme. My Wilson-inspired account is going to suggest something between the two extremes of a simple universal heuristic and the idea that each programme has its own. The general idea, inspired by what Wilson calls physics avoidance, can be captured in a slogan: heuristic is the art of model- making and improving. Since there are many ways to make models, there will be no simple, universal heuristic recipe, but there are some general strategies we can describe; and these strategies can be found at work throughout the sciences, both now and in the past. It is part of my aim not only to argue that this is a promising account of positive heuristic in general but that it is a plausible development of Lakatos's own brief remarks on the subject. I also

---

[5] For example: All polyhedra are Eulerian" (i.e., such that $V - E + F = 2$) becomes "All polyhedra with simply connected faces that are topologically equivalent to a sphere are Eulerian" through lemma-incorporation.

[6] This is the way Worrall (2002) and Hacking (1979) in his famous review of Lakatos's work understand him.

think that understanding Wilson's work in the context of MSRP can help us better appreciate the importance of his often difficult and complex writings.

I develop my argument in the following way. First, I sketch a rival view of heuristic offered by John Worrall. There aren't many detailed attempts to make better sense of heuristic in the literature, but Worrall's is an important one which is both clearly presented and has much in common with ideas of other students of Lakatos.[7] I claim it does not fit well with Lakatos's own remarks on heuristic or his broader philosophy of science. I illustrate my alternative, the idea of model making and improving, by sketching one physics avoidance strategy discussed in detail by Wilson, multi-scalar modelling. I then turn to Lakatos's own very brief discussion of the development of planetary models in Newtonian mechanics and show that it can be usefully thought of as illustrating another physics avoidance strategy. This, I take it, provides grounds for thinking that Wilson's work is capturing something important about Lakatos's idea. In Sections 10.4 and 10.5, I argue this more sophisticated account of heuristic can not only enrich our understanding of MSRP but help us see connections between Lakatos's work in the philosophy of science and mathematics and Wilson's work in the philosophy of language.

10.2 Worrall on Heuristic

Worrall, like me, finds Lakatos's account of positive heuristic "very sketchily presented" (Worrall, 2002, p. 88). He suggests we can improve on Lakatos by reconstructing "theoretical discovery ... as the result of a systematic argument from essentially uncontested or, at any rate, widely accepted premises." (Worrall, 2002, pp. 88–89) Worrall develops his ideas through several examples; let's consider a couple. According to Worrall (2002, pp. 91–92), Fresnel derives his transverse wave theory of light from some high-level theoretical knowledge, in this case, the general idea that light is some kind of wave, and some background data. If light is a wave, it must be either a longitudinal wave, like sound waves, or a transverse wave. Arago had shown that if the standard two-slit experiment is adjusted so that the light going through each slit is polarised in orthogonal directions, the interference pattern disappears. This is compatible with the transverse theory but incompatible with the longitudinal theory, and so Fresnel is able to *deduce* from the data that light must be a transverse wave.

Similarly, Worrall (2002, pp. 93–95) claims, the general idea that light is a wave was deduced by both Fresnel and Huygens from general principles of the mechanical philosophy and certain well-established data. The mechanical philosophy tells us light (and everything else) must either be matter in motion or a motion in matter. The passage of orthogonal beams of light through one another with no apparent effect beyond the point of crossing, and certain diffraction

---

[7] See in particular Zahar (1983), and Musgrave (1989).

phenomena rule out the first option. Hence, light must be some kind of motion through a material substance. Well-established observations like Newton's rings and interference effects demonstrate periodicities. So, light must consist of regular periodic motions through some material substance, the ether.

I don't doubt much of what Worrall says here. In fact, some of the ideas, like an emphasis on background knowledge, will also be important in the characterisation of heuristic that I present. My problem with his account is not so much that I think it is wrong about these particular historical episodes, I just don't think it fits very well with the few things Lakatos says about heuristic. There are two ways I think this is so. First, as Worrall admits, his examples seem quite different from the one case Lakatos does discuss in a bit more detail, Newton and later scientists' developing account of the motion of bodies in the solar system. Worrall says that this example is in "several ways unrepresentative" (Worrall, 2002, pp. 88).[8] But, I think, as we shall see in the next two sections, it is fairly typical of a certain sort of scientific growth, and if we want to capture what *Lakatos* means by heuristic, we need to be able to understand this example. One key element of Lakatos's idea, that I shall show is well illustrated by the Newtonian case, is that it is meant to provide a kind of road map or set of "instructions" (Lakatos, 1978, p. 50) for the working scientist. Worrall's account doesn't do anything like that. It simply reconstructs new developments as deductions from data and background knowledge. Each deduction like this seems independent from the last. The second problem with Worrall's account is that the examples he discusses seem to explain the wrong thing. Heuristic is meant to be something which drives a *particular* research programme forward. It is part of the story of intra-programme growth. The two examples above are really accounts of how parts of a new research programme, Fresnel's wave theory, come to be.[9] Worrall (2002, p. 93) is aware of this too but he considers it an advantage, not a drawback. It makes his account of the logic of discovery more general than Lakatos's. Again, I don't discount that he might be on to something here[10] if we want to characterise the growth of knowledge in general, but if our aim is to try to develop an idea of heuristic closer in spirit to Lakatos's work, and one which captures the examples he discussed, then some different ideas are needed.

---

[8] He doesn't elaborate how.

[9] Zahar's (1983) account of heuristic is also geared to explaining the emergence of radically new ideas like those of Einstein. Again, it is very plausible to think the features he picks out like the correspondence principle play an important role in the development of ideas in quantum theory and relativity but he is concerned, I would claim, with different issues than those that animate Lakatos.

[10] See the concluding section below.

10.3 Wilson on Multi-Scalar Modelling: An Example from Materials Science

We can find some of these ideas in Mark Wilson's work. He provides some wonderfully detailed illustrations of the slogan that heuristic is the art of model-making and improving. Consider a recent example of his that looks at a very concrete, practical problem. Say we wish to build a heavy structure on top of a large block of stone. We would obviously like to know if such a structure will be stable. Will the rock break, for example, under the stress of the building? This is a very hard problem to tackle. There are potentially many relevant variables, and certainly it would be impossible to arrive at an answer from the bottom up by considering the effects of the load on the atomic structure of our stone.

The engineer or the physicist needs to find a way into the problem. Wilson (2021a, pp. 97–103)[11] describes one way they can proceed. First, a highly simplified model of the block as a Hookean solid under stress from the weight of the structure is developed. In such a model the deformation of the rock depends upon just two empirically determined parameters, the Young's modulus, E, which measures the compressive stiffness of the material,[12] and the shear modulus, μ, which measures the shear stiffness.[13] A simple to write down but difficult to solve set of equations[14] can then be used to describe the rock.

This model by itself will not answer the question of whether the rock can bear the stress, it is in this regard too simple, treating the rock as homogenous and isotropic; but it can be used as a probe to answer that question. In combination with a suitable model of the forces the building will impress on the rock, we can use the model to investigate where there is likely to be high degrees of shearing stress. Having identified these areas of high stress, the second stage of modelling can begin. We must move from the macro to the meso scale and model the parts of the rock that interest us in a more complex manner to obtain (literally) some more fine-grained information.

One thing we know about the rock, which is obviously absent in the simple Hookean elastic solid model, is that it has a certain internal structure or grain. This we can see directly. In order to better understand the effect of the load on the rock, we need to know how these more complicated structures will behave under stress.

---

[11] See also Wilson (2021b: 490–491) and Wilson (2017: 203–205).

[12] That is the resistance the body has to bending and deformation from a force pushing down on it or squeezing it together.

[13] That is the resistance the body has to deformation by a shear force i.e., forces acting parallel or tangential to a face.

[14] For an equilibrium state the equation is: $\sigma_{ji,j} + f_i = 0$, where i and j represent spatial dimensions, σ_{ij} is the stress tensor, and $\sigma_{ji,j}$ is the divergence of the stress tensor. The system is further constrained by a compatibility equation: $\varepsilon_{ij} = 1/2 \; \partial u_i/\partial x_j + \partial u_j/\partial x_i$, where ε is the strain and **u** is the displacement. The shear and Young's moduli are buried in the stress tensor term. More detail can be found in materials science textbooks and lecture notes, e.g. Abeyaratne (1987).

We can do this, for example, by modelling the relevant part of the rock as consisting of a structure of individual grains, each orientated differently and in contact with one another. Using our macroscopic model to provide appropriate inputs to the stresses and strains on these structures, we can investigate how these grains will align themselves under stress.[15] If our model shows the macroscopically calculated stresses will cause the grains in the rock to align, then in this area, the rock will become damaged and weaker.[16]

This takes us to the third stage of modelling. The new information from our mesoscopic model shows our original simple Hookean model which assumes the solid to be isotropic is importantly inaccurate. We will need to input our new knowledge of grain alignment and weakening back into our macro model. According to Wilson:

> This is accomplished by inserting three or four additional constants (besides the original E and μ) into our Navier equation[17] modeling that reflect the fact that the rock is no longer isotropic within that little region due to its metamorphosized realignments. To estimate the new parameters required, we must homogenize (= average) our [mesoscopic model's] conclusions so that the results appear as simple elastic parameters within the [macroscopic model's] smoothed-over setting. In the case at hand, this adjustment only requires that the stress/strain matrix relationships within Navier's equations include a few more numbers than they did previously. (2021a, pp. 99–100)

Since our macromodel has altered, we must begin the process again, using our adjusted model to probe anew for areas of potential weakness; and then investigate those areas with our mesomodel to see if any new areas of weakness arise. If so, a new adjustment will be required to the macro-model, and we begin the process again. We keep looping through the models until we arrive at a stable (or approximately stable) model that requires no (or negligible) further corrections.

In some cases, it might be necessary or advisable to augment the structure with even more fine- grained models. We might add molecular models which describe important internal features of the grain, for example, if we are keen to investigate possible crack formations. But the basic procedure, although more complicated, is the same: lower level-models are used to provide corrections and improvements to higher-level models.[18]

Wilson's description of a multi-scalar architecture in a practical problem in materials science might seem quite specialised (and maybe a little dull!) but,

---

[15] Some kind of discrete element method might be used here with equations characterising the frictional, cohesive and other forces acting between the grains. As in the case at the macroscopic level, appropriate parametrizations require a great deal of empirical input.

[16] "Isotropic granite [becomes] metamorphized into anisotropic gneiss" as Wilson (99:2021) puts it.

[17] See fn 14 above for the details of the equations.

[18] As Wilson (2021a, 2021b, ch. 5, fns 6, 8 & 9) in fact notes, this example is in general way too simple. It ignores many factors, like temperature, which would be relevant in any real application. Since it is always possible that some ignored or unknown factor like this might be important all such modelmaking is, of course, fallible. I thank, a referee for asking me to clarify this point.

as he himself notes, this modelling strategy has application in many areas. For example, the same sorts of multi-scalar techniques are used in climate science, where lower-scale models of, for example, cloud-cover are used to improve the relevant parametrisations in global climate models (Tao et al., 2009); in biology and medicine in the investigation of cancer (e.g. Clarke et al., 2019); in biochemistry (Warshel, 2014) to combine quantum models of bond breaking in macromolecules like enzymes with a higher-level classical analysis of the molecule as a whole; and many, many other areas.[19] Different theories may be guiding model construction in each of these cases but the same basic structure as described by Wilson is in play. He sums up the core idea very well:

> The guiding strategy behind this computational policy recommends that we should remain content with our higher ΔL scale "dominant behavior" calculations until some warning criterion instructs us to consult a submodel associated with [a lower-level] scale length. (2021a, p. 92)

The idea is a nice concrete illustration of our slogan that heuristic is model-making and improving. We begin with a simple model which captures an important aspect of the dominant behaviour or behaviours of the system we are interested in. We improve that model *in this case* by making iterative use of the inputs of lower level-models and so we are able to apply our theories in a reliable way to a complicated situation. Knowledge grows, powered by heuristic.

10.4 Lakatos and Positive Heuristic in the Newtonian Research Programme

The multiscale modelling explosion is a recent, post-Lakatosian phenomenon, partly fuelled by recent improvements in computational capacity, but Lakatos's own brief remarks about positive heuristic, I claim, fit well the general structure of the model-making and improving idea outlined in the last section. As I said above, Lakatos doesn't say much about positive heuristic in MSRP. In fact, the term hardly features in his two worked out case studies, the research programmes of Prout and Bohr, but the few brief remarks he makes about heuristic when discussing Newtonian astronomy are revealing. According to Lakatos:

> The positive heuristic sets out a programme which lists a chain of ever more complicated *models* simulating reality: the scientist's attention is riveted on building his models following instructions which are laid down in the positive part of his programme.
> Newton first worked out his programme for a planetary system with a fixed point-like sun and one single point-like planet. It was in this model that he derived his inverse square law for Kepler's ellipse. Then he worked out the programme for more planets as if there were only heliocentric but no interplanetary forces. Then he worked out the case where the

---

[19] See Horstemeyer (2009) for some information on the growth of multiscale methods.

sun and planets were not mass-points but mass-balls. (Lakatos, 1978, p. 50. Italics in the original)

Very explicitly, then, we not only have the general idea that positive heuristic is a guide to model-making[20] but a (very brief) characterisation of how it works; and there are obvious similarities with Wilson's description of multiscale modelling. First, we begin with a simplified model: two point particles acting under a gravitational force. The motivations for choosing this model are again the same as in the multiscale case. The solar system is from the point of view of Newton's theory a complex, whirligig of mutual gravitational attraction. But what dominates the orbital behaviour of any particular planet is its gravitational interaction with the sun. If we model both the sun and planet as point particles that reduces the relevant degrees of freedom to a manageable handful and the problem of describing the motion of the planet becomes mathematically tractable. We know, of course, our model is incomplete and will be strictly inaccurate in all cases and in some, like the orbit of the moon, where multiple bodies have a large effect on its motion (the sun and earth in this case) utterly hopeless.

The next step is to improve the model. How do we do that? Again, a number of factors are in play in this case which are similar to the multiscalar example. First, we have background knowledge which guides us in looking for the appropriate sort of improvement. We know, in Wilson's example of multiscalar modelling, the rock we are interested in contains a grain, which we have ignored in our first stage of modelling. We know in the Newtonian case that the planets are not point particles, and from Newton's theory that the mass of the other planets must have some gravitational effect on all bodies in the solar system. We can consider this background knowledge part of the heuristic and that, I would suggest, is the most sensible way to interpret Lakatos's cryptic sounding remarks about heuristic being a metaphysical principle. It is not really that advocates of a research programme have some very general metaphysical ideas guiding their model-making, but rather they have a lot of low-level knowledge, like granite is composed of grains or the earth rotates on its own axis, which feed into our construction of more sophisticated models. The trick is, and this is where I believe heuristics develop over time, to work out what to do with that knowledge. Again, Lakatos's own brief remarks about the problem are insightful:

> Indeed, if the positive heuristic is clearly spelt out, the difficulties of the programme are mathematical rather than empirical. (1978, p. 51)

That is to say, the difficulty we face is knowing how to make a mathematically tractable model that incorporates some of our background knowledge which we know to be relevant to the behaviour of our target system. The use of numerical

---

[20] Lakatos's understanding of a model as "a set of initial conditions" (1978, p. 51) is a bit crude, especially if it is looked at through the explosion of model-talk in the philosophy of science in recent years but I don't think in any way this undermines the basic point that his understanding of heuristic is one of model-making and improving. Rather, the modern reader is in a position to improve Lakatos's account given our improved understanding of models and modelling.

methods and computational simulations are necessary to carry out any of the steps described in Wilson's multiscalar architecture, but the method also requires further clever mathematical elaboration, most importantly in finding techniques that allow lower-scale results to feed useful information back into higher-level macromodels.[21] The mathematical innovations of Newton are of course well-known, first and foremost the development of the calculus but equally important are model building techniques like the use of perturbation theory to deal with multi-body systems. These are both elements of the evolving Newtonian heuristic which continued to be refined throughout the eighteenth and nineteenth century by Laplace and Lagrange and many other giants of applied mathematics.

So, at a greater level of generality, we might say positive heuristic is something like the following:

1. Find a simple model which captures the dominant behaviour which interests us in a mathematically tractable fashion.
2. Seek to improve the model through a combination of background knowledge (knowledge of factors the model leaves out but ought to be relevant to the behaviour) and mathematical techniques (perturbation theory, multiscalar modelling,[22] for example).
3. Develop a theoretical account of the reliability of the mathematical methods employed at stage 2.

Wilson's work shows many of the factors, many of the strategies for physics avoidance,[23] that can go into finding a suitable model at stage 1. Sometimes we ignore the dynamic evolution of the system (which can often be very difficult to calculate) and focus on some appropriate equilibrium state. (This is in fact a feature of the elastic models briefly described above. The equations described in the footnotes model the equilibrium conditions of our rock.) Sometimes we can appeal to boundary conditions or fixed constraints to simplify our modelling. Modern physics provides some further interesting strategies. The huge complexity of the substances studied in condensed matter physics can be made into something tractable by redescribing the system in terms of quasi-particles like phonons or polaritons. Again, we reduce a problem with many variables (all of the elementary particles in the substance) to one in terms of just a few, the quasiparticle description;

---

[21] That is to say, appropriate ways to average out the results of the mesomodel so that it can be meaningfully articulated using the more coarse-grained parameters of the macromodel. At a more sophisticated level, relevant to point 3 below, we also need to show that there is good reason to think that this multiscale modelling will eventually converge to some fixed point.

[22] Interestingly Weinan (2011) characterises many strategies of physics avoidance as examples of multiscalar modelling *avant la lettre*. "[I]t is not a stretch to say that multiscale, multi-physics ideas played a role in most of the major developments in theoretical physics." (p. 18) Soon after listing the work of Planck, Cauchy and Boltzmann. To me, it does sound a bit of a stretch.

[23] Similar strategies can be found outwith physics. For example, Collin Rice (2015, 2019) gives some nice examples of equilibrium and optimization models used in biology, like Fischer's (1930) sex ratio model.

and again, our model building strategy will often start with the simplest case, independent quasiparticles, and, guided by our background knowledge of what a more realistic system would involve, add in extra structure. In this case, for example, some weak interaction between the quasiparticles, to arrive at a better model.

The third step is an important part of a flourishing heuristic and one emphasised by Wilson in his description of physics avoidance. As mathematicians develop techniques and approximations or more complex modelling methods, they investigate those techniques themselves. An example Wilson often gives involves the use of what he calls, "Eulerian marching methods"; types of numerical stepwise calculations of solutions to ordinary differential equations. This sort of simple method could be used, for example, to calculate the trajectory of a projectile. Applied mathematicians can specify conditions where this method can lead us astray. If the so-called Lipschitz condition[24] is not satisfied our approximate method can churn out answers that deviate wildly from the true path. (Wilson, 2017, pp. 395–397) At the fancier end of the spectrum, renormalisation group methods have been developed to justify or explain the use of renormalisation in perturbative calculations in quantum field theory.[25] This work helps to stabilise and reinforce the growing heuristic methods, aiding the working scientist in knowing the limits of various techniques.

This sketch of positive heuristic is appropriately general and, also, appropriately open-ended. There are useful, if abstract things to say in general about model-building and improving, my principles 1–3 capture some of that. We can learn more about heuristic by understanding some of these strategies in more detail, for example perturbative methods or multiscalar ones; and we can see that these methods are not at all research programme specific. Multiscalar modelling is used throughout the sciences; perturbative methods are used everywhere in physics. This Wilson inspired way of illuminating MSRP is genuinely insightful and an improvement on, and as we have seen in some ways, an elaboration of Lakatos's own ideas. I think too it helps reflect a little light back on Wilson's work. It is sometimes difficult to see in what he calls the "baggy expanses" (Wilson, 2011, p. 202) of his very long books exactly what it is we are supposed to learn. Placed in the context of MSRP, here is one way to think of some of what Wilson is doing: he is a masterful and detailed curator of heuristic strategies in the sciences, especially physics.

---

[24] For those who are interested: a function f (t, x) satisfies a Lipschitz condition in the variable x on a set D ⊂ R2 if a constant L > 0 exists with |f (t, x1) − f (t, x2)| ≤ L |x1 − x2|whenever (t, x1),(t, x2) are in D. L is Lipschitz constant. In other word small perturbations to the equation, only give rise to small perturbations of the solutions.

[25] See Ruetsche (2018) for a very nice explanation and some scepticism regarding the significance of these results for questions regarding realism in QFT.

10.5 Some Lessons: Refutation in MRSP and *Proofs and Refutations*

If we place this more sophisticated model of heuristic back into the quick sketch of MSRP I gave in Sect. 10.1, we can see the story I told contains certain misleading exaggerations. In the way I described MSRP, and the way many others do too, refutations are the first step in the growth of knowledge. Heuristic is the secret recipe which allows scientists to find appropriate auxiliary hypotheses to turn those refutations or anomalies into successes. When we look at the examples we've offered here, using our understanding of heuristic as model-making and improving, a different picture emerges. The real difficulty the working scientist encounters is not really how to deal with *refutations,* but how to find appropriate models so the theory can make empirical contact with reality; and these are often mathematical problems, as Lakatos says. Insight and advance come from the discovery of new mathematical techniques combined with our background knowledge.

Consider again the sequence of models that Lakatos takes to be part of the progressing Newtonian program, beginning with simple two-body point particle systems and steadily progressing to fuller, more realistic models. It is very odd to think of any particular problem with any particular model in this sequence as a refutation of some kind. Obviously, when presented these models are *known* to be incomplete and in the context of the general theory of Newtonian mechanics to omit factors which are *known* to be causally relevant. Lakatos is well aware of this: "Most, if not all, Newtonian 'puzzles', leading to a series of new variants superseding each other, were forseeable at the time of Newton's first naive model and no doubt Newton and his colleagues did forsee them." (1978, p. 51) (Although, he is inadvertently hilarious in where he thinks this knowledge of, for example, the shortcomings of simple point-particle models comes from: "infinite density was forbidden by an (inarticulated) touchstone theory, therefore planets had to be extended." (1978, p. 50). Maybe—but a quick peek out the curtains ought to be enough to suggest the Earth is not a point particle.[26])

Certain odd pronouncements aside, what is clear from both the examples and Lakatos's statements is that refutation is not essential here. What matters is that the research programme grows; that the positive heuristic guides the scientists to new successes (and that the heuristic grows adding new ways of model-making and improving). It is here we see the sense in which Lakatos's heuristic provides a set of instructions, something lacking in Worrall's account. The working scientist is not stumbling over new refutations which must be dealt with case by case but often is already well-placed, given their background knowledge, to know the likely deficiencies of their models. If the heuristic has some well-developed techniques for incorporating some of these known background factors, then it provides the scientist with the required guidance as to how those models can be improved.

---

[26] Here I think we see the unfortunate exaggeration of a theory dominated view of science. Not all relevant knowledge is in any serious sense theory dependent.

10 Heuristic, Physics Avoidance and the Growth of Knowledge

In fact, notwithstanding the titles of many of Lakatos's most famous writings, it is, I contend, a theme of his work that although refutations *may* play a role in the growth of knowledge, they are not *essential*. Consider this curious exchange from *Proofs and Refutations*, for example. After offering a summary of the eponymous method, Pi says this:

> [T]he power of the theory lies in its capacity to explain its refutations in the course of its growth. But there is a second main pattern of deductive guessing .... In this variation ... the growing theory not only explains but produces its refutations... *Not by extending a naive concept, but by extending the theoretical framework.* (Lakatos, 1976, p. 100)

And Alpha rightly objects:

> You now expand ['counterexample'] to cover heuristic counterexamples that never actually exist. Your claim that your 'second pattern' is full of counterexamples is based on the expansion of the concept of counterexample to counterexamples with zero life- time, whose discovery coincides with their explanation! (Lakatos, 1976, p. 101)

And when in Chap. 2, Poincaré's final proof is presented, refutation plays no substantial role. What matters in Lakatos's account of both mathematics and science is describing the growth of knowledge. It might be that a refutation or counterexample is the instigator of that growth, or it might be that our theories or proofs grow for other reasons, because we have a new mathematical method that allows us to apply our theories in new ways, or because as Pi says here, we extend our theoretical framework in some way by, for example, placing the Euler conjecture in the new, richer context of algebraic topology.

10.6 More Lessons: Concept-Stretching

One of the most interesting claims in *Proofs and Refutations* concerns the connection between the growth of mathematical knowledge and what Lakatos calls concept-stretching. As our proof of Euler's conjecture improves, we find the concept of polyhedron stretches beyond its intuitive limits. With Cauchy's proof in place, for example, we can see it applies just as well to shapes with curvilinear faces and so it becomes natural to expand the concept of polyhedron to include these shapes. In MSRP, concept stretching plays little explicit role.[27] But turning again to Wilson's work we can see physics avoidance or heuristic often involves concept-stretching or what Wilson (2006) calls wandering significance. As in the case of strategies of physics avoidance, Wilson has a superabundance of examples, but I will just sketch two.

---

[27] I can only find one explicit reference on p. 54 where Lakatos claims the concept element was stretched to include the idea of isotopes. The other references in the text are to *Proofs and Refutations*. However, he does say this: "The recognition that the history of science is the history of research programmes rather than of theories may therefore be seen as a partial vindication of the view that the history of science is the history of conceptual frameworks or of *scientific languages*." (Lakatos, 1978, p. 47 fn.1, my italics) It is not exactly clear what this means but it might be taken to gesture at the sort of concept-stretching that force and temperature enjoy described below.

Take an old physics textbook classic, a ball rolling down an inclined plane. There are in point of fact many complex interactions taking place between the surface of the ball and the plane here, too many to model directly. Following the guidance of our positive heuristic, we should seek out the dominant interactions. Obviously, the most important factor here is the gravitational force pulling the ball down the slope. It is convenient, then, to model the plane and ball as though both are rigid and to label the forces retarding the movement of the ball as frictional. But as Wilson (2011, p. 203) points out, when we do so we are in fact shifting the application of the term 'frictional force'. The real ball as it moves over the plane will cause a slight impression and so, there will be a slight lengthening of the journey. Since this effectively also retards the ball, the term 'frictional' force in our model will also pick out this effect and not just the intermolecular forces acting between the plane and ball. As Wilson says:

> "the seemingly innocuous act of evoking the constraint of rigidity silently operates as a kind of semantic switch that automatically adjusts the physical correlate that attaches to the term "frictional force" in a subtle manner." (2011, p. 204)

In other words, the concept frictional force gets stretched in this application to cover path elongation.

Consider now a second favourite example of Wilson's (2017, pp. 187–189): the temperature of rubber bands. On a standard way of thinking in thermodynamics, strictly speaking a body only has a well-defined temperature when at equilibrium. But most actual materials, including rubber bands, are not at thermal equilibrium at room temperature and would only be so in some distant future state when the band has dissolved into a liquid polymer mess. If we follow this strict conception of the concept, nothing rubber-band-like can have a temperature. Nevertheless, it still seems appropriate to talk of the temperature of a rubber band and to say, truly, that the temperature of the band goes up when stretched. How are we to make sense of this? Band stretching (and temperature change) require a little concept stretching.

The polymer chains that make up the band and allow it to stretch are pinned together at certain points and it is this pinning which makes the band stable over long periods of time. We can model the band as in a metastable state (when not stretched) free to wiggle in many ways except at these fixed points. By considering only configurations consistent with this metastable state, surrogates for temperature and entropy can be defined over these possible wiggle states using models analogous to those used to describe ideal gases in standard equilibrium thermodynamics. So, the concept temperature gets stretched to encompass not just substances in equilibrium but, through models analogous to ideal gas models, metastable structures like rubber bands.

If Wilson's analyses are right, concept-stretching is ubiquitous in the natural sciences. Just as Lakatos shows us that new proofs may extend our old concepts, Wilson shows that new applications, sometimes new applications making use of models used to describe other simpler systems, force our physical concepts to stretch. Frictional force is in part describing the effect of an elongated journey; temperature gets applied meaningfully beyond equilibrium states. That concept-stretching goes hand-in-hand with the kind of growth Lakatos and Wilson describe in mathematics and science should, on reflection, be no surprise. Both are anxious to show that simple-minded deductivist schemes cannot capture the important ways science and mathematics develop. Part of what makes such deductivist schemas inadequate is the realisation that with new growth, new theorems or new applications, comes old concepts deployed in new and unexpected ways.

10.7 Summing up

I have argued that the right way to think about positive heuristic is as a series of strategies for model-making and improving. We begin with simple models that capture the dominant behaviours of our target and that are mathematically tractable. Using our background knowledge and some mathematical tricks (some of which may have to be invented as the research programme develops) we can improve our models. Sometimes we adapt old models to new uses, as in the case of modelling the temperature of rubber bands using ideal-gas-like models. Sometimes we use other strategies. Ideally, the mathematical methods we develop should become the subject of independent mathematical investigation to ascertain the reliability and the limits of those methods. This general structure fits well with what Wilson calls physics avoidance and I have used his work to provide a more concrete account of heuristic, one which I hope begins to show how an account of positive heuristic could be as detailed as Lakatos's account of heuristic in the philosophy of mathematics. But I have also tried to suggest along the way that the account I am offering is close to what Lakatos had in mind all along and fits well with the very brief account he gives of heuristic in the Newtonian research programme.

The position we end up with corrects or, to use a Lakatosian term, sophisticates our initial sketch of MSRP. And I claim it also brings our understanding of MSRP a little closer in a lot of important ways to the Lakatos of *Proofs and Refutations*. Falsification and the hypothetic-deductive structure do not capture the real drivers of scientific growth; these are, as Lakatos says, often mathematical innovation: the imaginative application of old models to new cases or the careful extension and refinement of those models using new mathematical tools and our background knowledge. What we find then is not the simple deduction of predictions from well-chosen auxiliary hypotheses but, under the vicissitudes of application, the squashing and stretching of the concepts used in our successful models—something messier, something livelier and more dynamic. In *Proofs and Refutations*, Lakatos rails against the "deductivist style [that] hides the struggle, hides the adventure" (Lakatos, 1976, p. 151). Similarly, I think we should think of my Wilson-enriched Lakatos as railing against the hypothetico-deductivist style that ignores the dynamic way in which heuristic can guide the production of new models and stretch our theoretical concepts.[28]

---

[28] Part of Wilson's work is directed at showing how philosophy falls into error when it accepts a simplified hypothetico-deductive account of science, what he calls Theory-T thinking. This is not part of Lakatos's view, obviously, but I imagine he would have been sympathetic.

Let me end with a few final remarks about the limitations of what has been done here. I have sketched a general account of heuristic and shown how the details can be filled in with a few examples culled from Lakatos and Wilson. Obviously, there is work to do in showing other cases fit the general structure and finding other ways in which models get made and improved. Much of that is no doubt already implicit in Wilson's big books and the vast, recent literature on scientific modelling. But all of this is just an account of what I called earlier *intra*-programme growth. If MSRP or something like it is going to be an account of scientific growth in general, then we also need a better account of *inter*-programme growth and innovation. In particular, we need an account of something that Lakatos seems to have said nothing about at all: where new research programmes come from in the first place. I don't think anything I have said here or indeed anything we find in Wilson's work helps with that project. But some of the ideas sketched by Worrall and others seem like a promising start to that important task.[29]

Acknowledgements Many thanks to a referee for their helpful suggestions which greatly improved the focus of the paper. An ancestor of this piece was delivered as a talk at the Imre Lakatos Centenary Conference on the fourth of November at the LSE. Thanks to the audience for their useful comments and questions. Thanks also to the UCT URC that funded my trip to London. Special thanks to University of Barcelona and the LOGOS research group for providing such an excellent work environment while I was writing the paper.

References

Abeyaratne, R. (1987). Lecture notes on the mechanics of elastic solids. MIT. Retrieved January 3, 2023, from https://web.mit.edu/abeyaratne/ElasticSolids-Vol.1-Math.pdf

Chalmers, A. (1999). *What is this thing called science?* (3rd ed.). Open University Press.

Clarke, R., Tyson, J. J., Tan, M., Baumann, W. T., Jin, L., Xuan, J., & Wang, Y. (2019). Systems biology: Perspectives on multiscale modeling in research on endocrine-related cancers. *Endocrine-Related Cancer, 26*(6), R345–R368. https://doi.org/10.1530/ERC-18-0309

Fisher, R. A. (1930). *The genetical theory of natural selection*. Clarendon Press.

Glymour, C. N. (1980). *Theory and evidence*. Princeton University Press.

Godfrey-Smith, P. (2003). *Theory and reality: An introduction to the philosophy of science*. University of Chicago Press.

Hacking, I. (1979). Imre Lakatos's philosophy of science. *British Journal for the Philosophy of Science, 30*(4), 381–402.

Horstemeyer, M. F. (2009). Multiscale modeling: A review. In J. Leszczynski & M. Shukla (Eds.), *Practical aspects of computational chemistry* (pp. 87–135). Springer.

Lakatos, I. (1976). In J. Worrall & E. Zahar (Eds.), *Proofs and refutations: The logic of mathematical discovery*. Cambridge University Press.

Lakatos, I. (1978). *The methodology of scientific research programmes (philosophical papers: Volume 1)* (J. Worrall & G. Currie, Eds.). Cambridge University Press.

---

[29] Nancy Nersessian's (1989, 2008) work is clearly important here as a referee pointed out. Also van Fraassen's (2009) idea of empirical grounding in which data and experiment can be a guide to theory construction has many affinities with Worrall's work. (This is not surprising given their common source in Glymour (1980).) Unfortunately, I don't have the space to develop or explore any of these rich ideas and their connections.

Musgrave, A. (1989). Deductive heuristics. In K. Gavroglu, Y. Goudaroulis, & P. Nicolaccopoulos (Eds.), *Imre Lakatos and theories of scientific change* (pp. 323–334). Kluwer Academic Publishers.

Musgrave, A., & Pigden, C. (2023). Imre Lakatos. In E. N. Zalta & U. Nodelman (Eds.), *The Stanford encyclopedia of philosophy (spring 2023 Edition)*. Stanford University. Retrieved from https://plato.stanford.edu/archives/spr2023/entries/lakatos/

Nersessian, N. J. (1989). Scientific discovery and commensurability of meaning. In K. Gavroglu, Y. Goudaroulis, & P. Nicolaccopoulos (Eds.), *Imre Lakatos and theories of scientific change* (pp. 323–334). Kluwer Academic Publishers.

Nersessian, N. J. (2008). *Creating scientific concepts*. MIT Press.

Rice, C. (2015). Moving beyond causes: Optimality models and scientific explanation. *Noûs, 49*(3), 589–615.

Rice, C. (2019). Models don't decompose that way: A holistic view of idealized models. *British Journal for the Philosophy of Science, 70*(1), 179–208.

Ruetsche, L. (2018). Renormalization group realism: The ascent of pessimism. *Philosophy of Science, 85*(5), 1176–1189.

Tao, W., & Coauthors. (2009). A multiscale modeling system: Developments, applications, and critical issues. *Bulletin of the American Meteorological Society, 90*, 515–534.

van Fraassen, B. C. (2009). The perils of Perrin, in the hands of philosophers. *Philosophical Studies, 143*(1), 5–24.

Warshel, A. (2014). Multiscale modeling of biological functions: From enzymes to molecular machines (Nobel lecture). *Angewandte Chemie International Edition in English, 53*(38), 10020–10031. https://doi.org/10.1002/anie.201403689

Weinan, E. (2011). *Principles of multiscale modeling*. Princeton University Press.

Wilson, M. (2006). *Wandering significance: An essay on conceptual behavior*. Clarendon Press.

Wilson, M. (2011). Of whales and pendulums: A reply to Brandom. *Philosophy and Phenomenological Research, 82*(1), 202–211.

Wilson, M. (2017). *Physics avoidance: And other essays in conceptual strategy*. Oxford University Press.

Wilson, M. (2021a). *Imitation of rigor: An alternative history of analytic philosophy*. Oxford University Press.

Wilson, M. (2021b). Replies to critics. *Philosophy and Phenomenological Research, 103*(2), 488–498.

Worrall, J. (2002). "Heuristic power" and the "logic of scientific discovery": Why the methodology of scientific research programmes is less than half the story. In G. Kampis, L. Kvasz, & M. Stöltzner (Eds.), *Appraising Lakatos*. Kluwer Academic Publishers.

Zahar, E. (1983). Logic of discovery or psychology of invention? *British Journal for the Philosophy of Science, 34*(3), 243–261.

Open Access This chapter is licensed under the terms of the Creative Commons Attribution-NonCommercial-NoDerivatives 4.0 International License (http://creativecommons.org/licenses/by-nc-nd/4.0/), which permits any noncommercial use, sharing, distribution and reproduction in any medium or format, as long as you give appropriate credit to the original author(s) and the source, provide a link to the Creative Commons license and indicate if you modified the licensed material. You do not have permission under this license to share adapted material derived from this chapter or parts of it.

The images or other third party material in this chapter are included in the chapter's Creative Commons license, unless indicated otherwise in a credit line to the material. If material is not included in the chapter's Creative Commons license and your intended use is not permitted by statutory regulation or exceeds the permitted use, you will need to obtain permission directly from the copyright holder.

Chapter 11
Beyond Footnotes: Lakatos's Meta-philosophy and the History of Science

Samuel Schindler

Abstract In this chapter I revisit Lakatos's meta-philosophy concerning the use of historical facts for the purpose of philosophical theorizing about science. Despite Lakatos's bad reputation on that question—which mostly springs from his suggestion that the actual history could be detailed in the footnotes of texts of rational reconstructions of science—Lakatos in fact had quite reasonable things to say about the meta-philosophy of science. In particular, Lakatos's writings contain the idea that any philosophical methodology of science should aim at the maximization of rationally explainable facts, albeit without pretence to ever be able to explain *all* historical facts as rational. I will discuss this idea in the light of the contemporary meta-philosophical literature. Finally, I assess how Lakatos's own account of science – namely the methodology of research programmes – fares in the light of his meta-philosophical criteria, by comparing it to Kuhn's, account.

Keywords Lakatos · Metaphilosophy · History and philosophy of science · Rational reconstruction · Kuhn

11.1 Introduction

Lakatos famously advised to detail the actual history of science that "misbehaved" in the light of the rational reconstruction of science in the footnotes of philosophical texts (Lakatos, 1971, 107). Unsurprisingly, this did not bode well with historians (Kuhn, 1971; Holton, 1974, 1978; Kuhn, 1980; Arabatzis, 2017). It did not help much that Lakatos later called this an "'unsuccessful joke" and claimed never to have suggested in all seriousness that one should treat history of science like that (Lakatos, 1978, 192; see Nanay, 2010). The damage had already been done (see e.g., McMullin, 1970; Kuhn, 1971; Holton, 1974; Koertge, 1976; Laudan, 1977; Holton, 1978; Kuhn, 1980; Godfrey-Smith, 2009; Arabatzis, 2017). This is unfortunate in

S. Schindler (✉)
Centre for Science Studies, Department of Mathematics, Aarhus University, Aarhus, Denmark
e-mail: samuel.schindler@css.au.dk

many ways. Not only has the dismissive view of the history of science implied by Lakatos's "joke" given philosophers a bad reputation among historians, but it also has overshadowed some of the more reasonable things Lakatos had to say about historically grounded philosophy of science.

In this chapter I argue—contrary to the widespread cliché—that Lakatos actually did believe that good philosophy of science ought to accommodate as many historical facts as possible. My argument is based mostly on a close analysis of Lakatos's "History of Science and its Rational Reconstruction", which first appeared in the 1971 PSA proceedings (Lakatos, 1971), and later also in Lakatos (1978). This paper has often been misconstrued as giving advice to historians of how to do history of science (Holton, 1974, 1978; Arabatzis, 2017; Kuukkanen, 2017).[1] But even though Lakatos indeed seemed to address "historians" in his paper, he never really engaged with the work of historians nor was he really interested in their concerns (see also Dimitrakos, 2020).[2] Instead, he squarely focused on philosophy and how history could be used to support philosophical claims. As he put it himself, his arguments were "primarily addressed to the philosopher of science and aimed at showing how he can – and should – learn from the history of science" (122). Lakatos's project is thus better described as *meta-philosophical*.[3]

Lakatos, despite his bad reputation with the historians, actually can be seen as one of the main defenders of an integrated history and philosophy of science approach, as can be gleaned from his reformulation of Kant's famous dictum, with which he opened his metaphilosophical essay: "philosophy of science without history of science is empty; history of science without philosophy of science is blind" (Lakatos, 1971, 91).[4] Lakatos made clear that he opposed an "aprioristic philosophy" that is indifferent to empirical facts, and he also did not believe that there were any "immutable" scientific standards that could somehow be discovered by philosophers without analysis of the history of science (121). To argue that Lakatos had good grounds for making these claims—despite his notorious footnote quote—will be the goal of this chapter.

The structure of this chapter is as follows. Section 11.2 briefly introduces the basic ingredients of Lakatos's meta-philosophical account. Section 11.3 tracks more closely Lakatos's seemingly dismissive remarks about the history of science and whether there is anything to his claim that the famous footnote remarks was not meant seriously. Section 11.4 discusses several aspects of Lakatos's proposal as to how to test philosophical theories with historical evidence. There is one idea by Lakatos which I call the "maximization of rational facts" and which I will focus on

---

[1] Arabatzis criticizes Lakatos for not aiming to explain scientists' beliefs and judgments, which Arabatzis takes to be central to any historiographical project (Arabatzis, 2017, 72).

[2] Tellingly, the only historical work that Lakatos mentions (repeatedly), is by the *philosopher* Joseph Agassi.

[3] The meta-philosophy of history and philosophy of science has gained new momentum in recent years. See e.g., Schickore (2011), Kinzel (2015), Bolinska and Martin (2020), and Schindler and Scholl (2022). See also Sect. 11.6 of the current paper.

[4] The dictum goes back to Hanson (1962) and can also be found in Feigl (1970).

in Sect. 11.5. Section 11.6 discusses this idea in the context of the contemporary meta-philosophical literature. Section 11.7 focuses on the rationality of Lakatos's account by comparing it to Kuhn's account, which, as Kuhn pointed out, apparently very much influenced Lakatos. Section 11.8 concludes this chapter.

11.2 Rational Reconstruction, Methodology, and History

Lakatos thought that philosophy of science should be in the business of providing *rational reconstructions* of historical facts about science. The idea of rational reconstruction goes back to at least Carnap, for whom it was closely tied to concept explication (Carnap, 1950). Lakatos did not define what he meant by it, but he described the purpose of rational reconstruction as the "rational explanation of the growth of objective knowledge" (91). In particular, Lakatos was interested in making rational sense of historical facts concerning the appraisal of theories, or series of theories, which he also called "research programmes" (Lakatos, 1978). In other words, Lakatos saw the role of philosophy in explaining why it was rational for scientists to accept or reject research programmes.

The central notion in Lakatos's project of rational reconstruction is "methodology", which he also described as "theories of scientific rationality", "demarcation criteria", "definitions of science", or "logics of discovery" (Lakatos, 1971, 92). This is quite a mixed bag, but ultimately Lakatos meant to refer to philosophical theories about the scientific method. As examples of methodologies, Lakatos mentioned inductivism, conventionalism, falsificationism, and his own methodology of scientific research programmes. Each of these methodologies could be put to use in the analysis of the history of science and produce different "internal histories", i.e., histories that would make rational sense from the perspective of the very methodology that was used. External history was the history that could not be rationally reconstructed with the methodology at hand. Lakatos was happy to leave external history to "empirical psychology and sociology of discovery" (91).[5] For illustration, consider two of the methodologies discussed by Lakatos, namely inductivism and conventionalism.

In inductivism – or rather in Lakatos's caricature of it – scientific propositions are accepted only if "provenly true", which, according to Lakatos, is the case when they describe "hard facts" or when they are "infallible inductive generalizations" from the facts (92–93). Lakatos conceded that there are some episodes in the history of science which were consistent with inductivism, such as Kepler's conclusion that the shape of planetary orbits is elliptical and not circular, which he inferred from Brahe's exceptionally precise observations (92). One question that inductivism

---

[5] Obviously, Lakatos did not use internal and external history in their traditional sense. Traditionally, internal history of science is historiography concerned with the internal dynamics and developments of science, whereas external history of science is historiography concerned with science in relation to society. See e.g. Kuhn (1971).

cannot answer, according to Lakatos, is why scientists "select" certain facts and not others (93). That is, inductivism has not much to say about theoretical motivations that drive data collection and analysis. Such issues inductivism would then have to leave to external history.

Compare this to conventionalism. Conventionalism Lakatos defined as "pigeonholing" facts into "some coherent whole", whereby the resulting system is thought "true only by convention" (94–95).[6] Progress, for the conventionalist in Lakatos's portrayal, consists in a higher degree of simplicity of the classification of facts (96). Although Lakatos overall seemed more critical of conventionalism than inductivism, he was convinced that the Copernican revolution in astronomy could be described in conventionalist terms (96). As a central problem of conventionalism, Lakatos identified the question of why scientists chose to use certain theoretical classifications over others "at a stage when their relative merits were yet unclear" (96). Such questions, the conventionalist would thus have to leave to external history to sort out.

Thus, each methodology would result in a different external history, i.e., different historical facts that it could not explain. As we shall see later, in Sect. 11.4, Lakatos believed that one could assess different methodologies of science by the internal histories that they gave rise to. But first, let us turn to Lakatos's notorious footnote quote in the next section.

11.3 Of Footnotes and Distortions

Lakatos's notorious footnote quote pertains to the relation of internal and external history: "One way to indicate discrepancies between history and its rational reconstruction is to relate the internal history in the text, and indicate in the footnotes how actual history "misbehaved" in the light of its rational reconstruction" (107). This passage has drawn criticism for obvious reasons: it seems to prioritize rational reconstruction over the actual historical facts. But what good is a rational reconstruction of history when it distorts the historical facts? Rather shockingly, Lakatos explicitly embraced the distortion of facts: "Internal history is not just a selection of methodologically interpreted facts: it may be, on occasions, their *radically improved version*" (106; added emphasis).

The reaction to the disparaging view of history of science that seemed to be implied by quotes like these were understandably stark. Kuhn for instance commented: "What Lakatos conceives as history is not history at all but philosophy fabricating examples. ... Why is it ... that Lakatos feels the need to protect himself from real history? Why does he provide a parody in its place?" (Kuhn, 1971, 143). Similarly, Laudan wrote: "I object to the invention of historical figures and the fabrication of historical beliefs to score philosophical points or to teach philosophical lessons" (Laudan, 1977, 170).

---

[6] Lakatos mentions Duhem as a proponent.

As already mentioned in the introduction, Lakatos later retracted his footnote statement and called it "an unsuccessful joke" (Lakatos, 1978, 192). He even denied that he ever claimed that "this is the way in which history actually ought to be written and, indeed, I never wrote history in this way" (Lakatos, 1978, 192).[7] Can this be right at all? Note that in the notorious footnote passage, Lakatos indeed spoke of only *one way* of indicating discrepancies between internal and external history; he did not say that history of science *ought to* be written like that. Is it also true that, as Lakatos claimed, he never wrote history that way?

There are two main examples where Lakatos has been accused of distorting history. They concern Lakatos's discussion of the Bohr model and the Prout hypothesis. The former distortion concerns the fact that Lakatos himself describes the Bohr research programme as encompassing the introduction of electron spin. But Bohr never pondered the idea. This has been criticized by both Kuhn (1971), Holton (1974) and Kragh (2012). Holton describes this particular part of Lakatos's account as "ahistorical parody that makes one's hair stand on end" (Holton, 1974, 75). With regards to the Prout hypothesis, namely the idea that the atomic weight of all chemical elements are multiples of the atomic weight of hydrogen, the distortion consists in the fact that Lakatos depicted Prout as being aware of the real atomic weight of chlorine being 35.5 and that he, in spite of this, nevertheless held onto his belief that it ought to be 36. But that is not true. Prout was never aware what the real atomic weight of chlorine was; he just believed it was 36 (Koertge, 1976; Hacking, 1979). Lakatos did indeed mention this piece of "real history" in a footnote Lakatos (1978, 53). So at least on this one occasion, Lakatos did write history in way he later said he had not.

Neither of these two distortions strikes me as particularly severe. Surely, Lakatos would actually have helped his point if he had stated the facts regarding the history of the Prout hypothesis not only in a footnote (Hacking, 1979); yet the distortion seems minor. The other distortion concerns Lakatos's decision to *name* a research programme after Bohr. As long as there is no pretence that Bohr himself subscribed to every part of it, I think this "distortion" is actually also innocuous. And Lakatos made very clear that he did not think that philosophy was in the business of explaining the beliefs of individual scientists anyway (Lakatos, 1971, 106).

Of course, regardless of what he actually practiced himself, as a general recipe for philosophers doing history of science, Lakatos's recommendations regarding the distortion of history are unacceptable.[8] Unfortunately, they came to overshadow other parts of his view, which seem much more reasonable, and to which we shall now turn. Before doing so, however, let it be clear that the project of rationally reconstructing history is entirely distinct from distorting historical facts for philosophical means: the first does not imply the latter. One may very well try

---

[7] See Hacking (1979) and Nanay (2010). Lakatos exempted his *Proofs and Refutations* from this claim (Lakatos, 1978, 192).

[8] Hacking (1979) and Larvor (1998) have argued that Lakatos's idea of rational reconstruction can be traced back to Hegelian ideas.

to identify the logical structure underlying theory appraisal in science *while at the same time remaining truthful to the actual facts*. In what follows, this will be taken for granted.

11.4 Testing Philosophical Norms with History ... or How Not to

Lakatos considered the history of science as a "test" of philosophical theories about science (Lakatos, 1971, 108). Lakatos proposed two ways of going about this. The first one involved "falsificationism as a meta-criterion" and essentially consisted in seeking to falsify philosophical methodologies (in the sense introduced earlier) on the basis of historical facts, or rather on the basis of what he called "accepted 'basic value judgments' of the scientific elite" (Lakatos, 1971, 110). Lakatos does not motivate the introduction of this concept other than by stating that he had been inspired by Popper's contrast between the scientific general theory of relativity and the pseudo-scientific psychoanalysis in the context of his discussion of the demarcation problem (ibid.). But I think his reasons for introducing the concept are fairly obvious: methodologically it is prima facie problematic to test norms against facts. Some would even consider this a straight-out fallacy (Giere, 1973). Basic value judgments, on the other hand, are normative judgments, and thus allow Lakatos to get around the norm-fact divide. Furthermore, the basic value judgments of the *scientific elite*, if they could be had, should be fairly independent from the methodologies pondered by the philosophers. One would then have a testing procedure that would avoid charges of circularity (see also below). As an illustration of his proposal, Lakatos considers Popper's falsificationism inadequate as a scientific methodology, because contrary to what falsificationism would recommend, Newtonians did not reject classical mechanics, even when it could not account for the advance of Mercury's perihelion (Lakatos, 1971, 111–112). Lakatos concludes that Popper's falsificationism would "show up even the most brilliant scientists are irrational dogmatists" (112).

The proposal, despite its initial plausibility, must raise eyebrows. First of all, it is not so clear what exactly Lakatos might mean by "basic value judgments". Understood at face value, they seem rather elusive: scientists rarely make statements in print of the sort "this theory, but not that theory, is a good scientific theory". Instead, it seems, Lakatos views such judgments to be implied simply by the kinds of choices that scientists make in practice. For example, if the scientific elite decides to adopt theory A instead of theory B, then the implicit value judgment would be that A is a better theory than theory B. Furthermore, one could infer from the very fact that scientists chose theory A that *one ought to* choose theory A. But that seems to amount to pulling normative rabbits out of descriptive hats!

Lakatos did allow for the possibility that not all theory choices made by scientists are the right choices. He therefore proposed a "pluralistic system of authority" in which the judgments by philosophers may on occasion overrule the judgments of the scientific elite, particularly when research programmes degenerate (and scientists keep working on them) or when a "scientific school degenerates into pseudoscience" (Lakatos, 1971, 121). When the philosopher is supposed to step in, however, Lakatos left open (see also Sect. 11.7).[9]

Suppose, then, that we ignore the norm-fact divide and we do try to test philosophical theories on the basis of historical facts *directly*.[10] What kind of history would we then test our philosophical theories on? It cannot be external history because a particular methodology's external history is *by definition* inconsistent with that methodology. It is also not clear that we could test a methodology with the external history of *another* methodology; according to Lakatos *any* external history is supposed to be irrelevant to philosophical theories. It seems also obvious that we cannot test a methodology on the basis of the internal history that it led to, because that would be circular or, in the case of the use of internal history produced by another methodology, question-begging (McMullin, 1970; Kuhn, 1971). At least that would be so if the historical facts were distorted by the methodologies, which certainly should be concern if one were to follow some of the things Lakatos has said (see the previous section).

There is in fact a third category of history of science in Lakatos's writings, namely "actual history", which can be understood as the set of all historical facts from which internal and external history are constructed. Lakatos's most promising proposal is that we can compare and test competing methodologies by the number of historical facts they succeed in grouping under internal history. We will discuss this idea in more detail in the next section.

Before moving on, though, let us note that Lakatos proposed another way of testing methodologies on the basis of historical facts, namely by way of a "methodology of *historiographical* research programmes", which obviously was supposed to mirror his methodology of scientific research programmes (Lakatos, 1971, 116). Lakatos's reasons for switching from his first proposal to his second, are threefold, with the first of them being almost hilarious: (i) his own methodology of research programmes would also have to be falsified by the historical facts—obviously, that is not the best reason for the switch, (ii) the scientific community may want to reconsider their value judgment in case of a clash with methodology, (iii) if we give up falsificationism at the level of method, we should also give it up at the meta-level (Lakatos, 1971, 116).

The outline of the approach is clear enough, with Lakatos putting most emphasis on a methodology having to have predictive success: "We should, of course, insist that a good rationality theory must anticipate further basic value judgments unexpected in the light of its predecessors or that it must even lead to the revision of previously held basic value-judgements" (116–117). In case of a *progressive shift* we would then replace one methodology by another. When it comes to the details, though I think Lakatos's approach is underspecified. In particular, it remains rather

---

[9] See also related critiques by Feyerabend (1976) and Laudan (1986b).

[10] This is the approach recommended by Donovan et al. (1988).

vague what would count as novel success for Lakatos. For example, he considers the already mentioned value judgment that it was fine for Newtonians not to reject their theory in the face of the advance of Mercury's perihelion as "novel success" on his own methodology of research programmes, simply because scientists' behaviour can be accommodated on it, whereas it cannot be on Popper's falsificationism. It remains unclear, though, what is supposed to be "novel" or "anticipated" about this.

Although Lakatos's proposals are beset with difficulties, there is one element of his account, which I believe is much more promising. I call it the "maximization of rational facts".

11.5 Maximization of Rational Facts and Resistance to Falsification

A core idea of Lakatos's methodology of historiographical research programmes is that methodologies can be criticized by "criticizing the rational historical reconstructions to which they lead" (109). As already mentioned in the previous section, this statement must at first seem puzzling, because it would appear circular or question-begging. But if internal history is just guided, rather than distorted, by methodologies, and if the goal is to maximize the number of historical facts that can be accommodated as rational within internal history, then the proposal is in fact quite plausible. First consider Lakatos himself on the idea of *maximization of rational facts*:

> An 'impressive', 'sweeping', 'far-reaching' external explanation is usually the hallmark of a weak methodological substructure; and, in turn, *the hallmark of a relatively weak internal history* (in terms of which most actual history is either inexplicable or anomalous) *is that it leaves too much to be explained by external history*. When a better rationality theory is produced, internal history may *expand and reclaim ground from external history*. (Lakatos, 1970, 119; added emphasis)

This proposal provides an obvious way for conducting comparative tests of methodologies: if methodology A accommodates x historical facts as rational, and methodology B accommodates y historical facts as rational, then if $y > x$, we should prefer methodology B. Because Lakatos presumes that any facts that cannot be accommodated by a methodology will fall under external history, this would mean in this example that methodology B would minimize the number of historical facts that have to be treated as external 'irrational' history. In sum, the goal of philosophy of science on Lakatos's view must then be to construct methodologies that maximize the historical facts that come out as rational and to minimize the historical facts that come out as irrational.[11]

Consider once more (and for a final time!) Lakatos's own example of the advance of Mercury's perihelion, which was discovered in 1859. Newtonians could not

---

[11] I am certainly not the first to have noticed this idea of maximization of rational facts in Lakatos's account, although I perhaps give it more prominence here than others have before. See Kuhn (1980), Arabatzis (2017), Kuukkanen (2017), Schindler (2018), and Dimitrakos (2020).

account for it, but they did not reject their theory in the face of this anomaly. On the contrary, they kept their theory for another 56 years before Einstein provided them with a better alternative. As already mentioned, Popper's falsificationism has trouble dealing with such cases of "resistance to falsification", as one might call them. In contrast, Lakatos's account seems almost designed to deal with resistance to falsifiers: for Lakatos, it is rational to pursue a research programme so long as it is progressive and produces novel success. Empirical anomalies are being given much less weight in the development of the research programme than what Lakatos called "positive heuristic", that is, mostly theory-driven concerns (Lakatos, 1978, 151). Lakatos also rejected crucial experiments and took the Duhem thesis to heart. Lakatos concludes:

> what for the falsificationist looks like the (regrettably frequent) phenomenon of irrational adherence to a 'refuted' or to an inconsistent theory and which he therefore relegates to *external* history, may well be explained in terms of my methodology *internally* as a rational defence of a promising research programme." (Lakatos, 1971, 102; original emphasis)

Clearly, this third of Lakatos's proposals as to how to test philosophical theories is giving history a prominent role. It also avoids the charge of circularity or question-begging, because one can criticize a methodology on the basis of *the number* of facts it can explain rationally, in particular when there is another methodology that rationally explains *more* historical facts. At the same time, Lakatos was not naïve about rational reconstruction. He made clear that "the history of science is always richer than its rational reconstruction" (105) and that "no set of human judgments is completely rational and thus no rational reconstruction can ever coincide with actual history" (116). There could therefore never be a methodology that would render *all* historical facts rational.

The maximization of rational facts has another advantage that perhaps was not so apparent to even Lakatos himself: the norm-fact divide need no longer be bridged. That is because we are not directly testing philosophical norms against historical facts. Instead, we are asking what philosophical norms can account for the largest number of historical facts. There is thus no longer any need for the invocation of the problematic notion of basic value judgments (see previous section).

An objection one might raise against the maximization-of-rational-facts proposal is that it naively presumes an equality between all historical facts. But of course, some historical facts are more important than others. There will then surely have to be some kind of weighting of the historical facts and competing methodologies will have to be assessed on the basis of those. The weightings themselves may not be indisputable, so there is still room for disagreement, of course. And there might also be situations where one methodology accounts for one important set of facts and another does for another important set of facts, with only a slight overlap between the two. Even when there may be a numerical advantage for one methodology, it would not be clear that we should prefer the methodology with that numerical advantage. Of course, this need not be a devastating criticism of the maximization-of-rational-facts proposal: we simply should not expect ever to obtain some algorithmic procedure for arbitrating philosophical disputes anyway! The general idea of maximisation of rational facts seems sound enough.

There are more challenges to the maximization-of-rational-facts proposal, which have to do with the possibility that the historical facts may not be impartial and therefore not apt to distinguish between different methodologies in the way envisaged by Lakatos. To this and a related issue we will turn now in the context of current discussions in the meta-philosophy of science.

11.6 Lakatos's Meta-philosophy in Light of the Contemporary Literature

There are two major foci in contemporary meta-philosophical discussions concerning the use of historical case studies: one focus has been on the risk of case studies being used in tendentious ways when called upon to support philosophical claims (Pitt, 2001; Schickore, 2011; Kinzel, 2015; Chakravartty, 2017; Bolinska & Martin, 2020); another focus has been the issue of how the apparent particularity of historical cases can be reconciled with the aspiration of generality of philosophical claims (Pitt, 2001; Chang, 2011; Bolinska & Martin, 2020; Schindler & Scholl, 2022). In what follows I want to refer to the first problem as the *problem of impartiality* and to the second problem as the *problem of inductive warrant*.[12]

11.6.1 The Problem of Impartiality

There are three versions of the problem of impartiality: bias, distortion, and theory-ladenness. These three aspects should be kept apart, because they have very different implications. As we already mentioned in Sect. 11.3 of this chapter, distortion should have no place in any sound empirical approach, philosophical or otherwise. I do not think that respected philosophers using case studies have made themselves suspect of the charge, even though they have been accused of it. For example, Pitt writes in his highly influential dilemma for the case study approach:

---

[12] For completeness's sake, one should mention that there is another major meta-philosophical problem, which received some attention in the 1970s and 1980s, namely the problem of how to bridge the gap between philosophical *norms* and historical *facts*. Giere (1973) was the first to highlight this problem, but the current consensus seems to be that the problem disappears, either when we conceive of philosophical theories akin to scientific theories, that ought to be tested empirically (Giere, 1985; Donovan et al., 1988), or when we conceive of methodologies as *instrumental norms*, whereby empirical facts need to tell us whether the set goals are actually achieved by the used methods (Laudan, 1986a, 1990). For a criticism of the former option see Schickore (2011); for a criticism of the latter option see Schindler (2018).

> On the one hand, if the case is selected because it exemplifies the philosophical point being articulated, then it is not clear that the philosophical claims have been supported, *because it could be argued that the historical data was manipulated to fit the point*. On the other hand, if one starts with a case study, it is not clear where to go from there—for it is unreasonable to generalize from one case or even two or three. (Pitt, 2001, 373; added emphasis)

Possibly, Pitt means to say something weaker, namely that case studies are being cherry-picked by philosophers to fit their agendas. In that vein, Nickels once commented that "historical case studies can be too much like the Bible in the respect that if one looks long and hard enough, one can find an isolated instance that confirms or disconfirms almost any claim" (Nickles, 1995, 141).

Just like science, philosophy has safeguards against bias and cherry-picking. The better journals of the field have (pretty stringent) peer-review procedures that ensure that people not positively inclined to the author's thesis get a chance to criticize the author's bias. Even if peer-review fails to catch biases, there is a critical community of philosophers which is bound to either bring the bias to the attention of their peers or to present their own case studies challenging the thesis in question. And if peer-review and community criticism is apt to detect biases, then this should be even more the case for distortion. Of course, there is no guarantee that all biases and distortions will be caught by the community, but there is also no reason whatsoever to think that the very project of using history in support of philosophical theorizing is so skewed that all hope is lost.

A more serious concern than bias is theory-ladenness. Theory-ladenness is best understood not just as the theoretically biased selection of facts, but rather as something deeper, namely, as the impingement of theoretical presuppositions on the very way that the facts are described. Even when philosophers would take great care not to distort the facts, historical case studies could then never provide neutral support for any philosophical view, because the same historical facts could be seen in this or that light, depending on one's philosophical views (see Kinzel, 2015). In fact, that was a concern that Kuhn expressed in his review of a collection of essays discussing historical cases in the spirit of Lakatos's theoretical framework of research programmes (Howson, 1976):

> 'actual history' of the sort Lakatos requires is a myth. ... the data in most parts of the pool are not, until after much interpretation, the facts which appear in historical narratives. ... It is by no means clear, however, that proponents of those [compared] methodologies would accept the elements of his narrative as simply factual, and it is upon that agreement that his demonstration depends. History is interpretative throughout. (Kuhn, 1980, 184)

Interestingly, despite this concern, Kuhn concluded that "Lakatos is, I think, clearly right to suggest that improved historical narratives are often the ones that give a central role to a larger body of evidence" (Kuhn, 1980, 184–185). Kuhn just thought that the writing of history should be left to the historians, so as to ensure that philosophers' theories of science could not impinge on the writing of history in the problematic and distortive ways that Lakatos, regrettably, had suggested. Lakatos was not alive anymore, so he had not chance to respond this this suggestion by Kuhn. But I think we can be confident that he would have rejected this proposal, as he made clear already at the beginning of his meta-philosophical essay that he considered history without philosophy "blind".

How severe a threat to Lakatos's idea of maximization of rational facts is the theory-ladenness of historical facts then? How severe a threat is theory-ladenness to *any* philosophy seeking to use history as evidence? Perhaps not much more severe than theory-ladenness of observations in science. For example, Kinzel (2015) has argued that even when descriptions of observations are influenced by the vocabulary of the relevant philosophical theories, evidence can still be recalcitrant and act back against our theories. Kinzel is less optimistic that philosophical disputes can be arbitrated, because she believes that the very criteria for what may count as a good historical case study are likely to be theory-laden, for example, whether one thinks that social factors ought to be taken into consideration when constructing a case, or not (Kinzel, 2015, 55).

Perhaps argument criteria such as coherence and cogency can help us settle even such more fundamental disputes, at least in principle (Bolinska & Martin, 2020). In principle, it is also a matter of fact whether social factors are or are not required to explain a particular historical episode (ibid.). Furthermore, there are philosophical debates, in which the basic presuppositions are fairly widely shared. In such debates, the focus of discussion can even be put more squarely on the relevant historical facts. For example, in the realism debate, it is widely agreed between realists and antirealists that the success of scientific ought to be explained by epistemic factors and *not* by social factors. Now, of course, there is still room for disagreement, but I think it would be wrong to suggest that *therefore* the historical facts are evidentially toothless (Chakravartty, 2017). On the contrary, there are clear signs of progress that can be made by engaging with the historical details. For example, Psillos first argued that theorists of the caloric theory of heat did not embrace the reality of caloric as a substance when explaining heat phenomena and when deriving novel predictions (Psillos, 1999). But *on the basis of a further engagement with the historical facts* it has been shown that this is actually incorrect (Chang, 2003; Stanford, 2006). This is widely accepted.

There are other prominent examples of cases where concerted efforts by the community of philosophers have resulted in deeper and richer understanding of selected historical cases and arguably also in a sense of philosophical progress. Take for example the famous case of Semmelweis and his discovery of the cause of child bed fever. The case has been discussed by several philosophers, for several purposes. Hempel first took it to nicely illustrate the hypothetico-deductive method (Hempel, 1966). Lipton instead argued that Semmelweis in fact used the inference to the best explanation (Lipton, 1991/2004). Others disagree. Scholl (2013) argues that Semmelweis's method is best described in non-explanationist terms, namely Stuart Mill's method of agreement and concomitant variation. Superficially, it may look as though the history lends itself to almost any philosophical interpretation one pleases. And yet, the concerted study of the Semmelweis case has allowed philosophers to compare their abstract ideas in relation to a concrete case, with the succession of authors making arguments for their accounts to be a *better* representation of the actual case and to explain *more* historical facts (see Schindler and Scholl (2022) for a discussion). Even if one disagrees that we know more now about the case and about what philosophical method best represents the case, the historical case provides a focal point which has enriched the philosophical discussion, and without which the discussion may have remained rather sterile and removed from actual practice.[13]

Cases like these should give us confidence that there can be a fruitful interplay between philosophical theorizing and engagement with the historical facts, and that philosophical biases can be corrected by aiming for a fuller historical picture. This is not to say, though, that simply doing a rich historical analysis, involving analysis of contextual factors, will always give us a "truer" picture of science.[14] On the contrary, without a focus on a specific philosophical question, I believe with Lakatos (and Hanson) that we are bound to move philosophically "blindly" in historical territory: it is just much harder (and sometimes impossible) to address philosophical problems with history that is written without some philosophical question in mind. Hanson once put it, perhaps not fully charitably, but not entirely off target either: "To the philosopher, histories of science are often unilluminating because, as a result of their chaotic diffuseness, they never reflect monochromatically: only spectra of concepts and arguments result" (Hanson, 1962, 582).[15]

11.6.2 *The Problem of Inductive Warrant*

Let us now turn to the problem of inductive warrant: on what grounds are philosophers warranted to claim inductive support from just a few selected historical case studies (also see Pitt's dilemma cited in the previous section)? On the face of it, the problem seems much more challenging than the problem of impartiality: even when all the safeguards of the community work well, there is still an issue of how philosophers are licensed to infer from few cases to general philosophical claims about science.

Philosophy is not science. Philosophy thus cannot help itself to the same, established means that science can, such as statistics, for tackling inductive inferences. True, experimental philosophers are using statistics when collecting intuitive judgments from non-philosophers (Knobe & Nichols, 2017), but it is harder to see how this could work out with historical case studies. Obviously, not all cases are to be weighed equally, and constructing a single case in the first place, is very

---

[13] For another example of the same sort, see the discussion of the discovery of the weak neutral current in high energy physics (Schindler, 2014).

[14] This is what both Pitt himself, and Burian in his reply to Pitt's dilemma, seem to imply (Burian, 2001; Pitt, 2001).

[15] In all fairness, Hanson had similarly nice things to say about the work of philosophers: "To the historian such philosophy of science is often unilluminating because it does not enlighten one about any *thing*: nothing in the scientific record book is treated in such symbolic studies" (Hanson, 1962, 582).

time-consuming and dependent on choices that are subject to disagreement. As we have just seen, though, even when there is focus on just a few or even a single case, engagement with the history of science can be very fruitful for philosophical arguments and insight.

There is a less demanding look upon the kind of inductive support philosophy of science requires. Philosophical theories are often taken to be "generalizing", as in over-generalizing without inductive support (Pitt, 2001). But in reality, philosophers are not as naïve as they are sometimes made out to be by the critics. For example, in the literature on explanation, causal models of explanation have enjoyed an enormous popularity in the past few decades (Woodward, 2003). But even though proponents of these models obviously take causal explanation to be a widely used form of explanation in science, there is no pretence that causal explanation is the only explanation there is in science (Reutlinger & Saatsi, 2018). The same is true for the inference to the best explanation, which, as mentioned, has been motivated through historical case studies. Again, even though the inference to the best explanation is considered to be a widely used and important mode of inference, no philosopher would be silly enough to claim that it is the only kind of inference used in science.

Sometimes it is claimed that the history of science is so particular that it allows for no generalizations *whatsoever*.[16] But this is a claim that requires at least as much argument as the claim that there are at least *some* patterns and generalities that can be drawn from the history of science. Indeed, it would be surprising if the success of science was somehow based on complete happenstance and chance events. One is of course free to throw up one's hands and give up on the project of historically informed philosophy of science. But it is surely much more constructive to assume that the pessimistic view is false and to go to investigate philosophical theses about science under consideration of the history of science.

11.6.3 Lakatos and the Two Problems

Prima facie, Lakatos's core idea of the maximization of rational facts is threatened both by the problem of impartiality and the problem of inductive warrant, which both have been discussed extensively by the contemporary meta-philosophical literature. Yet, as we have seen in this section, bias is well-manageable by the profession of philosophers. It seems to have been Lakatos's view anyway, that bias could be controlled for by the number of facts that could be accounted for by a methodology's rational, internal history. Distortion, despite Lakatos's earlier slip of the tongue, should have no place in any respected academic discipline, as Lakatos later himself seemed to realize. Theory-ladenness is by far the most serious

---

[16] Pitt (2001) calls this the "Heraclitan view" of the history of science. See also Bolinska and Martin (2020) for a discussion.

threat not only to Lakatos's project of rational reconstruction, but to *any* meta-philosophical view that seeks to combine philosophical theorizing and historical facts. Yet, as we have seen, there is reason for optimism: if theory-ladenness in science is no reason for despair, then it should not be in philosophy either. And even if philosophy should somehow be worse off than science, there are still some examples where the study of history *has* given us a better understanding of the philosophical questions posed. Finally, the problem of inductive warrant seems more severe than the problem of impartiality, but it of course besets again not only Lakatos's approach, but *any* approach seeking to combine philosophy of science and history of science. There is also reason to think that philosophers are much more careful in their generalisations than the critics often have it.

11.7 Rationality in Lakatos and Kuhn

In the previous two sections we have been concerned with Lakatos's idea of the maximization of rational facts and how it compares to current work in the meta-philosophy of science. Throughout this chapter, we have been concerned with the question of how philosophical accounts of science may be assessed on the basis of historical facts. However, one may also ask about whether a philosophical account is internally coherent and what kind of rationality it entails. This is the kind of question we shall now turn to. To assess Lakatos's account with regard to this question, we shall compare it to Kuhn's account of science, which, as we shall see, Lakatos was quite substantially inspired by. Contrary to what Lakatos claimed, his account cannot be said to be an advance over Kuhn's in terms of rationality.

11.7.1 Resistance to Falsification, Again

In Sect. 11.5 we saw that what we referred to as "resistance to falsification" played a central role in Lakatos's own assessment of philosophical methodologies, and more specifically, both in his arguments against falsificationism and in his arguments favour of his own methodology of research programmes. It is interesting to note that the importance of resistance to falsification seems to have been impressed upon Lakatos by Kuhn. There is some circumstantial evidence for this, which can be found the precursor of his seminal "Falsification and the Methodology of Scientific Research Programmes" (Lakatos, 1970), namely a paper with the title "Criticism and the Methodology of Scientific Research Programmes", which was published in the *Proceedings of the Aristotelian Society* (Lakatos, 1968). In the first section of this precursor, titled "Popper vs. Kuhn", Lakatos contrasts Popper's account of science to Kuhn's. The former he describes thus: "Boldness in conjectures on the one hand and austerity in refutations on the other: this is Popper's recipe" (150). "Commitment" to a theory despite counterexamples, on the other hand, Lakatos describes as being "an outright crime" for Popper (ibid.). Kuhn, on the other hand, took commitment (to a paradigm) to be central to science. Lakatos explains that for Kuhn "the transition from criticism to commitment marks the point where progress -and 'normal' science-begins" (ibid.).[17] Lakatos concludes that the debate between Popper and Kuhn is "not merely over a technical point in epistemology" but rather "over our central intellectual values" (151).

In the second section of the precursor to the *Methodology* paper, Lakatos sought to develop less "naïve" versions of Popper's account (which he characteristically referred to with subscripts: $Popper_1$, $Popper_2$, $Popper_3$), ultimately resulting in his methodology of research programmes. As the "main difference" between Popper's falsificationism and his own account Lakatos tellingly described that the "criticism [of theories] does not – and must not – kill as fast as Popper imagined" (Lakatos, 1968, 183). Later, in his final version of his *Methodology* paper, Lakatos would attribute to Kuhn the insight that falsificationism fails as a "rational account of scientific growth" (Lakatos, 1970, 93).

There is even further, more direct evidence that Lakatos was inspired by Kuhn's highlighting of resistance to falsification in the history of science. In the unpublished precursor to the precursor of the *Methodology* paper, dated 1967 and stored in the Lakatos Achieve at the LSE, Lakatos speaks of "Kuhn's thesis that 'theories are born refuted'" (8). He also notes that if the thesis is correct, "then refutations play no dramatic role in science ... The [Popperian] slogan 'make sincere attempts to refute your theories', falls flat if any new theory emerges in an ocean of counterexamples" (8).

11.7.2 Other Parallels

Arguably, resistance to falsification was not the only inspiration that Lakatos took from Kuhn. Kuhn in fact once noted, in his comments on Lakatos's meta-philosophical paper—which has been the main subject of the current chapter—that "I have read no paper on scientific method which expresses opinions so closely paralleling my own" (Kuhn, 1971, 137). In particular, Kuhn believed that Lakatos' hard core, protective belt, and a programme's "degenerative phase" were analogous to his notion of a paradigm, normal science, and crisis, respectively (Kuhn, 1970, 256), and he voiced his frustration that Lakatos was "so unable to see" these parallels (Kuhn, 1971, 139). Let us have a closer look at these points in turn.

First, Kuhn thought that Lakatos's choice of basic analytical unit of a research programme was not so different at all from his unit of a paradigm (Kuhn, 1971, 138), which is fair because both are more encompassing than theories or hypotheses. Kuhn also believed that Lakatos's notion of a hard core of research programs really differed little from how he had described paradigms, namely as containing

---

[17] However, see my Schindler (2024).

"elements which are not themselves subject to attack" (Kuhn, 1971, 138). Second, although Kuhn did not mention this, Lakatos's notion of positive heuristic bears some similarities to Kuhn's notion of normal science: the former helps to develop a research programme by guiding scientists in their articulation of the 'protective belt'. In normal science, practitioners strive to increase the scope and the precision of a paradigm. Both of these activities are carried out while taken as given the hard core / basic assumptions of the paradigm, and neither activity is unseated by the discovery of anomalies.

Third, with regard to the phenomenon of resistance to falsification, Lakatos thought that as long as research programmes are "progressive", and as long as no better alternatives are available, scientists do not and should not abandon them (Lakatos, 1970, 100, 130, and 176–177). How does Lakatos compare here to Kuhn? Kuhn offers *two* (albeit closely related) rationales for why it might be rational for scientists to resist falsification during periods of normal science: (i) scientists would not get much work done if they constantly let themselves bogged down by anomalies, rather than addressing those that the paradigm can solve (Kuhn, 1962/1996, 82) and (ii) since, according to Kuhn, *all* theories are contradicted by *some* data, we could never embrace *any* theory if we would view a theory to be falsified whenever the theory's predictions did not match the evidence (Kuhn, 1962/1996, 146). We may refer to those two rationales as "efficiency" and "pragmatism", respectively. Lakatos clearly also adopted the second rationale (Lakatos, 1970, 120, fn. 1). He does not say anything resembling the first rationale; but he does not say anything that would contradict it either.

Despite these apparent parallels between Lakatos's and Kuhn's accounts of science, Lakatos did not have much positive to say about Kuhn. The few comments that Lakatos makes about Kuhn in his *Methodology* paper are negative and focused on theory change. For example, Lakatos described paradigm change as "a mystical conversion which is not and cannot be governed by rules of reason ... [It is] a kind of religious change" (Lakatos, 1970, 93). Furthermore, Lakatos noted that "there is no particular rational cause for the appearance of a Kuhnian 'crisis'. 'Crisis' is a psychological concept; it is a contagious panic. ... Thus in Kuhn's view scientific revolution is irrational, a matter for mob psychology" (Lakatos, 1970, 187).

Even though Lakatos's charge of irrationality concerning paradigm change was perhaps not entirely off target, Kuhn vehemently protested against what he took to be a caricature of his views. First of all, he argued that paradigm crisis was in fact not at all dissimilar to degenerative phases of research programmes, because also for paradigm crisis to occur, there needs to be a piling up of anomalies (Kuhn, 1971). Second, Kuhn pointed out that, although he indeed did not believe that theory-choice was governed by rules, he believed that theory-choice was guided by standard criteria of theory choice, and based on the value judgments by scientists (Kuhn, 1970). Kuhn also pointed out that Lakatos's account required several 'decisions' (in the parlance that Popper and Lakatos preferred) by the scientific community, which he took to be similar to the value judgments in theory choice (Kuhn, 1970, 238–240). For example, what are the assumptions that belong to the hard core and which assumptions are part of the protective belt? When is a research programme progressive and when is it degenerating? The second question is particularly problematic, because, even though Lakatos presented a *prima facie* clear-cut criterion for progress in the form of novel success, good research programmes can be degenerating, and thus fail to generate novel success, for a long time (see e.g., Feyerabend, 1976). For example, one of the earliest, most impressive scientific achievements, namely the Copernican research program, was degenerating for no less than a hundred years (!) according to Lakatos's own criteria of progress, before it would be salvaged by the likes of Galileo and Newton (Lakatos & Zahar, 1978).

Being aware of such cases, Lakatos insisted that his account did not offer "instant rationality" and that research programs must be treated "leniently" in their early stages (Lakatos, 1970, 179). He also suggested that "it is perfectly rational to play a risky game: what is irrational is to deceive oneself about the risk" (Lakatos, 1971, 104, footnote). In other words, it is not irrational for scientists to pursue a research programme, even when it is degenerative, because the programme might ultimately turn out to be progressive. But one should be aware that one is taking a risk, particularly when another research programme is available that is progressive. This all seems reasonable, but Lakatos thus fails himself to supply the "rules of reason" that he criticized Kuhn for not providing.[18]

In recent years, Lakatos's preferred criterion of measuring success of research programmes, namely novel success, has come under attack. Even Lakatos—perhaps through the works of Zahar on the topic—was not committed to research programmes necessarily having temporarily novel success. But even a weaker criterion like "use-novelty" (Worrall, 1989b, 2014), according to which a piece of evidence is novel if it was not used in the construction of a theory, comes with all kinds of problems (Schindler, 2018). So it is not clear whether novel success can at all serve the meta-philosophical purpose assigned to it by Lakatos in the assessment of research programmes.

All in all, apart from Kuhn's highly controversial notion of incommensurability of paradigms, I think there is not a single dimension where one could say that Lakatos offered a more rational reconstruction than Kuhn did. There do not seem to be any historical facts that Lakatos, but not Kuhn, managed to explain rationally either. Most notably, both Kuhn and Lakatos rationally explained the resistance to falsification. By his own account of rationality, Lakatos thus failed to provide a better "methodology" than Kuhn.

---

[18] Like Popper, Reichenbach and others, Lakatos thought the context of discovery was the domain of psychology and sociology, but not of the philosophy of science. Instead, he thought philosophers ought to be concerned with finding normative rules for the "appraisal of ready, articulated theories" (Lakatos, 1971, 92). Confusingly, though, he spoke of proposals for those rules as "logics of discovery" (ibid.).

11.8 Conclusion

This chapter argued that despite his notorious footnote quote, Lakatos actually had quite reasonable things to say about the meta-philosophical question of how to use historical facts in philosophical theorizing. In particular, the idea of the maximization of rationally explainable facts strikes me as a decent project for a historically informed philosophy of science. This kind of philosophy of science has admittedly gone out of fashion. One can only speculate about the reasons why. One concern Lakatos and his contemporaries shared was the demarcation problem: what is it that characterizes science? Popper obviously had his answer, and so did Kuhn and Lakatos (Lakatos, 1978). The consensus today appears to be in many quarters that solving the demarcation problem is hopeless and misconceived: there are just too many dissimilarities between the sciences for one to look for an answer that "fits all".[19] This concern is related to another common contemporary view, namely that the methods used by science are very diverse *even within a single discipline*, so it may appear hopeless to assess the rationality of "the" scientific method (Laudan, 1977).[20] Second, philosophers of science seem to have largely lost interest in the diachronic dimension of science, to which the study of history is indispensable.[21] Whether one thinks that these trends are to be welcomed or regrettable, Lakatos's work still provides a rich resource for thinking about how the philosophy of science could fruitfully be engaged with the history of science.

References

Arabatzis, T. (2017). What's in it for the historian of science? Reflections on the value of philosophy of science for history of science. *International Studies in the Philosophy of Science, 31*(1), 69–82.

Bolinska, A., & Martin, J. D. (2020). Negotiating history: Contingency, canonicity, and case studies. *Studies in History and Philosophy of Science Part A, 80*, 37–46.

Burian, R. M. (2001). The dilemma of case studies resolved: The virtues of using case studies in the history and philosophy of science. *Perspectives on Science, 9*(4), 383–404.

Carnap, R. (1950). *Logical foundations of probability*. Chicago University Press.

Chakravartty, A. (2017). *Scientific ontology*. Oxford University Press.

Chang, H. (2003). Preservative realism and its discontents: Revisiting caloric. *Philosophy of Science, 70*(5), 902–912.

Chang, H. (2011). Beyond case-studies: History as philosophy. In S. Mauskopf & T. Schmaltz (Eds.), *Integrating history and philosophy of science* (pp. 109–124). Springer.

---

[19] But see Pigliucci and Boudry (2013) and Schindler (2018) for recent revival attempts.

[20] But see the intriguing discussion between Laudan and Worrall, who defends the view of "fixed methodology" (Worrall, 1988; Laudan, 1989; Worrall, 1989a).

[21] One exception to this trend is the realism debate, and in particular the pessimistic meta-induction, where philosophers of science have quite closely engaged with the history of science (although, again, without much focus on the diachronic element of theories).

Dimitrakos, T. (2020). Reconstructing rational reconstructions: On Lakatos's account on the relation between history and philosophy of science. *European Journal for Philosophy of Science, 10*, 1–29.

Donovan, A., Laudan, L., & Laudan, R. (1988). *Scrutinizing science: Empirical studies of scientific change* (Vol. 193). John Hopkins University Press.

Feigl, H. (1970). Beyond peaceful coexistence. *Minnesota Studies in the Philosophy of Science, 5*, 3–11.

Feyerabend, P. (1976). On the critique of scientific reason. In C. Howson (Ed.), *Method and appraisal in the physical sciences: The critical background to modern science, 1800–1905* (pp. 309–339). Cambridge University Press.

Giere, R. N. (1973). History and philosophy of science: Marriage of convenience or intimate relationship. *British Journal for the Philosophy of Science, 24*, 282–297.

Giere, R. N. (1985). Philosophy of science naturalized. *Philosophy of Science, 52*(3), 331–356.

Godfrey-Smith, P. (2009). *Theory and reality: An introduction to the philosophy of science*. University of Chicago Press.

Hacking, I. (1979). Imre Lakatos's philosophy of science. *British Journal for the Philosophy of Science, 30*(4), 381–402.

Hanson, N. R. (1962). The irrelevance of history of science to philosophy of science to philosophy of science. *The Journal of Philosophy, 59*(21), 574–586.

Hempel, C. G. (1966). Philosophy of natural science. *British Journal for the Philosophy of Science, 18*(1), 70–72.

Holton, G. (1974). On being caught between Dionysians and Apollonians. *Daedalus, 103*(3), 65–81.

Holton, G. (1978). *The scientific imagination: Case studies*. Cambridge University Press.

Howson, C. (1976). *Method and appraisal in the physical sciences: The critical background to modern science, 1800–1905*. Cambridge University Press.

Kinzel, K. (2015). Narrative and evidence. How can case studies from the history of science support claims in the philosophy of science? *Studies in History and Philosophy of Science Part A, 49*, 48–57.

Knobe, J., & Nichols, S. (2017). Experimental philosophy. In N. Edward (Ed.), *The Stanford encyclopedia of philosophy (winter 2017 edition)*. Zalta.

Koertge, N. (1976). Rational reconstructions. In R. S. Cohen, P. Feyerabend, & M. W. Wartofsky (Eds.), *Essays in memory of Imre Lakatos* (pp. 359–369). Reidel.

Kragh, H. (2012). *Niels Bohr and the quantum atom: The Bohr model of atomic structure 1913–1925*. Oxford University Press.

Kuhn, T. S. (1962/1996). *The structure of scientific revolutions* (3rd ed.). University of Chicago Press.

Kuhn, T. S. (1970). Reflections on my critics. In I. Lakatos & A. Musgrave (Eds.), *Criticism and the growth of knowledge, proceedings of the international colloquium in the philosophy of science* (pp. 231–278). Cambridge University Press.

Kuhn, T. S. (1971). Notes on Lakatos. In R. C. Buck & R. S. Cohen (Eds.), *Psa (1970): Proceedings of the biennial meeting of the philosophy of science association* (pp. 137–146). Springer.

Kuhn, T. S. (1980). The halt and the blind: Philosophy and history of science. *The British Journal for the Philosophy of Science, 31*(2), 181–192.

Kuukkanen, J.-M. (2017). Lakatosian rational reconstruction updated. *International Studies in the Philosophy of Science, 31*(1), 83–102. https://doi.org/10.1080/02698595.2017.1370930

Lakatos, I. (1968). Criticism and the methodology of scientific research Programmes. *Proceedings of the Aristotelian Society, 69*, 149–186.

Lakatos, I. (1970). Falsification and the methodology of scientific research Programmes. In I. Lakatos & A. Musgrave (Eds.), *Criticism and the growth of knowledge* (pp. 91–196). Cambridge University Press.

Lakatos, I. (1971). History of science and its rational reconstructions. In R. C. Buck & R. S. Cohen (Eds.), *Psa (1970): Proceedings of the biennial meeting of the philosophy of science association* (pp. 91–136). Springer.

Lakatos, I. (1978). *The methodology of scientific research Programmes: Volume 1: Philosophical papers*. In J. Worrall (Ed.), *And Gregory Currie*. Cambridge University Press.

Lakatos, I., & Zahar, E. (1978). Why did Copernicus' research program supersede Ptolemy's? In J. Worrall & G. Currie (Eds.), *The methodology of scientific research Programmes: Volume 1: Philosophical papers* (pp. 168–192). Cambridge University Press.

Larvor, B. (1998). *Lakatos: An introduction*. Routledge.

Laudan, L. (1977). *Progress and its problems: Towards a theory of scientific growth*. University of California Press.

Laudan, L. (1986a). *Science and values: The aims of science and their role in scientific debate*. University of California Press.

Laudan, L. (1986b). Some problems facing intuitionist meta-methodologies. *Synthese, 67*(1), 115–129.

Laudan, L. (1989). If it ain't broke, don't fix it. *The British Journal for the Philosophy of Science, 40*(3), 369–375.

Laudan, L. (1990). Normative naturalism. *Philosophy of Science, 57*(1), 44–59.

Lipton, P. (1991/2004). *Inference to the best explanation* (2nd ed.). Routledge.

McMullin, E. (1970). The history and philosophy of science: A taxonomy. *Minnesota Studies in the Philosophy of Science, 5*, 12–67.

Nanay, B. (2010). Rational reconstruction reconsidered. *The Monist, 93*(4), 598–617.

Nickles, T. (1995). Philosophy of science and history of science. *Osiris, 10*, 139–163.

Pigliucci, M., & Boudry, M. (2013). *Philosophy of pseudoscience: Reconsidering the demarcation problem*. University of Chicago Press.

Pitt, J. C. (2001). The dilemma of case studies: Toward a Heraclitian philosophy of science. *Perspectives on Science, 9*(4), 373–382.

Psillos, S. (1999). *Scientific realism: How science tracks truth*. Routledge.

Reutlinger, A., & Saatsi, J. (2018). *Explanation beyond causation: Philosophical perspectives on non-causal explanations*. Oxford University Press.

Schickore, J. (2011). More thoughts on Hps: Another 20 years later. *Perspectives on Science, 19*(4), 453–481.

Schindler, S. (2014). A matter of Kuhnian theory choice? The Gws model and the neutral current. *Perspectives on Science, 22*(4), 491–522.

Schindler, S. (2018). *Theoretical virtues in science: Uncovering reality through theory*. Cambridge University Press.

Schindler, S. (2024). Normal science: Not uncritical or dogmatic. *Synthese, 203*, 1–22.

Schindler, S., & Scholl, R. (2022). Historical case studies: The "model organisms" of philosophy of science. *Erkenntnis, 87*, 933–952.

Scholl, R. (2013). Causal inference, mechanisms, and the Semmelweis case. *Studies in History and Philosophy of Science Part A, 44*(1), 66–76.

Stanford, P. K. (2006). *Exceeding our grasp: Science, history, and the problem of unconceived alternatives*. Oxford University Press.

Woodward, J. (2003). *Making things happen: A theory of causal explanation*. Oxford University Press.

Worrall, J. (1988). The value of a fixed methodology. *The British Journal for the Philosophy of Science, 39*(2), 263–275.

Worrall, J. (1989a). Fix it and be damned: A reply to Laudan. *The British Journal for the Philosophy of Science, 40*(3), 376–388.

Worrall, J. (1989b). Fresnel, Poisson and the 'white spot': The role of successful prediction in theory-acceptance. In D. Gooding, T. Pinch, & S. Schaffer (Eds.), *The uses of experiment* (pp. 135–158). Cambridge University Press.

Worrall, J. (2014). Prediction and accommodation revisited. *Studies in History and Philosophy of Science Part A, 45*, 54–61.

Open Access This chapter is licensed under the terms of the Creative Commons Attribution-NonCommercial-NoDerivatives 4.0 International License (http://creativecommons.org/licenses/by-nc-nd/4.0/), which permits any noncommercial use, sharing, distribution and reproduction in any medium or format, as long as you give appropriate credit to the original author(s) and the source, provide a link to the Creative Commons license and indicate if you modified the licensed material. You do not have permission under this license to share adapted material derived from this chapter or parts of it.

The images or other third party material in this chapter are included in the chapter's Creative Commons license, unless indicated otherwise in a credit line to the material. If material is not included in the chapter's Creative Commons license and your intended use is not permitted by statutory regulation or exceeds the permitted use, you will need to obtain permission directly from the copyright holder.

Chapter 12
Cholesterol and Cardio-Vascular Disease: Degenerating Research Programmes in Current Medical Science

John Worrall

Abstract This paper argues that, when examined from a Lakatosian perspective, the (mini-) research programmes that have been built to defend two important theories in modern medicine show all the marks of consistent empirical degeneration. Yet those two theories remain enormously influential—underpinning, as they continue to do, advice and treatment given to millions of people worldwide. As Lakatos and others have shown, theories in successful sciences, such as physics and chemistry, whose associated research programmes have degenerated have invariably been rejected. This is an area, then, in which Lakatosian ideas might have enormous impact—if the verdict of degeneration is correct and if it were accepted by the medical community, it could lead to change of medical treatment for millions of people. The final part of the paper looks at reasons why recommended treatment has so far *not* changed despite the apparent degeneration of the supporting research programmes. It therefore takes us into another area of Lakatosian scholarship—his much-discussed distinction between 'internal' and 'external history of science'.

Keywords Ad hoc · Independent testability · Degenerating research programmes · Cholesterol · Cardio-vascular disease · Statins · Internal and external history

J. Worrall (✉)
Department of Philosophy, Logic and Scientific Method, London School of Economics and Political Science, London, UK
e-mail: j.worrall@lse.ac.uk

As noted in the entry on his work in *The Stanford Encyclopedia of Philosophy,* Imre Lakatos was "very much more than a philosopher's philosopher".[1] In particular, researchers in a variety of fields—Biology, Psychology, International Relations and Management Science amongst them—have found it enlightening to conceptualise pieces of theorising in their subjects as research programmes and to assess them for progress or degeneration in Lakatosian terms.

This paper concerns a *potential* impact of Lakatos's ideas outside of philosophy—in fact in contemporary medicine. I will show that a whole series of steps in two "mini-programmes" built to defend influential medical claims constitute clear-cut cases of Lakatosian degeneration. While I cannot here survey all the evidence for all the very many saving hypotheses potentially involved in these two mini-programmes, I do hold that I make a *prima facie* case for the overall degeneration of those mini-programmes.

Assuming that a full analysis of all the evidence would bear out this judgment, then one would have expected both of the hypotheses involved to have been firmly rejected. But the reality is very different. The second hypothesis in particular continues to be very widely accepted in medicine and to form the basis for advice to, and treatment of, millions of people worldwide. As for the first, the consensus that it attracted for decades has recently shown some signs of breaking up, but it continues to have a firm hold on public opinion and certainly remained accepted as true and as the basis for dietary and medical advice long after the research programme built to defend it had shown clear signs of degeneration.

I hope that by characterising the situation in explicitly Lakatosian terms and hence relating it to cases in "harder" sciences such as physics, where such degeneration has historically always led to the rejection of the theories/research programmes at issue, that this will strengthen the hand, and hence the influence, of those few within medicine who have been and remain sceptical about the hypotheses concerned.[2] And hence that it will have an impact, both on the science and its application (in terms of approved advice and treatment).

In the final section of the paper, I will address the conflict between the judgements arguably supplied by Lakatos's methodology and what has actually happened—and is actually happening—in medicine. This will point us in the direction of "group think", vested interests and vast amounts of money via Lakatos's famous (some might hold, infamous) distinction between "internal" and "external history" of science.

---

[1] My academic career would never even have started without the inspiration, guidance and support of Imre Lakatos; and so, it was a special pleasure and honour for me to present a plenary address at the "Lakatos @100" Centenary conference held at LSE in November 2022. This paper is a revised version of that presentation—some revisions having been made in response to helpful criticisms from a referee.

[2] These include Dr. Malcolm Kendrick, to whose 2007 and 2018 books this paper is greatly indebted.

12.1 Methodological Preliminaries: Adhocness, Independent Testability and Degeneration

I will turn to the medical examples very shortly, but first some preliminary clarificatory remarks about degeneration and its relationship to adhocness. Many—Paul Feyerabend amongst them[3]—interpret Lakatos as identifying the two notions: as in effect claiming that

- A shift in theory constitutes degeneration just in case the new theory is an ad hoc response to some experimental difficulty or anomaly for its predecessor theory.

If it were committed to that identification then of course Lakatos's position would be refuted (just as Feyerabend claimed) by instances of theories that, while definitely ad hoc, were also clearly scientifically valuable. But endorsing that identification would be a mistake—not one that Lakatos in fact made. On the contrary, had he still been alive to hear it, Imre would have fully agreed with a talk given at the Popper Seminar at LSE in the 1970s a few years after his death. The talk was by the experimental physicist and historian of science, Allan Franklin, and was entitled "Ad hoc is not a four-letter word". Franklin's message was not, of course, the trivial literal one, but instead a much more systematic version of Feyerabend's thesis that there are many theories in science that were produced only as ad hoc responses to some difficulty for a predecessor theory but should clearly count as good progressive science.[4]

Some theories are *both* ad hoc *and also* clearly scientifically unacceptable. My favourite example was provided by Philip Henry Gosse. In his book *Omphalos: an attempt to untie the geological knot* (1857), Gosse defended what later came to be called Young Earth Creationism (the view that the Universe was created in 4004 BC or thereabouts) against the evidence that many parts of the Earth's furniture seem to be much more than 6000 or so years' old by shifting to the theory that, just as God had created Adam with a navel despite this being an unnecessary, even misleading embellishment in Adam's case ("Omphalos" is Greek for "navel"), so God had created the universe in 4004 BC or thereabouts with many aspects of the Creation looking already very old. (Gosse never, it seems, made it clear why he believed he knew that Adam had a navel—I could find no mention of this aspect of Adam's anatomy in the *Book of Genesis*.) Gosse's hypothesis is both patently ad hoc and patently unscientific, but it is unscientific, not because it is ad hoc, but rather

---

[3] See for example Feyerabend (1975).

[4] Several different notions of adhocness can be found in the subsequent literature—many of them with automatic negative ("four-letter word") overtones. It is important to emphasise, therefore, that this Lakatos-Feyerabend-Franklin debate makes sense only if 'ad hoc' is understood, as it is throughout the current paper, strictly in the literal sense of 'being introduced to deal with some particular difficulty as opposed to planned in advance'. (See for example *The Cambridge Dictionary*.) In the case of theories, this means introduced purely to deal with some difficulty in the form of an experimental anomaly for an earlier accepted theory.

because it is totally untestable independently of the phenomena it was constructed to explain. (Indeed, it is constructed precisely to guarantee that there is no independent testability.)

Consider, by contrast, the theorizing of Adams and of Leverrier that resulted in the discovery of the planet Neptune. Herschel had earlier discovered the planet Uranus simply through careful observation of the night sky. When Uranus's orbit was calculated using Newton's theory, the calculations were in significant disagreement with the observational results concerning that orbit. Adams and, independently, Leverrier produced a clearly ad hoc defence of Newton's theory. They took it that that theory had to be correct in view of all the other evidence in its favour. But in effect made the Duhemian point that no testable prediction about Uranus's orbit follows deductively from Newton's theory taken in isolation. Amongst other assumptions, some hypothesis about the total gravitational force acting on Uranus is clearly needed: there might, Adams and Leverrier each suggested, be a still further planet which was so far unknown and hence whose gravitational influence had not yet then been taken into account. And, working back from the assumption that Newton's theory was correct, they calculated what that extra gravitational influence had to be in order to yield correct predictions about the orbit of Uranus. Those calculations amounted to the prediction of the existence of a hitherto undiscovered planet—subsequently observed and named Neptune. Clearly, the Adams-Leverrier hypothesis was ad hoc: the postulation of the extra planet was motivated solely by the desire to defend Newtonian physics against the initially anomalous data concerning Uranus. But it led to a verifiable prediction, independent of the now correct "predictions" about Uranus, and that independently testable prediction was confirmed (Neptune really exists and can be observed). No wonder this is so often cited as one of the great success stories in the history of science: a great success for ad hocness!

So, the key question so far as the progressiveness of a theory-shift is concerned is *not* whether or not that shift was an ad hoc response to experimental difficulties encountered by the earlier theory (the theory shifted from). Instead, the key issue is *independent testability*.

- A research programme is **progressive** if and only if its successive theories are always independently testable in principle, sometimes independently testable in practice *and* confirmed in (at least some of) those independent tests.
- A research programme is, on the contrary, **degenerative** if and only if each new theory explains only the evidence that was anomalous for its predecessor and has no *independent* success: meaning *either* that the new theory is not independently testable at all *or* that it does make independently testable predictions but those predictions are themselves falsified—requiring a further shift that in turn has no independent predictive success *etc*.

Here 'independent' always means: different from any data that were anomalous for the predecessor theory and were worked into the later theory. The modified system of classical physics created by Adams and Leverrier was bound to entail the correct orbit of Uranus—it was specifically engineered to do so. The surprise and therefore the confirmation comes from its correct prediction of the hitherto unknown planet Neptune.

I now turn to the medical examples. As we will see, all the theory-shifts involved in these examples of "mini-research programmes" that I shall cite are patently ad hoc; but, as we have just noted, that in itself is not necessarily a scientific defect. The key question is always whether or not the theories shifted to are independently testable and independently confirmed.

12.2 Cholesterol and Coronary Vascular Disease

A number of relationships between diet (specifically foods high in cholesterol and/or saturated (animal) fats), "blood cholesterol level" and Cardiovascular Disease (CVD) have been alleged to hold over the years since the "Diet-Heart hypothesis" was first publicised by the nutritionist Ancel Keys in the 1950s. I will concentrate on two. They are.

- Theory 1: A diet high in saturated fats causes (i.e., is a positive risk factor for) CVD—via its effect on "blood cholesterol".[5]
- Theory 2: A "high" blood cholesterol level—independently of how it got to be high, whether through dietary or other reasons—causes (i.e., is a positive risk factor for) CVD.

The two main forms of CVD are heart attacks (myocardial infarctions) and ischaemic strokes.

The story of the overall "Diet-Heart Hypothesis" is full of twists and turns, involving several changes in the meanings of key terms. In order to avoid over-complicating matters, I restrict myself to one preliminary clarification. Since cholesterol is not soluble in blood, you, strictly speaking, cannot have a blood cholesterol level whether high or low. Instead, cholesterol is carried round in the blood as a component, along with some fatty acids, of a lipoprotein. These lipoproteins come in various forms and sizes and, when not ingested from food, are manufactured in the gut or (mainly) in the liver—they range from VLDLs (very low density lipoproteins), also sometimes categorized as triglycerides, to IDLs (intermediate density lipoproteins, formed from VLDLs when they lose triglycerides to fat cells), these in turn may shrink to form LDL (low density

---

[5] Defenders of the 'Diet-Heart Hypothesis', like Keys, initially stressed the role, not of saturated fats but of dietary cholesterol (from, for example, egg yolks and avocados) in (allegedly) causing high blood cholesterol and hence (allegedly) CVD. However, even its most fervent initial advocates, including Keys himself, soon found intolerable the degeneration involved in defending the dietary cholesterol part of the hypothesis. So nowadays (almost) no one mentions dietary cholesterol and the emphasis is (almost) exclusively on saturated (animal) fats. Despite its interest, I omit the part of the evidential story about the demise of the dietary cholesterol hypothesis in the interests of brevity.

lipoproteins) and finally the smallest lipoprotein is HDL (high density lipoproteins). It was LDL that was eventually identified as the alleged bad guy in terms of increased risk of CVD. A later twist—one that we will eventually consider in some detail—saw the emergence of the theory that, while a high level of LDL is a cause of CVD, a high level of HDL is, on the contrary, *protective against* CVD. I shall from hereonin follow the now usual (though distinctly odd) practice of talking about 'LDL-cholesterol' (so called "bad cholesterol") as opposed to 'HDL-cholesterol' ("good cholesterol"). Hence the two claims whose evidential status we will investigate read:

Theory 1: A diet high in saturated fats causes CVD, via its effect on LDL-cholesterol.
Theory 2: A high LDL-cholesterol level in the blood (independently of how it got to be high, whether through dietary or other reasons) causes CVD.

Both of these claims should, I believe, be rejected as false on the basis of all the evidence. As noted earlier, I shall not pretend to show this fully here. A full demonstration would in any case involve a number of elements—especially the logic of the confirmation of hypotheses that are "causal" but non-deterministic—to which Lakatos, in common with all the other philosophers of science of his era, gave scant attention at best. However, one central plank of the case for a negative evidential judgment about theories 1 and 2 is also a central notion in Lakatos's methodology of scientific research programmes: namely degeneration. I shall show that the development and defence of both theories have been beset by several instances of classic Lakatosian degeneration.

12.2.1 *A Problem for Theory 1: The "French Paradox"*

One objection to theory 1 (that a diet high in saturated fats causes CVD) was raised long ago, has been much discussed and is generally referred to as "The French Paradox". Compared to people from the UK, the French—on average of course—consume considerably more saturated fat as a proportion of their total diet (they also smoke more and exercise less), and have a (fractionally) higher average LDL-cholesterol level; but, despite the higher fat consumption and the (slightly) higher LDL level, the French rate of CVD and of death from CVD is not just lower than the UK rate, it is around *one quarter* of the UK rate. So, higher saturated fat consumption, yet strikingly lower rate of CVD and of CVD deaths. This looks like a problem for theory 1. In fact, the French have the highest rate of saturated fat consumption and the lowest rate of CVD in Europe. (Incidentally, the second highest in the fat consumption stakes is Switzerland which also has the highest average LDL level in Europe but the second lowest CVD rate after France. The country with the lowest rate of saturated fats as a proportion of overall diet is Russia, which happens to have the highest rate of CVD and CVD deaths. So quite a lot of initial "paradoxicality" surrounds Theory 1!)

This fact about the French compared to the UK diet has long been known and so, unsurprisingly (and of course quite justifiably), there have been responses to it from those who continue to defend Theory 1. One response was that the recorded lower rate of CVD in France was not real, but rather a reflection of some difference between the criteria applied in France for counting a death as a death from CVD, compared to the criteria applied in the UK and elsewhere. (What counts as 'cause of death' on a death certificate is by no means always a straightforward matter.)

Well, this hypothesis is certainly ad hoc, but, as noted, ad hoc is not a four-letter word; and the real question is whether or not the hypothesis is testable. And it clearly *is* testable—French practices of classifying deaths as from CVD or otherwise can be checked. The WHO (World Health Organisation) recognised this and sent a team to make exactly that audit: the result was that the French doctors were classifying CVD deaths in precisely the same way as those from the UK. Ad hoc response, testable but no confirmation equals one form of degenerative step.

A second response to the French Paradox was the suggestion that the French have not yet had a relatively high saturated fat diet for long enough for the (alleged) effects to be felt in terms of increased CVD and CV mortality.

> "We propose that the difference is due to the time lag between increases in consumption of animal fat and serum cholesterol concentrations and the resulting [sic] increase in mortality from heart disease – similar to the recognised time lag between smoking and lung cancer." (Law & Wald, 1999)

This "time lag" theory clearly requires the identification, and dating, of some major change in the French diet toward greater consumption of animal products and that is by no means straightforward. But assuming this problem to have been solved, the theory is testable provided that some sort of time period is specified at which changes in CVD rates will start to become visible. (If defenders of this hypothesis are allowed to wait forever for the change then that hypothesis is almost Gosseian in its untestability. Popper's favourite category of unfalsifiable hypotheses was, remember, the "purely existential" hypothesis.) Suffice it to say that this saving hypothesis was first advertised in 1998 (published 1999) since when the French diet has been essentially unchanged and its CVD rates have gone downwards not upwards (see European Heart Network and European Society of Cardiology, 2012).

The most popular response to the "French Paradox", however, is a different one and in fact amounts to a response-schema: a diet high in saturated fats does indeed make CVD more likely, even among the French, *ceteris paribus*, but some other factor X in the French diet (or perhaps in their way of life more generally) intervenes to make other things in fact *un*equal via X's having a contrary and positive effect— one that more than compensates for the negative effect of the saturated fats. Left with X unspecified, this is again untestable, but there have in fact been many attempted specifications on offer. And, so long as some confounding factor is specified, there is nothing unscientific about responding to difficulties for an initial hypothesis in this way. There is, after all, no reason why the impact of diet, or indeed any lifestyle factor, on some disease should not be simply part of the picture—multiple factors may be involved and may be" mixed": some conducing toward the disease and others protective against the disease). Indeed, many diseases are known to be multifactorial in this way. The issue, as always, is testability and success in tests: if the "French Paradox" is to be resolved scientifically in this way, then the extra, allegedly protective factor has to be specified, tested and the evidence provided by the test should support the claim that the specified factor is indeed protective against CVD.

As indicated, there has been no shortage of contenders for dietary factors (allegedly) found more often in the French than in the UK population and (allegedly) protective against CVD: extra garlic consumption, extra consumption of red wine, and more lightly cooked vegetables amongst them. All of these have been tested—by the obvious method of a controlled trial in which some participants are given a diet high in, say, garlic and the others form a control group with no garlic in their diet—and those tests have generally been failures. Of course, you would have to run the RCT for many years in order to compare rates of CVD and CVD deaths in the two groups, so these investigators use a proxy outcome in the form of lowering of blood cholesterol, (that is, in effect, they assume that theory 2 is correct). Serious studies find no difference even in this proxy marker (see Kendrick, 2007, Chap. 6.) There are some studies that claim to find a small effect of "garlic supplements" in reducing moderately raised cholesterol levels. (See, for example, Ried, 2016.) But these studies are invariably sponsored by the "natural foods" lobby.[6] So far as I can tell, there is no evidence at all that garlic consumption, in any form, affects the variables of real interest: cardiovascular mortality and all-cause mortality.

As for red wine, some studies have endorsed a small negative correlation between moderate alcohol consumption and CVD, but even if it is true that there are more French than UK moderate alcohol drinkers, the effect is much too small to account for the observed France/UK difference in CVD rates. Other attempts to specify the factor X have been even less successful empirically. So again ad hoc but testable theories have been proposed to defend theory 1, but garnered no independent confirmation.

12.2.2 Other Problems for Theory 1

One trial performed as part of the Framingham project (which has been running since 1948—see www.framinghamheartstudy.org) found that eating a high-fat diet was associated with a *decreased* rate of (ischaemic) stroke. Given that ischaemic

---

[6] The list of organisations involved in sponsoring the Ried (2016) research, just cited, is impressively long and includes the American Botanical Council; the American Herbal Products Association; Bionam; Eco-Nutraceuticos; Healthy U 2000 Ltd.; Nature's Farm Pte. Ltd.; Nature Valley W.L.L.; Organic Health Ltd.; Purity Life Health Products L.P.;Vitae Natural Nutrition; Wakunaga Pharmaceutical Co., Ltd.; and Wakunaga of America Co., Ltd. Wakunaga of America, Co., Ltd., for example, describes itself as "a privately held, family-owned health and wellness company dedicated to offering high-quality dietary supplements."

stroke is one important form of CVD, this again looks like a direct problem for our theory 1. As always, ad hoc responses are, however, available. For example: ah! but strokes primarily affect the elderly; no doubt their fatty diet is causing many to die of heart disease before a stroke gets the chance to despatch them.

An ad hoc hypothesis, but again clearly testable and those involved in the Framingham sub-study just mentioned had in fact already tested *and refuted* it:

> "This hypothesis, however, depends on the presence of a strong direct association of fat intake with coronary heart disease. Since we found no such association, competing mortality from coronary heart disease is very unlikely to explain our results." (They are being polite!)

The Women's Health Intervention USA trial whose result was published in 2006 involved 48,835 women studied over 8.1 years. It was a randomised intervention study with those in the experimental group receiving intensive counselling to reduce their fat intake (they were also counselled to increase their intake of fruits and vegetables to at least 5 servings daily and to increase grain consumption to at least 6 servings daily). By the end of the sixth year those in the experimental group were on average consuming 29% of their calories as fat (9.5% saturated fat) compared to 37% fat in the control (uncounselled) group (12.4% saturated fat). The result was no significant difference between experimental and control group in any of: Coronary Heart Disease or Stroke incidence, Coronary Heart Disease or Stroke Mortality or Overall Mortality (for references see Nabel, 2006).

The mainstream reaction to this result brings us to a *ne plus ultra* in ad hoc responses: the promissory note—give us time and we promise that we'll come up with something to explain these negative results. Dr. Elizabeth Nabel the head of the Heart Section of the US National Institutes of Health (which managed the Women's Health Initiative and hence this trial) said (*op cit.*) "There may have been some 'disappointment' that the studies didn't always give clear answers [in fact they gave clear answers, just not the ones that she and her colleagues expected/wanted]. The findings are what they are ... Now we are in a second wave of putting the findings into perspective [i.e. of trying to dream up some specific ad hoc response]."

In the meanwhile, despite the fact that the trial raised serious questions about the evidential basis of the NIH's dietary advice, the advice must it seems remain in force. Nabel pronounced *ex cathedra*: "The results of this study do not change established recommendations on disease prevention. Women should continue to ... work with their doctors to reduce their risks for heart disease including following a diet low in saturated fat ... ".

Before moving on to the different, though, of course, related theory 2, here for luck is just one more "paradox" facing theory 1: the "Japanese paradox". Japan is often described as having been the initial "poster boy" for the diet-heart hypothesis (essentially theory 1). In the late 1950s/early 1960s when the nutritionist Ancel Keys was first championing the hypothesis, Japan stood out as having the lowest animal fat intake of any country for which there were figures, the lowest average cholesterol level (it was 3.9 millimoles per litre compared to 5 mmol/L in the UK and 5.2 mmol/L in the USA) and by far the lowest rate of CVD and CVD mortality.

How much "proof" of the link between saturated fats and CVD could you want? Certainly, Keys himself needed nothing more.

Since that time however there have been significant changes in the Japanese diet involving a 400% increase in animal saturated fat intake. The average cholesterol level in Japan is now the same as in the USA (5.2 mmol/L). Theory 1 therefore seems to predict that the rate of death from heart disease in Japan will have risen since the early 1960s. In fact, it has *fallen* by 60%.[7]

One reaction to this was to claim that the Japanese have some special genetic feature that protects them against heart disease. This is again certainly ad hoc, but again testable; indeed, in this case, somewhat surprisingly, testable even in the absence of any specification of what particular genetic feature that might be. It predicts that Japanese émigré populations will have lower rates of CVD and CV deaths than the host populations. But again, this independently testable prediction is refuted: the Japanese community in the USA, for example, exhibits the same CVD and CV mortality rates as the US population as a whole.[8]

12.2.3 A Problem for Theory 2: Low *Cholesterol Causes CVD*

We now come to some "paradoxes" for, i.e. seeming refutations of, theory 2; which states, remember, that a high level of LDL-cholesterol in the blood, no matter how it got to be high, causes CVD.

A 2001 paper in *The Lancet* by a group of researchers from the University of Hawaii, Honolulu reported a study which found "increased mortality in elderly people [not with high but rather] with *low* serum cholesterol" (Schatz et al., 2001); emphasis supplied). Their data showed "that long term persistence of *low* cholesterol concentration actually increases the risk of death [by a whopping 65%!]. [Moreover], the earlier that patients start to have lower cholesterol concentrations, the greater the risk of death. These data cast doubt on the scientific justification for lowering cholesterol to very low levels." (You can say that again—though few in medicine have taken notice!) As these researchers pointed out, far from constituting

---

[7] In the same period, the rate of ischaemic stroke in Japan has plummeted by seven-fold. In the early 60s, Japan had the highest rate of strokes of any country (so, since ischaemic stroke is the other form of CVD alongside heart attacks, a little thought would have taken the sheen off its diet-heart poster boy image from the outset). So overall a 400% increase in saturated fat intake in Japan was associated with a near six-fold fall in overall CVD. (See Kendrick, 2007).

[8] Ueshema presents another possible ad hoc explanation for the "Japanese paradox"—that although cholesterol levels have risen in the Japanese population as a whole, they are still lower in the Japanese elderly than they are in the elderly in, for example, the US; and CVD, especially CVD mortality primarily of course afflicts the elderly. But this is hopeless: it would at best predict that the rate of CVD and CVD mortality would have remained roughly the same despite the overall increase in cholesterol levels in Japan, but not the actual fact that it fell dramatically. Moreover the explanation is in clear conflict with a series of studies that we will come to next, all showing that lowered cholesterol levels in the elderly is associated with an *increase* in CVD.

an outlier, their "data accord with previous findings of increased mortality in elderly people with *low* serum cholesterol". The Honolulu study was in fact the culmination of a series of studies, including reports from the Framingham project (which is generally seen as providing the initial basis for the high LDL-cholesterol/CVD link but many of whose original findings have been reversed by later research based on a much enlarged data set). All of these studies found that it was *low* cholesterol levels, rather than high ones, that were predictive of CVD in the elderly.

This certainly seems like a problem for theory 2 but there is an obvious escape route: maybe the elderly who have low cholesterol are a special case; maybe there is some further factor that affects them and which is the real cause of the higher rate of CVD and CVD mortality, where that factor also happens independently to cause a lowering of the LDL level. Again, although patently ad hoc, there is nothing inherently unscientific about this suggestion. On the contrary, a standard way of testing whether an observed correlation between factors A and B is genuinely causal is by checking that the correlation still holds when A is conditionalized on further factors C_n which might plausibly also be causes of B. So, for example, Hill and Doll provided strong evidence that smoking cigarettes causes lung cancer not simply by showing that there is an observable correlation between smoking and lung cancer, but by showing further that this correlation continues to hold when independent possible causes of cancer (such as, for example, living in an area with heavy air pollution) are conditionalized on. If, on the contrary, the correlation "disappears" conditional on C (that is, A and B are probabilistically independent given C) then that is evidence that the correlation between A and B is "accidental" and that, rather than A causing B (or vice versa), A and B are two separate effects of some underlying "common cause". (In the standard example, there is a definite positive correlation between having yellowed fingers and developing lung cancer but that probabilistic correlation "disappears" upon conditionalization on cigarette smoking: despite the strong probabilistic correlation, there is (of course) no causal connection between yellowed fingers and lung cancer, instead they are separate effects of the common cause: cigarette smoking.)

So, given the finding of a correlation between low cholesterol level and increased risk of CVD in the elderly, there is certainly nothing automatically unscientific about reacting by postulating that some other factor afflicts the elderly that "explains away" the observed low cholesterol/CVD link. And there's a fairly obvious candidate: the elderly often have comorbidities—maybe they come into these trials with some other (non-CVD) illness which *both* lowers their LDL-cholesterol *and also* independently predisposes them to develop CVD. This is undeniably ad hoc but we are learning that ad hoc is not a four-letter word. And indeed this suggestion is plainly testable: there should be a higher rate of comorbidities in the experimental arms of the trials that showed a correlation between low LDL-cholesterol level and high rates of CVD or CVD mortality. Moreover, it is known that certain diseases, – for example, advanced cancer and liver diseases such as chronic hepatitis B— *are* indeed associated with low LDL-cholesterol levels. However, people with comorbidities were excluded from all these trials—exclusions were based not just on cancer and hepatitis but on any significant comorbidity. Testability but again refutation rather than confirmation.

Undeterred, Iribarren and colleagues sought to continue to defend theory 2 by in effect pointing out that any comorbidities had to be overt if elderly people were to be excluded from the trials on that ground. Perhaps, Iribarren postulated,[9] covert or subclinical illness was the common cause of low LDL-cholesterol and high rates of CVD—perhaps even as much as a decade (or more) before the illness became overt. Again ad hoc, again testable but again in conflict with the data. The Honolulu study, for example, reported (Schatz et al., 2001) that: "[in the light of our data] Iribarren's hypothesis is implausible and unlikely to account for the adverse effects of low cholesterol levels." (They too were being polite!)

Iribarren and colleagues also suggested that simple frailty (strongly associated with old age of course) might be a hidden common cause of low LDL-cholesterol and CVD. But a large Austrian study in 2004 found that the low LDL-cholesterol/CVD link is not in fact restricted to the elderly: "For the first time, we demonstrate that the low cholesterol effect occurs even amongst younger respondents, contradicting the previous [theories] ... that this is a proxy or marker for frailty occurring with age." (Ulmer et al., 2004).

Attempts to defend theory 2 are indeed looking like a degenerating research programme, but, in the (expressive if strictly logically ill-informed) words of the song, "you ain't seen nothing yet".

12.2.4 Another Problem for Theory 2: The "Female Paradox" and the Sex Hormone Hypothesis

Right from early on, it was recognised that the claim that high levels of LDL-cholesterol cause CVD faces a problem from facts about women. In general (though with some, independently interesting exceptions), across various populations, women have much lower rates of CVD than men; while—again in general though with some exceptions—women have much *higher* LDL-cholesterol levels.

Obviously, theory 2 predicts to the contrary that, given their higher LDL-cholesterol level, then, ceteris paribus, women ought to exhibit a higher CVD rate. This is a well-known problem for theory 2 with, however, you might think, a well-known solution: there must be some other difference between men and women that means that women are protected against CVD despite their higher cholesterol level; and the most obvious suggestion for that role, the most obvious biochemical difference between women and men, is their sex hormones.[10] So this is another

---

[9] Iribarren et al. (1995).

[10] It is true that women smoke less than men, but if you compare men smokers with women smokers or men non-smokers with women non-smokers you still generally find higher levels of LDL but lower rates of CVD in the women's groups.

of those "conflicting causes" saving hypotheses: it still may be correct that high LDL-cholesterol causes CVD and so women with their higher cholesterol level *would have* in general a higher CVD rate than men if other things were equal, but in fact at the same time, their distinctive sex hormones make other things in fact unequal, by somehow operating physiologically to lower CVD rates so as to more than compensate for the (supposed) effect of the high LDL level. Again: undeniably an ad hoc attempt to save the initial hypothesis, but again it is plainly testable.

The sex hormone hypothesis predicts, for example, that amongst women who have had hysterectomies, those who had their ovaries removed at the same time as their womb will, since they will no longer produce any sex hormones, have a higher rate of CVD than those women who only had their womb removed. But in fact, a 1963 study of several hundred woman already found no difference in the prevalence of coronary heart disease between those women who had had their ovaries removed as well as their womb and those who had had only their womb removed—both groups exhibiting a rate of 8% CVD some 15 to 20 years after their operation.[11]

A second clear prediction of the sex hormone hypothesis is that women who have been through the menopause should lose the protection allegedly afforded by their sex hormones, and so older women's rate of developing CVD should start to move up toward that found in males. Although this is widely believed to be true, scientific studies belied it. As early as 1987 a study found that "The normal menopause, which causes a gradual decrease in oestrogen production, was not associated with any increase in the risk of coronary heart disease."

A third prediction is that those women who take the contraceptive pill—which of course contains female sex hormones—should have a still lower rate of CVD than equivalent women not taking the pill. But the evidence is that women taking the pill in fact have *a greater* rather than reduced risk of dying from coronary heart disease (see, for example, Tanis et al., 2010)—even when other possible confounders, notably smoking, are controlled-for.

Perhaps the most famous prediction, however, made by the sex hormone hypothesis is still a fourth one: that women receiving Hormone Replacement Therapy (HRT) to counteract the negative effects of the menopause (or for some other reason) should exhibit lower rates of CVD (and CVD death) than equivalent women who are not on HRT. And, indeed, this looked for a while like being a first instance of Lakatosian *progress:* a 1983 observational study showed a 42% reduction in strokes and heart attacks in a cohort taking HRT compared to the average CVD rates amongst women of the same age not on HRT (Bush et al., 1983). This 42% is a *relative* risk reduction and so not as impressive as it might sound, but is still fairly substantial. And indeed, HRT became recommended treatment in the US on the basis of this study. However, as we will reflect in a moment, the result of this study was later completely overturned by a couple of large, reasonably high-quality randomised controlled trials which yielded an estimate of a 29% *increased* risk of

---

[11] Ritterband et al. in *Circulation,* 1963, **27**, 237 (reported in *The British Medical Journal*, Dec 14 1963, 1487).

CVD in those undergoing HRT. The official guidelines were promptly changed to recommend against HRT as a treatment aimed at reducing the risk of CVD.

This is one of the turnarounds (42% *protection* yielding to a 29% *increased risk*) that are often cited as showing that you can "never trust" an observational study. The correct view is, however, surely that you shouldn't trust the outcome of an observational study if a moment's reflection would suggest, on the basis of background knowledge, that the study was likely be multiply confounded. The women who were involved in the "treatment arm" of the observational study and therefore had chosen themselves to take HRT before it became mainstream treatment formed a self-selected and very special group: particularly fitness- and health-conscious, predominantly middle class and well-educated, containing very few smokers and so on; and hence should never have been thought of as representative of the general female population.

In any event, our latest ad hoc hypothesis was indeed eventually subjected to rigorous tests via a couple of large randomised trials. One of these was the **H**eart and **E**strogen/Progestin **R**eplacement **S**tudy (**HERS**). HERS ran for 6 years from 1998 to 2004 (various interim results were published during that period) and ended up involving 2763 women all of whom had a history of heart disease. Those women were randomised to either HRT or placebo. The outcome variables were either non-fatal MI (myocardial infarction) or death from CHD (coronary heart disease). The outcome was 172 fatal or non-fatal cases of heart disease in the HRT group compared to 176 in the placebo group. Of course, this is a tiny difference in such a large sample; and, moreover, there was actually a 24% *increase* in fatalities in the HRT group compared to placebo (compensated for by a 9% decline in the HRT group in the more numerous non-fatal events to produce the final barely distinguishable overall numbers). The study concluded "Based on the finding of no overall cardiovascular benefit [combined with notably more negative side effects in the HRT group] ... the investigators do not recommend starting this treatment for the purpose of ... prevention of [CVD]." (Hulley et al., 1998). Ad hoc theory (female sex hormones protect against CVD and therefore cancel out the (alleged) effect of the higher average LDL-cholesterol amongst women); is testable (women taking HRT should have a lower heart disease rate than equivalent women not taking HRT); but the test result is entirely negative; and so again a case of degeneration.

Another even larger randomized controlled trial on the effects of HRT ran for over 5 years and was published in 2002. This was a further trial under the auspices of the US Women's Health Initiative and looked at the effect of HRT not just on heart disease (as HERS did) but on CVD more generally—so including fatal or non-fatal ischaemic strokes. The result of this larger trial could hardly have been more definite: "[A]fter 5.2 years, there was a 29% *increase* in coronary heart disease risk, including an 18% increase in risk of CHD (coronary heart disease) mortality and a 32% increase in risk of nonfatal myocardial infarction in the HRT group. There was a 20% *increase* in the risk of fatal stroke and a 50% increase in the risk of nonfatal stroke in women assigned to HRT." (Writing Group for the Women's Health Initiative Investigators 2002; emphases supplied.)

12.2.5 The Switch to the Theory that "Good Cholesterol" Protects Against CVD

So, lots of ad hoc but testable hypotheses aimed at saving theory 2 but all of them immediately refuted. However, defenders of the theory were a far from fainthearted bunch and were certainly not ready to roll over just yet. Some of them began, for example, to argue that the sex hormones response to the "female paradox" had always been the wrong response. While it is true that women generally have higher levels of either LDL or total cholesterol and yet lower CVD levels, perhaps they also have higher levels of HDL-cholesterol. Several researchers had suggested that, in complete contrast to LDL ("bad cholesterol"), high rates of HDL may actually be protective against CVD (and hence count as "good cholesterol"). So perhaps it is their higher levels of HDL, rather than their sex hormones that reduce women's rates of CVD.

Well, same story—certainly ad hoc, but definitely testable. So, for example, since in the HERS study the average level of HDL-cholesterol was observed to be higher in the HRT group compared to placebo, this new HDL hypothesis predicts that CVD rates should have fallen in that group. Instead, as noted earlier, the rate of CVD mortality actually increased.[12] There was similar lack of confirmation in other studies. A large Russian study published in 1994, for example, reported "... there was no association of HDL cholesterol with mortality in Russian women." Despite the fact that the name "good cholesterol" somehow lives on (as the ghost of what ought to be a departed theory?) there seems to be no serious evidence that high HDL is protective against CVD.

Attempts to provide such evidence have unsurprisingly been made alongside attempts to develop drugs that raise HDL-cholesterol levels and so, if the HDL hypothesis were correct, would reduce the risk of CVD. The main group of such drugs to be tested were the "rapibs". The first of these was Torcetrapib. In tests, Torcetrapib raised HDL levels by around 60%; sadly, it also raised overall morality by almost 50% and was never approved for use. Delcetrapib had no effect on either LDL level or CVD. Tests on Anacetrapib provided evidence of a small positive effect but so small that Merck decided not to market it. (http.//www.pmlive.com/pharma_news/cetp_inhibitor_class_finally_dies_as merck_abandons_anacetrapib_1208239—see Kendrick 2018, p. 107).

However, the most interesting case is that of Evacetrapib. A very large study showed that this drug managed to more than double HDL levels (120% increase), it also lowered LDL by 37%—significantly more than statins manage. So, here's the next blockbuster, right? Unfortunately not: in tests Evacetrapib had zero effect on CVD.[13] As Steve Nissen, a celebrated cardiologist and head of The Cleveland

---

[12] See again Hulley et al. (1998).

[13] https://www.forbes.com/sites/matthewherper/2015/10/12/eli-lillys-good-cholesterol-goes-bad/#47d83c527de8.

Institute in the US, wrote "the results can't be explained because the study was too small or because too few heart attacks and strokes occurred. The drug didn't work." This looks like a severe blow both to the hypothesis that raising HDL protects against CVD and also to the hypothesis (our theory 2) that lowering LDL protects against CVD.

Nissen was, however, firmly attached at least to the latter hypothesis. Having decided that the negative test result concerning Evacetrapib could not be questioned, Nissen therefore had to find another explanation of the fact that LDL levels had gone down substantially but CVD rates were not affected. "There are" he wrote "two hypotheses to explain the results". One of these was that "lowering LDL cholesterol was beneficial but something else evacetrapib did causes toxicity [so as to outweigh the supposed good effect of the lowered LDL]". The other was that "it matters *how* you lower LDL cholesterol". (I perhaps do not need to point out that there is a third explanation: viz. that lowering LDL-cholesterol has no effect on CVD. But, as noted, Nissen could not bring himself to countenance this possibility.) So CVD risk *is* lowered by lowering LDL levels, but some ways of lowering LDL are ineffective even though that is not because they trigger some other mechanism that outweighs the alleged benefit of the lowered LDL.[14] This seems to be another maximum in untestable adhoccery. Since there are no signs to pick up of interfering processes (that's the first possible hypothesis which Nissen dismisses), presumably the only way to tell if someone's cholesterol has been lowered "in the right way" is by seeing if they develop CVD; if they don't develop CVD then their LDL was lowered in the right way; if they do develop CVD, it was lowered in the wrong way.

Finally, turning back from theory 2 to the HDL/"good cholesterol" hypothesis generated to protect it, a further interesting twist in the story originated in the picturesque Italian lakeside village of Limone sul Garde. A family living there was identified, all descendants of one man—Giovanni Pomarelli—born in the village in 1780. Both the family history and the current family (consisting of some 40-odd souls) had been extensively studied as the family exhibited amazing longevity and, especially, exceptional immunity to heart disease. The cholesterol levels—in particular the HDL levels—of the current members of the family had been carefully measured. Despite their immunity to heart disease, their average HDL level (and HDL is, remember, supposed to be protective against heart disease) was remarkably low—*much* lower than the Italian population average. Ah!, but what is special about this family is that all carry a genetic mutation, inherited from Giovanni Pomarelli, which produces a distinctive form of the apolipoprotein that holds the HDL and the fatty particles together to form the HDL-cholesterol lipoproteins that circulate in the blood. This distinctive form of the HDL apolipoprotein was dubbed "ApoA-1 Milano". (It was first identified and analysed in laboratories in Milan.) Perhaps, although standard HDL-cholesterol is indeed the more protective against CVD the higher its level, ApoA-1 Milano HDL is, by contrast, a very special case and

---

[14] https://mdedge.com/ecardiologynews/article/108182/lipid-disorders/accelerate-evacetrapibs-clinical-failure-sinks-lipid.

is protective no matter what its level—maybe even the lower its level the more protective it is.

So we have layers of adhoccery here. The theory that high levels of HDL are protective against CVD was originally produced as an ad hoc response to the "female paradox" difficulties for theory 2 (and the lack of success of dealing with those anomalies via the sex hormone route). Then that HDL theory was itself put into empirical difficulties and the ApoA-1 Milano hypothesis was an ad hoc response to those difficulties. But layered or not, the hypothesis is again testable—at least in principle. The most direct test would be via genetic engineering: give initially normal people the Apo-A1 Milano variant HDL and see if they exhibit lower rates of CVD and CVD mortality. But this was something, if at all, for a date far in the future. Perhaps, it was conjectured, if people were injected with Apo-A1 Milano HDL, they would exhibit at least *some* reduction in CVD risk.

A pharmaceutical company called Esperion Therapeutics obtained a patent on the production of cloned Apo-A1 Milano HDL; and some small initial trials, using the proxy marker of reduction in the volume of arterial plaque, along with some animal experiments were trumpeted as 'amazingly' positive. However, it was clear that producing convincing evidence of an effect of these injections on CVD rates in humans would require a very large trial—one affordable only by a BigPharma giant. And Pfizer in fact duly bought out Esperion with the sole motive of thereby acquiring the patent on Apo-A1Milano HDL and of running that trial (and of course with a view to reaping the financial benefits if the trial was positive and Apo-A1 Milano injections became the new blockbuster treatment).

We do not know what the outcome of that large trial was (despite some regulatory efforts, pharmaceutical companies are able to keep very tight control on the data from trials that they fund and they release results only when convenient for them). However, I think we can infer just how negative that test outcome must have been: Pfizer paid $1.25 billion to buy Esperion Therapeutics and hence obtain the patent on Apo-A1 Milano; 5 years later, after running the trial, Pfizer sold the patent for $ 10 million. (This represents a loss of greater than 99% on their initial investment.)

The patent was bought by a company called The Medicines Company—a start-up that specialises in picking up treatments for which there was as yet no solid evidence, but which they judged still to be somewhat hopeful. The Medicines Company ran a further trial on Apo-A1 Milano—though, since they did not have the financial clout of Pfizer, this trial was again on a proxy marker (again atherosclerotic plaque volume) rather than on CVD itself. They did publish the result of this trial: "Percent atheroma volume decreased 0.94% with placebo and 0.21% with [ApoA1- Milano]". So Apo-A1 Milano was actually outperformed by placebo—albeit fractionally and on a by no means clearly meaningful proxy outcome.[15]

Unsurprisingly, nothing has been heard of Apo-A1 Milano since.

---

[15] For details of this whole story see https://www.science.org/content/blog-post/long-saga-apo-a1-milano.

In sum, then, the theory that a high rate of HDL-cholesterol is protective against CVD was born as an ad hoc response to a difficulty for theory 2, the difficulty being that women generally have higher rates of LDL-cholesterol and yet at the same time lower rates of CVD than men; the initial ad hoc attempts to explain this difficulty, via the sex hormones hypothesis, produced significant degeneration, so there was a switch to the theory that it is HDL, not female sex hormones, that is protective; that theory is itself testable in a number of ways—all of them in fact being met by immediate refutation. Degeneration piled on degeneration.

12.3 The Clash Between What Ought to Have Happened and What Actually Happened

To reiterate my earlier concession: I do not claim to have looked at all the different responses that have been made in the research literature to difficulties for theories 1 and 2 and how those responses have fared evidentially—let alone, of course, at all possible responses, of which there are clearly indefinitely many. A series of degenerative steps does not entail that a programme overall has degenerated beyond hope of redemption; and Lakatos, remember, was always keen (I believe, too keen) to stress that a degenerating programme, no matter how degenerate, might "always stage a comeback". Nonetheless, the above does, I suggest, form a reasonably telling case that both of the mini-programmes at issue have degenerated sufficiently to call for the rejection of the two hypotheses in whose defence those mini-programmes were built. Especially since there seem to be no instances of empirical progress produced by either programme to balance against the degenerative steps.

However, instead of being rejected on the basis of this degeneration, theory 2 is still very much enshrined in medical orthodoxy: everyone is urged by the medical profession to 'know (and frequently check) their number', i.e. their LDL-cholesterol level and immediately treat it—by taking statins—if it is "high", in the expectation that by reducing their LDL-cholesterol level, the statin will, in accordance with theory 2, in turn reduce their risk of developing CVD. Millions and millions of people worldwide are taking statins life-long in the firm, and medically endorsed, expectation that it will reduce their chances of suffering from cardiovascular disease. (And the cholesterol level that counts as 'high' keeps on being lowered.) As for theory 1, there was until recently a similarly firm consensus that the uniquely healthy diet is one low in saturated fats; and hence that reducing the amount of saturated fat in your diet and replacing it with "healthy" carbohydrates and unsaturated fats was a sure way to reduce your risk of developing either a stroke or heart attack. The programme to defend theory 1 had definitely begun to degenerate long before this consensus began to be (rather reluctantly and very patchily) questioned.[16]

---

[16] See, for example, Harcombe et al. 2016.

This clash between what you might expect to happen on the basis of the Methodology of Scientific Research Programmes (MSRP) and what has actually happened (and is happening) brings us to another aspect of Lakatos's thought.

12.3.1 Internal and External History

In a much-discussed 1971 paper, Lakatos introduced the idea of "internal" and "external history". As Lakatos saw it, each methodology or philosophy of science endorses a narrative of how the history of science *ought* to have gone in terms of the acceptance or rejection of the available hypotheses at a given time, depending on the evidence available at that time. This is "internal history" which Lakatos famously also called a "rational reconstruction" of the history of science. In case the actual history differs from its rational reconstruction, the methodology supplying that rational reconstruction is, he went on to claim, obliged to provide an "external" historical explanation of the difference, where that external history should of course itself be empirically testable and empirically confirmed: "... when history differs from its rational reconstruction, [external history] provides an empirical explanation of why it differs." (Lakatos, 1971, p. 118). The underlying idea (Lakatos's "meta-methodology" for the appraisal of rival philosophies of science), then, was that a philosophy itself gets confirmational brownie points from the acceptance (or rejection) of a theory at a particular time if it *either* delivers the judgment that the acceptance (or rejection) of that theory at that time was rational (scientific/evidence-based) *or* it entails that the acceptance/rejection of the theory was not rational but there is a supplementary "external" historical account of the divergence between what ought to have happened and what actually did happen; where that "external" account is empirically confirmable and empirically confirmed.

While Lakatos argued that there have been any number of clashes between actual history and, for example, its naïve-falsificationist-reconstruction, the only clear example of a clash between actual history and its rational reconstruction in the light of Lakatos's MSRP that any of us could come up with at the time he was writing in the early 1970s was the "Lysenko affair" in Stalinist Russia. This involved the endorsement by some Russian scientists of Lysenko's half-baked neo-Lamarckian views about genetics (more specifically, plant genetics) over the orthodox neo-Mendelian approach of Lysenko's original mentor, Vavilov (along of course with that of all competent geneticists from the West). (See, for example, https://en.wikipedia.org/wiki/Trofim_Lysenko) This case, however, was not entirely satisfactory: in part because the "external factor" was so singular and obvious—Stalin took the view that Lysenkoism was altogether the "more Communist" approach and of course was in a position to ensure that his opinion was "influential"; and (mainly) because it is for that reason unclear how many of the Russian endorsements of Lysenkoism were genuine, rather than feigned with a view to political convenience and/or personal safety.

The medical example we have been considering, however, constitutes a further, and altogether more challenging case. There are, as we have seen, in this case sharp divergences between the internal history endorsed by MSRP and real history. And there is no serious doubt that the real opinions of medics and dieticians generate these clashes. The defender of MSRP must, therefore, provide an "external history" to explain those clashes; and that external history should of course be firmly empirically supported.

I shall not attempt to develop here anything resembling a full external history of the attitudes toward theories 1 and 2. Instead, I shall just point to four factors that were clearly involved—introducing them in an acknowledgedly preliminary and sketchy way but saying enough, I hope, to indicate that they are all firmly based on evidence. I will deal with them in order of increasing importance.

First, the intuitive, one might almost say emotional, appeal of theory 1 is undeniable. It is difficult not to be repelled by the sight of large amounts of solid fat. All recoil from images of "fatbergs" blocking the London sewers. The physiological counterpart seems so natural: despite our better selves, we eat fat, and so have fat in our blood stream; fat can get deposited on artery walls and eventually block them. Eating animal fats *must* be doing us harm—it "stands to reason". (This despite the facts that: (i) the atherosclerotic plaques involved in CVD contain cholesterol (amongst other things) rather than fats; (ii) dietary cholesterol was very quickly abandoned as a cause of the plaques, even by the most vocal supporters of the Diet-Heart hypothesis; *and* (iii) it is not plaque blocking the artery that causes the MI or ischaemic stroke, but rather a blood clot that breaks away from that plaque and blocks an artery closer to the heart or brain than the artery on which the plaque was formed!). The intuitive appeal of the theory certainly seems to account for its uncritical acceptance by the general public and for how quick some were to accept guidelines for "healthy" (low fat) eating despite the lack of anything remotely resembling telling evidence. And it also seems to have played some role even amongst scientists.

Secondly, it is a well-known and often recurring phenomenon that scientists who have become associated with a particular hypothesis go on to defend it against attack almost as if they were being attacked personally—especially in "softer" sciences where effects are generally multifactorial and the impact of evidence therefore less direct. As Malcolm Kendrick puts it in his 2014 book *Doctoring Data* (p. 141):

"When an expert is wrong, he, or she is far less able to change their mind than you. Because it matters so much more to them than anyone else. Their entire reputation, status and income may be built on the hypothesis they ... support."

A few charismatic individuals who fit this description and so are determined to defend a hypothesis at all costs may exert an inordinate influence on the attitudes of others. In the case of theory 1, the nutritionist Ancel Keys from the University of Minnesota was such an individual. Keys became the embodiment of the heart-diet hypothesis, becoming widely known as "Mr Cholesterol" and, for example, appearing as such on the front cover of *Time* magazine. Keys became very attached indeed to theory 1. He had, it seems, a very charismatic personality and consequently attracted many followers. Having become famous as the champion of the theory, the length to which Keys was willing to go to defend its public standing is vividly illustrated by an episode that came to light only after his death.

Keys first became famous for his "Seven Countries Study" of 1957 which, for the seven countries he considered, pointed to a straight-line relationship between percentage contribution of saturated fat to the diet and incidence of heart disease: for those countries, the higher the fat intake, the higher the rate of CVD. Whether or not Keys consciously selected his seven countries to support his favoured theory, objectively speaking his study suffers from the worst sort of selection bias: it is not difficult to select a different set of seven countries for which the relationship goes in exactly the opposite direction—the higher the animal fat consumption, the lower the rate of CVD (see, for example, Kendrick, 2007, p. 63). It is unsurprising, therefore, that within scientific circles at least, Keys' study received at best a patchy, in fact predominantly cool reception and that the influential American Heart Association (AHA) refused to follow Keys' suggestion that it issue strong advice to adopt a low-fat diet as a means of reducing the risk of CVD.

Keys reacted to this initially cool reception of his work in two ways: first by getting himself and a co-defender of theory 1 elected to the relevant AHA Committee with a view to changing the judgement about "what the evidence shows" concerning fat in the diet—independently, if necessary, of any change in the evidence itself. (In this, he actually succeeded. We will consider the role of committees, guidelines and government directives very shortly.) Keys' other reaction to his initial disappointment was to begin to plan a large randomized blinded trial which, he assumed, would provide "gold standard" evidence that he had been right about saturated fat and CVD all along. He conducted the trial together with his Minnesota colleague Ivan Frantz. It involved a treatment group whose diet was modified to replace saturated fats with food items that have naturally high or artificially raised content of linoleic acid—an allegedly healthy polyunsaturated omega-6 fatty acid. Although the trial was completed in 1973, only a few snippets of the results were published (by junior members of the research team, including PhD students). Until, that is, in 2013 a group of Australian researchers discovered all the raw data and the analysis of that data in a set of cardboard boxes in the garage of the son of Keys' principal co-investigator, Ivan Frantz. The newly discovered results showed (a) a statistically significant lowering of serum cholesterol levels in the intervention (polyunsaturated) group; *but* (b) no effect of the cholesterol-lowering on either mortality from coronary heart disease or all-cause mortality. (In fact, and to the contrary, the trial found a 22% *higher* risk of death for each 30 mg/dL reduction in serum cholesterol (mg/dL is milligrams per decilitre, the preferred unit in the US—30 mg/dL is equivalent to 0.78 mmol/L (millimoles per Litre) in European units). Keys was the lead investigator and must surely have been complicit in the "burying" of these results (which of course tell strongly against theory 2 as well as theory 1). Keys was indeed *very* determined not to see his favoured hypothesis undermined in scientific and public estimation.

A few charismatic individuals with total devotion to a theory can, it seems, persuade surprisingly many others.

An important *third* element of our "external history" is the role played by advice from influential professional bodies and more especially guidelines issued by governmental bodies. As just noted, in 1957 after the publication of Keys' "Seven Countries Study", the American Heart Association (AHA) resisted strong pressure from Keys and his supporters and found that "The evidence does not convey any specific implications for dietary changes." (quoted from Le Fanu, 2018, p. 66) Keys responded by getting himself, and also Jeremiah Stamler his great ally in the fight for the acceptance of theory 1, elected to the relevant Committee of the AHA. That committee fairly promptly recommended a reduction in saturated fat in the diet with a view to (allegedly) reducing the risk of heart disease-, while admitting that there was, as yet, "no final proof". Thereafter the AHA has continued to play a major role in propounding ever stronger advice to the US population to avoid saturated fats and replace them either with carbohydrates or unsaturated fats—despite the fact that the prospect of "final proof" receded further and further in the light of negative results.[17]

As for theory 2, the AHA also, over the years, issued ever stronger advice to be aware of your cholesterol level and, if that level was "high", treat it by either dietary change or by taking statins. And a similar and still more influential role was played by a US government body. The National Cholesterol Education Program (NCEP), a programme managed by the National Heart, Lung and Blood Institute, itself a division of the National Institutes of Health was set up in 1985 and continues to operate. Its goal, according to Wikipedia, is "to reduce increased cardiovascular disease rates due to hypercholesterolemia (elevated cholesterol levels)" in the USA—a goal which, of course, inextricably ties the program's very existence to the truth of theory 2. The guidelines the NCEP supplies for correct medical practice— fundamentally the LDL-cholesterol level at which to institute statin treatment—in effect have the force of law: a medical practitioner can be successfully sued for malpractice in the USA if s/he contravenes those guidelines. It is difficult for a medical practitioner to question the evidential basis of a claim if by questioning it they might end up in court.

The NCEP's guidelines amount to interference in the practice of medicine to encourage (really mandate) application of theory 2. Such interference is by no means confined to the USA. In the UK, for example, the QOF (Quality Outcomes Framework) was introduced in the NHS in 2004. Regarded by many acute observers as having proved to be a major mistake, the QOF makes general practitioners' pay dependent on the extent to which they meet certain outcome targets. One of these is the number of patients whose cholesterol level they have measured and (if regarded as high) treated. Des Spence, a Scottish GP and regular writer for the *British Medical Journal,* estimates that over the first 10 years of its existence, the QOF had produced 3 million extra statin users (without any discernible effect on CVD rates). (See Spence, 2013). Again, it is not easy to be analytical about the real evidential basis of

---

[17] See the study headed by Salim Yusuf reported in https://www.medscape.com/view-article/884937#vp_3, as well as Harcombe et al. 2016.

a theory, if that claim is governmentally endorsed and your income depends (albeit in part) on applying it.

So far, our (outline) external history has been very much Hamlet without the Prince. The *fourth* and overwhelmingly most powerful influence in promoting theories 1 and 2, despite what I claim is their extremely poor evidential record, has been money—massive amounts of money from the Sugar lobby as regards claim 1 and even more massive amounts from Big Pharma regarding claim 2. Here I concentrate just on the latter.[18]

In 1976, Henry Gadsden, the Chief Executive Officer at Merck, then the world's largest pharmaceutical company, gave an interview to *Forbes* magazine in which he bemoaned the fact that his company only sold to those who were sick. He wanted instead to be able to sell to everyone: sick or well—"just like Wrigley's sell chewing gum". It might be thought that Gadsden was joking, but the fact is that in the 40 years between his making that remark and 2016, pharmaceutical company profits increased by 40fold—most of that due to the vast increase in the sale of drugs to treat risk of developing an illness rather than an actually developed illness (Le Fanu, 2018).[19] Of course, by far the main risk treated was the risk of developing CVD and the (allegedly) risk-reducing drugs were statins. Within a couple of years after its approval for use in 1987, lovastatin (sold under the trade name 'Mevacor' in the USA and the first statin to be given FDA-approval) was generating as much revenue as Merck's entire drugs portfolio a decade earlier. Statins have been the blockbuster drugs to beat all blockbusters. At peak (things have quietened down a little of late as the statins progressively come off patent), statins as a whole were bringing in an estimated $5 billion per annum in profits and it is estimated that total profit from all statins has been upwards of $1 trillion. The selling point of statins is of course that they reduce cholesterol levels; and, crucially, that by reducing cholesterol they reduce CVD risk. It would therefore be difficult to overstate the extent to which "Big Pharma" has been dependent for its profitability on the acceptance of the truth of theory 2.

So, over the past several decades, Big Pharma has certainly wanted theory 2 to be accepted as true. And the extent to which "What Big Pharma wants, Big Pharma gets" can be gauged by reading Ben Goldacre's 2012 *Bad Pharma. How Drug Companies Mislead Doctors and Harm Patients*, and, especially, Marcia Angell's 2004 *The Truth about the Drug Companies: How they deceive us and what to do about it*.

---

[18] The Sugar lobby (fruit juice and soft-drink manufacturers as well as table sugar), along with manufacturers of "healthy fat" products, such as Unilever who manufacture Flora margarine—an allegedly more healthy replacement for saturated-fat-replete butter—have of course a vested interest in seeing saturated fat branded unhealthy since it keeps sugar out of the frame. A good place to start reading about their influence is Chap. 12 of Kendrick 2018.

[19] See also Part Three of Greene [2007]—Greene does not question the medical orthodoxy (that high LDL-cholesterol causes CVD) but nonetheless supplies much fascinating detail of the history of statins.

Sadly, medical research largely exemplifies "The Golden Rule" ("He who has the gold, makes the rules"): The Pharmaceutical Companies have an enormous amount of control over what medical research gets done, how it gets done, which results get published and which do not, and even what the results of the research are declared to be to an extent that is both staggering and frightening. Marcia Angell—no radical, anti-establishment figure but instead a former Editor-in-Chief of *The New England Journal of Medicine* (the world's most prestigious medical journal)—wrote "It is simply no longer possible to believe much of the clinical research that is published, or to rely on the judgment of trusted physicians or authoritative medical guidelines. I take no pleasure in this conclusion, which I reached slowly and reluctantly over my two decades as an editor of *The New England Journal of Medicine*." The main reason for this was the influence of Big Pharma, not only in controlling research and publication of research results, but also in affecting medical opinion; and, as Angell points out, it exerts this influence largely through *money:*

> "No one knows the total amount provided by drug companies to physicians, but I estimate from the annual reports of the top 9 U.S.-based drug companies that it comes to tens of billions of dollars a year in North America alone. By such means, the pharmaceutical industry has gained enormous control over how doctors evaluate and use its own products."[20]

Here is just one illustrative example directly relevant to our cholesterol case: The decision taken by the National Cholesterol Program in 2004 to lower the point at which a person's LDL-cholesterol level starts to count as high, and therefore at which statin treatment is recommended/mandated, meant that millions more Americans were declared to have a high LDL-cholesterol level and were duly prescribed statins. This in turn of course resulted in tens of billions of extra profit for the drug companies. The financial interest statement for the 8 members of the Committee who made this decision (aside from the Chair who was employed by the National Institutes of Health and not allowed any overt connection with the drug industry) showed just short of 70 individual financial conflicts of interest—in terms of research and travel support, honoraria, consultancy fees and the like paid to them by companies directly involved in manufacturing and selling statins.[21]

To summarize: this section of the paper has considered the *prima facie* surprising clash between what Lakatosian methodology would have predicted would happen to theories 1 and 2 and what has in fact happened to them in contemporary medicine. I hope that I have at least indicated that a plausible, external, evidence-based explanation of that clash can be constructed. This in turn lends weight to the view that the negative appraisal of the evidential basis of theories 1 and 2 that I have argued is supplied by MSRP is correct; and that therefore both currently

---

[20] I should add that Angell does not accuse the medical experts of being directly bribed into endorsing claims that they do not really believe. (This is not a rerun of the Lysenko Affair with Big Pharma playing the role of Stalin!) The influence is much more covert and subtle—akin to the way that advertising clearly works (why else would financially successful companies spend so much on it?) despite the fact that (nearly) everyone insists they take no notice of it.

[21] See Kendrick (2014) pp. 160–161.

accepted official dietary advice and currently accepted medical practice in this area are in urgent need of reform. If this view were to be absorbed, it would be a truly significant case of Lakatosian thought having an impact outside of philosophy.

References

Bush, T. L., et al. (1983). Estrogen use and all-cause mortality. Preliminary results from the Lipid Research Clinical Program follow-up study. *JAMA, 249*(7), 903–906.

European Heart Network and European Society of Cardiology. (2012). *European Cardiovascular Disease Statistics 2012 Edition* September 2012. https://www.bhf.org.uk/plugins/PublicationsSearchResults

Feyerabend, P. K. (1975). *Against method*. New Left Books.

Gosse, P. H. (1857). Omphalos: An attempt to untie the geological knot. *BioScience, 48*(10), 848–850.

Greene, J. A. (2007). *Prescribing by numbers. Drugs and the definition of disease*. Johns Hopkins Press.

Harcombe, Z., et al. (2016). Evidence from randomized controlled trials does not support current dietary fat guidelines: A systematic review and meta-analysis. *Open Heart, 3*, e000409.

Hulley, S., et al. (1998). Randomized trial of estrogen plus progestin for secondary prevention of coronary heart disease. Heart and estrogen/progestin replacement study (HERS) research group. *JAMA, 280*(7), 605–613.

Iribarren, C., et al. (1995). Low serum cholesterol and mortality: Which is cause and which is effect? *Circulation, 92*(9), 2396–2403.

Kendrick, M. (2007). *The great cholesterol con. The truth about what really causes heart disease and how to avoid it*. John Blake.

Kendrick, M. (2014). *Doctoring data. How to sort out medical advice from medical nonsense*. Columbia.

Kendrick, M. (2018). *A statin nation. Damaging millions in a brave new post-health world*. John Blake.

Lakatos, I. (1971). History of science and its rational reconstructions. In R. C. Buck & R. S. Cohen (Eds.), *PSA 1970, Boston studies in the philosophy of science* (Vol. 8, pp. 91–135). Reidel. (reprinted as chapter 2 of Imre Lakatos: *The methodology of scientific research Programmes. Philosophical papers, volume 1* edited by John Worrall and Gregory Currie, Cambridge university press, 1978).

Law, M., & Wald, N. (1999). Why heart disease mortality is low in France: The time lag explanation. *British Medical Journal*, 1999. http://www.xcbi.nlm.nih.gov/pmc/articles/PMC1115846/

Le Fanu, J. (2018). *Too many pills. How too much medicine is endangering our health and what we can do about it*. Little, Brown.

Nabel, E. G. (2006). The Women's health initiative. *Science, 313*(5194), 1703.

Ried, K. (2016). Garlic Lowers Blood Pressure in Hypertensive Individuals, Regulates Serum Cholesterol, and Stimulates Immunity: An Updated Meta-analysis and Review. *The Journal of Nutrition, 146*(2), 389S–396S. https://doi.org/10.3945/jn.114.202192

Schatz, I. J., et al. (2001). Cholesterol and all-cause mortality in elderly people from the Honolulu heart program: A cohort study. *The Lancet, 358*(9279), 351–355. https://ncbi.nlm.nih.gov/pubmed/11502313

Spence, D. (2013). Kill the QOF. *The British Medical Journal, 346*, f1498. http://www.bmj.com/content/346/bmj.f1498

Tanis, B. C., et al. (2010). Oral contraceptives and the risk of myocardial infarction. *New England Journal of Medicine, 345*(5), 1787–1793.

Ulmer, H., et al. (2004). Why eve is not Adam: Prospective follow-up in 149650 women and men of cholesterol and other risk factors related to cardiovascular and all-cause mortality. *Journal of Women's Health, 13*(1), 41–53. https://www.nvbi.nlm.nih.gov/pubmed/15006277

Writing Group for the Women's Health Initiative Investigators. (2002). Risks and benefits of estrogen plus progestin in healthy postmenopausal women. Principal results from the Women's Health Initiative Randomized Controlled Trial. *JAMA, 288*(3), 321–333. https://jama.jamanetwork.com/article.aspx?articleid=195120

Open Access This chapter is licensed under the terms of the Creative Commons Attribution-NonCommercial-NoDerivatives 4.0 International License (http://creativecommons.org/licenses/by-nc-nd/4.0/), which permits any noncommercial use, sharing, distribution and reproduction in any medium or format, as long as you give appropriate credit to the original author(s) and the source, provide a link to the Creative Commons license and indicate if you modified the licensed material. You do not have permission under this license to share adapted material derived from this chapter or parts of it.

The images or other third party material in this chapter are included in the chapter's Creative Commons license, unless indicated otherwise in a credit line to the material. If material is not included in the chapter's Creative Commons license and your intended use is not permitted by statutory regulation or exceeds the permitted use, you will need to obtain permission directly from the copyright holder.

Chapter 13
Trade-offs and Progress in Cancer Science

Anya Plutynski ⓘ

Abstract There are a variety of trade-offs involved in biomedical research on cancer. This chapter focuses on reductive heuristics (Wimsatt, 1994, *Canadian Journal of Philosophy Supplementary, 20*, 207–274, 1997, *Philosophy of Science, 64*(S4), S372–S384, 2007, *Re-engineering philosophy for limited beings: Piecewise approximations to reality*. Harvard University Press), and their role in the cancer genomes projects. The advantage of reductive heuristics is that they yield "solvable" problems and can scaffold future research; the disadvantage is that they can lead to relative conservativism. I consider this picture viz. Lakatos's picture of progress in science and argue that the import of enterprises such as TCGA needs to be assessed viz. a larger set of heuristics than the ones Lakatos imagined, including attention to practical applications, and questions of value or social aims, as well as potential future import for scientific inquiry.

Keywords Cancer · Genomics · TCGA (Cancer Genome Atlas Project) · Pan-cancer atlas project · Research program · Progress in science

13.1 Introduction

One of Lakatos's central concerns was to characterize progress in science. What exactly does progress involve? Very briefly, on Lakatos's view, assessing progress in science requires attending to not a single theory in isolation, but series of theories, embedded within a larger "research program." A research program includes a negative heuristic (consisting of principles that cannot be violated), and a positive heuristic that enables and guides future research. A research program is "theoretically progressive" if its growth is "continuous"—each new theory in the sequence must not only accommodate previous empirical content, but also predict excess empirical content over its predecessor (Lakatos, 1970, 33), and it is

---

A. Plutynski (✉)
Washington University, St. Louis, MO, USA
e-mail: aplutyns@wustl.edu

"empirically progressive" if the new empirical predictions are in fact corroborated. A progressive program *anticipates* novel auxiliary theories or has "heuristic power." A "degenerating" research program, in contrast, either makes no predictions at all, or is a "patched up, arbitrary," or "disconnected" series of theories, involving increasingly more ad hoc adjustments made in the face of anomalies (Lakatos, 1970, 1978).

Lakatos left us with several puzzles: How do we demarcate research programs? When exactly should we give up an unproductive research program? When is a research program degenerating?[1] These puzzles have generated decades of debate among historians and philosophers of science. At issue also was whether Lakatos's "rational reconstructions" of history of science insufficiently attended to the complex social, political and economic factors that shape the course of scientific inquiry. Arguably, any reconstruction of the history of a science in service of philosophical analysis trades off between two risks: "unwarranted generalizations from historical cases" on the one hand, and "entirely "local" histories with no bearing on an overall understanding of the scientific process," on the other (Chang, 2011, p. 110).

One of my goals here is to try to strike this balance. This is particularly difficult in applied sciences like biomedicine. Not only does such research often seem to lack (explicit) unified theoretical commitments, but the biomedical sciences are also very difficult to "rationally reconstruct" as an inquiry with a strictly "internal" logic, unaffected by practical concerns. Cancer research is a diverse, heterogeneous domain of study; cancer scientists are engaged in a wide array of research projects, drawing upon a wide range of disciplinary specializations—some closer to "basic" science (genetics, cell and molecular biology, genomics, bioinformatics), and some more "applied" (epidemiology, clinical oncology, pharmaceutical research). As I've argued (Plutynski, 2018), most such research seems focused on *local* puzzles, rather than broad, unifying theories. It is far from clear what count as general benchmarks or measures of success across such varied domains, given such varied goals, questions, and practical concerns.

Moreover, particularly since WWII, there has been an increase in stratification of scientific communities, ever greater competition for limited resources, and thus, pressure for rapid publication. This has shaped scientists' choices and the direction of research. Particularly in cancer research, there has been an increasing focus on what some dub "doable" problems—clearly defined research projects using well-established experimental methods, in service of research questions that are likely to be well-funded and to generate publishable results (Fujimura, 1988, 1996). Since the 1980s, in particular, investment in such research has tended to shift toward "big science" projects—multi-institution, multi-year projects aimed at harnessing federal

---

[1] For discussion of these and related questions, see, e.g., Cohen, R.S., P.K. Feyerabend, and M.W. Wartofsky (eds.), (1976); Musgrave, A. and C. Pigden, (2023 Edition); Kampis, G., L. Kvasz, and M. Stöltzner (eds.), (2002); and more recently, Havstad, J. C., & Smith, N. A. (2019).

support, involving hundreds if not thousands of researchers (Leonelli, 2019; Ankeny & Leonelli, 2016).

A vivid example is the cancer genome projects: the Cancer Genome Atlas Project (TCGA), and its legacy project, the "Pan-Cancer Atlas" project. These multi-year, multi-institutional projects were in service of an array of distinct goals, and it's far from clear whether generating theoretical novelties was one of them. As I've argued (forthcoming), most such research was "scaffolding" future research. Thus, rather than concern myself with the question of whether cancer research is (overall) is "progressive," following Chang (2011), I will focus on the specific "systems of practice," or sets of "epistemic activities" involved in TCGA and the Pan-Cancer Atlas Project. I'll argue that while they were not "theoretically" progressive in Lakatos's sense, they nonetheless "scaffolded" future progress, insofar as they enabled further development of instrumentation, data gathering, and tools for data analysis. Moreover, each distinct "epistemic activity" was progressive, at least in light of the projects' highly specific aims, and thus criteria of success. I use this case to illustrate how scientific activities may be successful in one respect, or given one set of aims, but less so in others.

A central trade-off in play in the cancer genomes projects was that between "reductive" heuristics, and what I'll call "integrative" research. "Reductive heuristics" (Wimsatt, 1994, 2007) are a set of tools commonly used in approaching complex phenomena in biology; they involve breaking down a complex system into parts and studying their activities in relative isolation. As Wimsatt argued, such approaches risk the entrenchment of reductive perspectives on the systems of interest, which can slow progress toward more "integrative" multiscale research in the same domain. Below, I show how this trade-off was in play in the TCGA, reinforced by institutional, economic, and social factors.[2]

Scientists face a trade-off between risky, complex problems, and relatively "doable" projects that promise quick success in a competitive research environment. In choosing the latter, they trade off one scientific aim and set of values for another— something that may well serve their ends in the short term but may be detrimental in the longer term. Attention to such trade-offs is necessary for a "thick description" of biomedicine (cf. Currie, 2019a, 2019b). Any questions about "progress" versus "degeneration" in science today cannot avoid situating these questions within the larger social, economic, and institutional contexts that constrain how "big" science (such as the genomes projects) moves forward. As I illustrate here, choice of research project is all too often a compromise between various competing ideals, given institutional and economic constraints. While this may seem a critique of Lakatos, in one sense, arguably this historical episode only reinforces Lakatos's admonition that "Philosophy of science without history of science is empty; history of science without philosophy of science is blind." (Lakatos, 1978).

---

[2] As has been argued at length by Elliott (2011, 2017), Brown (2020), Douglas (2009) and many others, there are multiple decision points over the course of scientific inquiry informed by values, or normative judgments about which ends to prioritize, e.g., false positives versus false negatives.

13.2 Situating TCGA in Historical Context: The Rise of Cancer Genomics

The rise of cancer genomics can be traced to a variety of different beginnings, but one of the most formative ones was Bishop and Varmus's (1989) Nobel Prize for Physiology and Medicine for their "discovery of the cellular origin of retroviral oncogenes." The two showed that near identical versions of certain genes associated with cell division were carried by retroviruses (viruses that integrate themselves into the DNA of infected cells), and present in the genome of normal cells in a wide range of species. They called these "proto" onco-genes (or, oncogenes). Mutations to these genes—either inherited or acquired over the course of a lifetime of cell division—are associated with the disorderly behavior of cancer cells. These genes typically play key functional roles in regulation of cell division. When they are mutated in specific ways, cells may grow without limit, and in some cases, eventuate in an invasive tumor. After Bishop and Varmus's discovery, there was a rush to identify yet more "proto-onco" genes, what some historians have called the "molecular biological bandwagon" in cancer research (Fujimura, 1988, 1996).

This rush to jump on the bandwagon was promoted in part by institutional and economic incentives, growing in part out of a shift in federal funding for scientific research, starting in 1980. The Bayh-Dole Act encouraged greater partnering of scientific researchers and private industry in the U.S., shifting institutions toward privately funded research that might result in patentable products and applications. One such application that has become a focus of investment is genome sequencing technologies. Such technologies have been touted as a potential solution to the great ills of cancer, by enabling more precise diagnosis, prognosis, and treatment. At the event celebrating the completion of the human genome project (or, HGP) then President Bill Clinton promised that our "children's children" would know cancer only as a "constellation of stars." Clinton was instrumental in the appointment of Francis Collins to head up the National Human Genome Research Institute, under the auspices of which, along with the NIH, the cancer genome projects were funded. The Cancer Genome Atlas was advertised as the first step toward greater precision in diagnosis, screening, treatment of cancer.

The longest serving director of NIH (1993–2008), Francis Collins has been instrumental in a shift to significant investment in genomics research in biomedicine as a whole. Dr. Collins is a physician-geneticist noted for his landmark discoveries of disease genes and his leadership of the international Human Genome Project, which culminated in April 2003. Not surprisingly, then, he was a great booster for cancer genomics, and for developing the technologies to enable faster, more efficient, and less expensive sequencing. In part due to investments promoted by Collins and his group at the National Human Genome Research Institute (NHGRI), cancer genomics became a central focus of research in the last decades of the 20th, and first decades of the twenty-first Century, dwarfing investments in almost every other domain funded by the NHGRI. The Cancer Genome Atlas Project was a massive, multi-institutional research project, spanning 10 years (2005–2015), and involving thousands of researchers and research subjects, and resulting in hundreds of publications. Notably, it resulted in published sequences of 33 major cancer types, "marker papers," part of what a 2011 National Research Council report called "a knowledge network for biomedical research and a new taxonomy of disease." (NRC, 2011)

The initiative was promoted with the idea that this research would yield more effective, "targeted" therapies. In defending and explaining his motivation for the project, Collins often cited Tamoxifen and Herceptin as examples of drugs that could assist patients based on genetic or molecular features associated with their cancer type or subtype (Collins & Barker, 2007). Though both drugs were developed a decade prior to the genome projects (and so their discovery was in no way informed by cancer genomics, per se), examples of such "targeted" forms of treatment were used to promote cancer genomics, with the view that understanding the genetic basis of various cancers would lead to innovative diagnostic tools, prognoses, and treatments. "Targeted" drugs would intervene directly on the molecular pathway affected in disease, preventing both side effects, and cancer recurrence. Despite such promises, 10 years out from completion of TCGA, most cancers today are still treated with surgery and chemotherapy, and recurrence is common for many new targeted therapies (Prasad, 2020; Tabery, 2023).

Once TCGA was underway, it acquired a momentum associated in part with Clinton's romantic promise of cancer becoming known only as "a constellation of stars." Echoing similar sentiments, President Obama, at the State of the Union Presidential Address in 2015, announced his "Precision Medicine Initiative." This initiative devoted substantial funds to "ramping up" cancer genomics, so that each patient would receive "the right drug, at the right time." This, and Biden's "Moonshot" for cancer, led to yet more continued funding for TCGA and its downstream "Pan-Cancer" Atlas project, two of the most well-funded multi-year scientific investments in U.S. history. Together, they have funded more researchers, for far longer, than all research on environmental epidemiology of cancer together in the same span of time. The TCGA arguably was a continuation of the "bandwagon" effect already underway in the 90 s (Fujimura, 1996), but the pile on to cancer genomics was many orders of magnitude more significant, in terms of sheer numbers of participants, funding prioritization, and institutional support. So, what has been the upshot of this tremendous investment?

13.3 Trade-offs and Progress

Was the TCGA an instance of "progress" in Lakatos's sense? Some might argue that TCGA was not science, at all. On such a view, applying Lakatos's criteria may be a category error. The central goals of these investments were only indirectly epistemic: the creation of a research infrastructure, bringing down the cost of sequencing, training future researchers in genomic data analysis, and the creation of a curated public database of mutations to genes associated with cancer, which could be mined for future research (Hutter, 2018). With respect to these goals, the research project was a success. Indeed, it was engineered to succeed: project managers were deliberate in their choices of funding, whether of technologies, institutions or researchers, based on relative speed and efficiency. They set up a competitive, market based framework for research, aimed at both building alliances with industry, and speeding innovations in technology. They set deadlines to speed up efficient sample collection and analysis of the data; there were competitions among algorithm developers for fast, accurate genomic data analysis. While allowing for self-governance within subgroups, they also put institutional subgroups into competition for grant funding, to speed production of high-quality results. In other words, purely "scientific" concerns were not the only (or even primary) concerns of project managers, in the sense that they were not promoting the development or testing of *novel* hypotheses, per se. By the point the genome projects were underway, there was no question in researchers' minds regarding whether mutations played a causal role in cancer; the matter in question was which and how many mutations were associated with which cancers. Drawing upon their successes and failures with the Human Genome Project, the program managers at the NIH and NHGRI reasoned it would be effective to put researchers into competition in the race for documentation of ever more such mutations.

Thus, one could argue that it is simply a mistake to ask whether the TCGA made "progress" in the senses Lakatos intended. The effort was a strategic, technological and institutional shift of resources into building a scaffold for future research. Indeed, one could argue that talk of discovery of new targeted drugs or new cancer classifications was merely a front; if they were being frank, most researchers acknowledged that they did not think that investing in cancer genomics was likely to generate groundbreaking understanding of cancer as a disease, at least immediately. Most granted this project was instrumental to future insights, (potentially) down the line; one researcher remarked on how they simply "did not know what we did not know." (Govindan, 2018) While TCGA did generate empirical discoveries (we know far more now about the extent and nature of diversity of cancer genomes), these empirical discoveries were intended to create scaffolding for future research. Rather than conclusive, "complete" genomes, these early marker papers were known to be incomplete "drafts," that would require revisiting. Most grant that larger samples of tumors need to be gathered than were initially, and the initial marker papers and data should be re-analyzed with better techniques, to remove batch effects, and enable more precise predictions.

By most practical measures, however, TCGA achieved its goals: it significantly reduced the cost and improved the speed of genome sequencing technology, leading to advances in efficiency in analysis of genetic data. It led to the creation of public databases of mutations to genes affected in cancer, and a network of researchers well trained in both genomic sequencing and analysis. As such, the projects have put in place a scaffold for future research. However, the value of creating such scaffolding may seem opaque when we turn to a critical assessment of theoretical or predictive outcomes Lakatos focuses on in his analysis of progress. While researchers learnt a good deal about the complexity and heterogeneity of cancer genomics, most cancer researchers today acknowledge that this data is only the "tip of the iceberg" in gaining a comprehensive understanding of the *causal role* of various genes in cancer initiation, progression, recurrence, or responses to treatment. Having a catalogue of mutations associated to cancer is only a first step—indeed, some scientists refer to this as a kind of "dictionary" of cancer biology, but we are still far from reading the book. If what we wish to understand is how these mutations act and interact in cancer, we must turn to the "downstream" projects of TCTA—the Pan-cancer Atlas Project, or the "next stage" in cancer bioinformatics: "proteomics," "epigenomics," and "transcriptomics." These are investigations into how mutations to cancer cells are expressed, differentially, in different tissues and organs, tracing the pathways associated with these molecules in the cell, in service of identifying "biomarkers" for both better screening, prognosis, and targeted treatments.

Many critics of TCGA and affiliated projects from the beginning were skeptical on these very grounds. Critics argued that this information was only a very small piece of a comprehensive understanding of cancer as a disease. Cancer initiation and progression is shaped by the tissue microenvironment, the structural organization of tissue and organs, and activity of the endocrine and immune system, as well as the larger environment. Moreover, as it emerged that the mutational burden in healthy somatic tissues is relatively high, it has increasingly become evident that mutation alone is not sufficient for cancer; thus, many researchers have urged greater attention to mechanisms apart from those that were the central focus of the genomes projects that both prevent and promote cancer (Balmain, 2023). The danger of *reductive heuristics* is that they can lead to myopic perspectives on the system of interest. TCGA focused attention on the genome, reducing the problem of investigating cancer's causes to one specific unit and scale of analysis. This type of reductive research can generate results much faster than interdisciplinary, multi-year studies of environmental carcinogens. Such studies are expensive and time consuming, in that they require identifying risk factors, isolating their effects from the multitude of potential confounders, and investigating how such factors interact over the course of a lifetime. The rapid production of "marker" papers published in *Nature* and *Cell*, by comparison, appears a triumphant success. However, one conclusion of this research, echoed by many different participants, was that they are only *beginning* to understand the diversity and complexity of the causal pathways yielding cancer, even 50 (plus) years after Nixon declared his "war" on cancer.

There were many more mutations involved in cancer initiation and progression than anticipated, and understanding how these genes act and interact with one another, in various cancer types and subtypes, is the kind of complex problem that will take some decades to uncover (Ding, 2018; Govindan, 2018). Very few cancers are driven by a small handful of four or five mutations that can be used in service of generating "magic bullets," or specific interventions on specific pathways that lead to lasting remissions (Plutynski, 2023). Thus, on the one hand, the research led to a great deal of new information: the sheer scale of genomic information available now about cancer dwarfs, for instance, astronomical data by orders of magnitude (Stephens et al., 2015). This new data has *potential* to change both how we diagnose, and treat, cancers. However, the project also changed the landscape of cancer research, in some ways for the worse.

The sheer size and institutional inertia behind the cancer genome projects shifted many young scientists' careers toward "dry" lab research, and away from the "wet" lab, leading to a shift in the culture of research. Given that the definitive measure of success in a young scientist's career is—currently—continued grant funding and H-Index, young cancer researchers would do well to choose data analysis or informatics over wet lab research. Unlike wet-lab research, which can take decades to get off the ground, genomic data analysis leads to rapid publications with high rates of citation. Unlike the tedious "bench" lab research that preceded it, which requires dedicated time and technique, bioinformatics is portable and can rely public databases. Yet, how "significant" such research is seems to be far from straightforward. Critics worry that unreflective reliance on big data, or use of machine learning in cancer screening, diagnosis, and treatment carries risks of harm—misdiagnosis, overdiagnosis, and overtreatment, as well as compromise of privacy (Ngiam & Khor, 2019; Vogt & Green, 2020).

The investment in genomics, and the celebration of these achievements, has arguably led to a lower estimation of the variety of scientists at the "margins" of cancer research—those investigating causal factors apart from genetics. Yet investigation into these causes have the potential to significantly reduce cancer incidence and mortality (Brennan & Davey-Smith, 2022). In my (Plutynski, 2018), I give several examples of how exclusive focus on the cell and molecular level in cancer research is less than ideal. By way of examples:

- It can lead to an overly simplistic picture of how best to classify cancers (Chap. 1)
- It can prevent us from developing novel forms of treatment, or lead us to fail to attend to other disciplines of potential relevance (e.g., immunology, evolution, development) (Chaps. 4 and 5)
- It can lead us to ignore or fail to investigate the role of complex environmental, economic, social and institutional factors that place some vulnerable populations at greater risk (Chap. 3)
- It can lead to overly simplistic explanations of patterns of incidence, in turn informing unwise public health policies, especially with respect to screening (Chap. 2).

While I cannot review all these concerns here, I will briefly consider a counterfactual history—or a way this research might have gone differently (see, e.g., Radick, 2023)—one proposed by a cancer epidemiologist, Richard Peto, in the 1980s.

13.4 An Alternative Path? Progress and the Scientific Imagination

How could cancer research have gone differently? One possibility was proposed by Richard Peto, an epidemiologist who did important work on cancer in the 1970s and 1980s. In an insightful paper published in the *Encyclopedia of Medical Ignorance* in 1984, Peto made a plea for "ignorance" in cancer research.[3] The paper's definition of "ignorance" was somewhat broad, but he called for greater curiosity, imagination, and open-ended questions.[4] Peto was resisting the very tendencies in cancer research that Fujimura called the "molecular bandwagon." He urged attention to questions that required that researchers step back and look at epidemiological patterns (e.g., of incidence and mortality, over the course of a lifetime), as well as evolutionary history, in service of gaining greater insight into cancer's causes. He proposed a "paradox," or "puzzle," that seemed to require attention to this "longer view". The question that has been dubbed "Peto's paradox": Why—given that cells divide over the course of a lifetime, and mutations arise as they divide—don't larger animals get cancer more often than we do? This question eventuated in a small but important body of research about evolutionary and developmental bases of patterns of differential cancer incidence within and across taxa (for a review, see Nunney & Muir, 2015).

Peto also defended what he called "black box," rather than "mechanistic," approaches to understanding and preventing cancer. He suggested that the "mechanistic" approach was more likely to be favored by "pure" scientists, because it allowed for experimental manipulation on causal processes that yield disease (at least in cells and cell culture). In other words, mechanistic research deploys "reductive" heuristics –heuristics that led to the rapid advances in cell and molecular biology of cancer, and exciting, novel technologies that were so widely hailed as promising in the 1980s. However, Peto was concerned that this enthusiasm for reductive heuristics would take away resources from the important work of epidemiologists. He pointed out how "black box" epidemiology had led to far greater reductions in cancer mortality than any research in molecular biology so far. Work by Doll, Hill, Wydner, and Graham on the link between smoking and lung cancer provided the best grounds for recommending smoking cessation decades before any molecular or genetic factors had been identified that linked smoking and lung cancer. Thus, he argued that environmental epidemiology should not be set aside in favor of more "mechanistic" approaches, but that both should be pursued simultaneously.

What might Lakatos make of Peto's plea for curiosity, open ended questions, and attention to the diversity of causes and disciplinary perspectives on cancer? If we had

---

[3] See also Firestein, S. (2012). *Ignorance: How it drives science*. OUP USA. Thanks to a reviewer for this suggestion.

[4] See also Gero's (2007) "Honoring Ambiguity/Problematizing Certitude" *Journal of Archaeological Method and Theory*, Vol. 14, No. 3 pp. 311–327.

invested equally as much in cancer epidemiology and public health in the 1980s, the landscape of cancer research (and perhaps also, cancer incidence and mortality) might look rather different today. Today, cancer is primarily spoken of and understood as a "genetic," or "genomic" disease. The standard premedical undergraduate curriculum tends to emphasize reductive approaches, big data, and experimental replication, over epidemiological or public health research. Yet, emphasis on these styles of scientific inquiry in cancer research—while certainly important—can carry the false implication that "black box" methods are not legitimate forms of arriving at causal knowledge. Arguably, both modes of investigation are necessary for understanding and preventing cancer. Reductive approaches are important tools of inquiry, but so too are approaches that attend to patterns of association of the sort investigated by Doll, Hill and others in the mid-twentieth Century (cf. Plutynski, 2023). That is, the speed and (apparent) effectiveness of reductive heuristics at generating empirical results has led to an increase in focus on cancer causation at the cell and molecular level, overshadowing the value of multiscale perspectives on the problem of cancer, and underestimating the importance of public health and epidemiology. Yet, arguably, such perspectives are required, if we wish to intervene successfully in service of better cancer prevention.

13.5 Conclusions

There are a variety of trade-offs involved in biomedical research on cancer that can both lead to local success, and to "degenerating" research. My focus here has been on reductive heuristics—narrowing one's focus to a specific temporal and spatial scale. On the one hand, as Wimsatt (1994, 1997, 2007), (see also: Bechtel & Richardson, 2010), argued, reductionism as a methodological strategy in biology is an effective tool. Carving off subparts of a system, investigating these parts and their causal interaction in isolation, can help us identify core mechanisms at work in the production of complex diseases. This is a specific version of a general principle, which seems to be illustrated in a variety of examples of scientific progress: part of science involves suitably constraining the problem to be solved. This is what the advocates of the "molecular bandwagon" were doing. Schaffner (2006) calls the product of these forms of inquiry "partial" reductions. Such partial reductions can look like spectacular successes; indeed, as he points out, they can be Nobel prize winning discoveries.

On the other hand, however, these partial reductions only give insight into a small part of a complex, dynamic process. So, measures of progress of this sort, involving generation of ever more, or "big," data, seem somewhat myopic. More seriously for our purposes, the institutional organization of science is such that the successes of reductive heuristics can lead scientists to "pile on" one very narrow avenue of inquiry. Such practices can be reinforced by other practical trade-offs involved in scientific decision making—trade-offs facing scientists' career choices, such as how to choose projects that speed publication of "smallest publishable units." There's a trade-off between the professional advantages associated with "conservatism," or entrenched research ("sticking with what we know") and risky, innovative, or "creative" research (cf. Sonnenschein & Soto, 2018, 2020).

Indeed, there seems to be a consensus that the current organization of scientific research by and large has led to less innovative science (Stanford, 2019; see also, Luukkonen, 2012; O'Connor, 2019; Wu et al., 2023; Currie, 2019a, 2019b; Schneider, 2021; Peacock, 2009). On the one hand, there are several advantages to well-entrenched scientific research. Young scientists can be trained in specializations that give them a clear path for generating knowledge. Focus on very specific, well-defined problems allows for incremental advances along a very well-marked path. A scientist choosing such a path has the advantage of joining a large network of scholars working on the same solvable problems. Given such well-defined problems have widely agreed upon shared standards for publication and funding, the "rules of the game" (or, how best to advance knowledge, as well as advance in one's career) are clearly laid out. On the other hand, this approach leads to research that may be less genuinely innovative. How might one reframe this in Lakatos's terms?

First, in my view, it is a mistake to read this history as a matter of "bad theory," or a "degenerating research program." The participants in TCGA were not attempting to "save" a false theory with "ad hoc" accommodations; mutations do play a causal role in cancer. As a result of TCGA, the number, type, and role of mutation in cancer is better understood. The disagreement is with respect to prioritizing investment in scientific resources currently (see, e.g., Brennan & Davey-Smith, 2022). Moreover, cancer research is by and large not "theory-driven," but problem-driven (Plutynski, 2018, 2019). To be sure, theoretical presuppositions inform these problems, but as the case above illustrates, it is perhaps equally as important to (most) cancer scientists that the problems they seek to solve are "doable," and "fundable." This has led to a highly specialized, "siloed" form of scientific research.

How might one integrate this picture into Lakatos's framework? Perhaps one might say that while scaffolding and specialized "problem-based" knowledge is useful, its value in advancing *scientific* progress is indirect at best, and perhaps can lead to slowing of progress in the worst-case scenario. Perhaps it is better to say that the import of these enterprises need to be assessed viz. a larger set of heuristics than the ones Lakatos imagined, including attention to practical applications, and potential future import for scientific inquiry. By this measure, TCGA may indeed have been quite successful; it served to launch many hundreds if not thousands of downstream research projects into everything from the diversity of mutational signatures in cancer and their association to environmental exposures (Alexandrov et al., 2020), to how metabolic disorders promote mutant expansion in the oesophagus (Herms et al., 2024) to the relationships between cancer and heart disease (Heyde et al., 2021). That said, perhaps by investing in TCGA, opportunities for more integrative, interdisciplinary research were lost.

That said, as this case illustrates, there is a trade-off between scaffolding future research and doing the science that the scaffolds enable—both are required for scientific progress, and so part of any complete picture of a scientific "research program." Perhaps Lakatos might include such "scaffolding" projects as part of the positive heuristic of a research program. This part may be less concerned with generating *theoretical* or *(wholly unexpected) empirical* novelties than building infrastructure that makes future discoveries possible. This heuristic might include generation and testing of novel technologies, the creation of coordinated data platforms, and the training of scientific workforces with the skills necessary to use and refine these tools. These are legitimate scientific goals, but rather different ones from those philosophers have traditionally taken to be central to science—namely, generating theoretical and empirical novelties. Arguably, both kinds of research are essential components of scientific progress. But, just as with reductive heuristics, there can be diminishing returns, or over-investment in one kind of research, at the expense of the other.

Perhaps significantly unlike the picture of science Lakatos suggested, each of the above trade-offs can be more broadly characterized as trade-offs in values.[5] Which sorts of research to prioritize, and thus how to assess progress, depends on which sorts of outcome we value, a matter that is not reducible to the epistemic aims of science. Incorporating such values into an assessment of progress of TCGA would require considering how and which kinds of information, technology, and downstream research matters most, over the short and long term—no small feat! On the one hand, we know a great deal more about the genetics and molecular biology of cancer than we knew even a decade ago. We can sequence a human genome for a few hundred dollars. Cancer patients' tumors are now more or less routinely tested for the presence of mutations that could affect their metabolization of a drug that might otherwise be ineffective. This is no small achievement. On the other hand, however, as illustrated above, such focus on "solvable" problems can lead to research silos, or myopic views of the subject matter and limits to our imagination as to what count as interesting scientific questions.

Scientists—even those in the very same discipline—can seem to be working on isolated peaks, at a great distance from neighboring peaks in the landscape of research. This siloing can make science relatively predictable. When scientists' work is measured by "productivity indexes" (such as the number of publications or citations), the risk we run is emphasizing this measure over and above novel theoretical innovations, predictions, or explanations, let alone consideration of the larger social value of the research.[6] Taking seriously the question of "progress" in science requires we attend to institutional factors such as how scientists are trained and rewarded, as well as considerations of the relative value of outcomes, over and above mere number of papers published or cited. The social and economic organization of science tends to isolate scientists from addressing such questions, but in a field such as cancer research, it seems such questions are unavoidable.

---

[5] See also, e.g., Kuhn, T. (1977) *The Essential Tension*, particularly his chapter on "Objectivity, value judgment, and theory choice"; Douglas (2009); Elliott (2011, 2017), and Brown (2020).

[6] According to Aksnes et al. (2019) "there is no evidence that citations reflect other key dimensions of research quality. Hence, an increased use of citation indicators in research evaluation and funding may imply less attention to these other research quality dimensions, such as solidity/plausibility, originality, and societal value."

One important perspective cancer researchers ought to attend to more closely is the perspective of the patient. Consider recent critiques of how new cancer drugs are studied and approved (Prasad, 2020). Many clinical trials of new drugs fail to take into consideration questions such as: How does this drug improve *quality* as opposed to mere quantity of life? What are the side effects that are of greatest concern to patients? Are the outcomes we are measuring in clinical research good proxies for outcomes that genuinely matter to patients and families? In other words, assessing progress in biomedical science seems to require that scientists devote some time and attention to assessing and critically reflecting on how to prioritize and assess value-laden outcomes. The values that have largely shaped our current model for funding policy, as well as our presuppositions about what counts as progress in science, seem to focus too narrowly on the magnitude of empirical results generated, without a sense of *which* results *matter, to whom, and how*. Answering the question of what it might mean for a research program to be progressive in the context of biomedicine requires that we engage these questions about trade-offs in value, and genuinely reflect on the kind of progress or success we might hope for. Perhaps, the question of progress is more normatively loaded than Lakatos understood.[7]

The promise of basic science is that it will (eventually) serve our larger social as well as epistemic aims. In many ways, it has. However, research can become entrenched, not only because of "ad hoc" adjustment, but because of prioritizing certain kinds of scientific ends, and values, over others. Thus, perhaps scientists need to be more aware of and open to readjusting focus along the way, engaging and considering other, competing aims. This may require replacing Lakatos's picture of progress in science with one that recognizes the prevalence of trade-offs intrinsic to the culture of science, and the role of values in prioritizing different goals or research questions with different import. If a central goal of cancer science is to not only to understand cancer, but reduce the suffering caused by cancer, we need to broaden our appreciation of the variety of measures of progress, and work to traverse the divides both between scientists, and between scientists and the patients they hope to benefit.

References

Aksnes, D. W., Langfeldt, L., & Wouters, P. (2019). Citations, citation indicators, and research quality: An overview of basic concepts and theories. *SAGE Open, 9*(1), 2158244019829575.

Alexandrov, L. B., Kim, J., Haradhvala, N. J., Huang, M. N., Tian Ng, A. W., Wu, Y., et al. (2020). The repertoire of mutational signatures in human cancer. *Nature, 578*(7793), 94–101.

Ankeny, R. A., & Leonelli, S. (2016). Repertoires: A post-Kuhnian perspective on scientific change and collaborative research. *Studies in History and Philosophy of Science Part A, 60,* 18–28.

Balmain, A. (2023). Peto's paradox revisited: Black box vs mechanistic approaches to understanding the roles of mutations and promoting factors in cancer. *European Journal of Epidemiology, 38*(12), 1251–1258.

Bechtel, W., & Richardson, R. C. (2010. Originally, 1993). *Discovering complexity: Decomposition and localization as strategies in scientific research.* MIT Press.

---

[7] See also: Currie, 2018a, 2018b, postscript & Brown, 2020.

Brennan, P., & Davey-Smith, G. (2022). Identifying novel causes of cancers to enhance cancer prevention: New strategies are needed. *JNCI: Journal of the National Cancer Institute, 114*(3), 353–360.

Brown, M. J. (2020). *Science and moral imagination: A new ideal for values in science.* University of Pittsburgh Press.

Chang, H. (2011). Beyond case-studies: History as philosophy. In S. Mauskopf & T. Schmaltz (Eds.), *Integrating history and philosophy of science: Problems and prospects* (pp. 109–124). Springer Netherlands.

Cohen, R. S., Feyerabend, P. K., & Wartofsky, M. W. (Eds.). (1976). *Essays in memory of Imre Lakatos, Boston studies in the philosophy of science* (Vol. 39). Reidel.

Collins, F. S., & Barker, A. D. (2007). Mapping the cancer genome. *Scientific American, 296*(3), 50–57.

Currie, A. (2018a). Introduction: Creativity, conservatism & the social epistemology of science. Studies in History and Philosophy of Science A.

Currie, A. (2018b). *Rock, bone, and ruin: An optimist's guide to the historical sciences.* MIT Press.

Currie, A. (2019a). Existential risk, creativity & well-adapted science. *Studies in History and Philosophy of Science Part A, 76*, 39–48.

Currie, A. (2019b). Creativity, conservativeness & the social epistemology of science. *Studies in History and Philosophy of Science Part A, 76*, 1–4.

Ding, L. (2018). *Personal interview with Anya Plutynski.* Washington University in St. Louis.

Douglas, H. (2009). *Science, policy, and the value-free ideal.* University of Pittsburgh Pre.

Elliott, K. C. (2011). *Is a little pollution good for you? Incorporating societal values in environmental research.* OUP USA.

Elliott, K. C. (2017). *A tapestry of values: An introduction to values in science.* Oxford University Press.

Firestein, S. (2012). *Ignorance: How it drives science.* OUP USA.

Fujimura, J. H. (1988). The molecular biological Bandwagon in cancer research: Where social worlds meet. *Social Problems, 35*(3) Special Issue: The Sociology of Science and Technology, 261–283.

Fujimura, J. H. (1996). *Crafting science: A Sociohistory of the quest for the genetics of cancer.* Harvard University Press.

Gero, J. M. (2007). Honoring ambiguity/problematizing certitude. *Journal of Archaeological Method and Theory, 14*(3), 311–327.

Govindan, R. (2018). *Personal interview with Anya Plutynski.* Washington University in St. Louis.

Havstad, J. C., & Smith, N. A. (2019). Fossils with feathers and philosophy of science. *Systematic Biology, 68*(5), 840–851.

Herms, A., Colom, B., Piedrafita, G., Kalogeropoulou, A., Banerjee, U., King, C., et al. (2024). Organismal metabolism regulates the expansion of oncogenic PIK3CA mutant clones in normal esophagus. *Nature Genetics, 56*(10), 2144–2157.

Heyde, A., Rohde, D., McAlpine, C. S., Zhang, S., Hoyer, F. F., Gerold, J. M., et al. (2021). Increased stem cell proliferation in atherosclerosis accelerates clonal hematopoiesis. *Cell, 184*(5), 1348–1361.

Hutter, C. (2018). *Co-director of TCGA for the last five years of its tenure.* Personal interview, NHGRI.

Kampis, G., Kvasz, L., & Stöltzner, M. (Eds.). (2002). *Appraising Lakatos.* Kluwer.

Kuhn, T. (1977). *The essential tension.* University of Chicago Press.

Lakatos, I. (1970). Falsification and the methodology of scientific research Programmes. In Lakatos & Musgrave (Eds.), *Criticism and the growth of knowledge* (pp. 91–195). Cambridge University Press. Reprinted in Lakatos, 1978: 8–101.

Lakatos, I. (1971). The history of science and its rational reconstructions. In R. C. Buck & R. S. Cohen (Eds.), *PSA 1970: Boston studies in the philosophy of science* (Vol. 8, pp. 91–135). Reidel.

Lakatos, I. (1978). In J. Worrall & G. Currie (Eds.), *The methodology of scientific research programmes: Philosophical papers* (Vol. 1). Cambridge University Press.

Lakatos, I., & Musgrave, A. (Eds.). (1970). *Criticism and the growth of knowledge*. Cambridge University Press.
Leonelli, S. (2019). *Data-centric biology: A philosophical study*. University of Chicago Press.
Luukkonen, T. (2012). Conservatism and risk-taking in peer review: Emerging ERC practices. *Research Evaluation, 21*, 48–60.
Musgrave, A., & Pigden, C. (2023). "Imre Lakatos", The Stanford Encyclopedia of Philosophy (Spring 2023 Edition). https://plato.stanford.edu/archives/spr2023/entries/lakatos/
National Research Council. (2011). *Toward precision medicine: Building a knowledge network for biomedical research and a new taxonomy of disease*. National Academies Press.
Ngiam, K. Y., & Khor, W. (2019). Big data and machine learning algorithms for health-care delivery. *The Lancet Oncology, 20*(5), e262–e273.
Nunney, L., & Muir, B. (2015). Peto's paradox and the hallmarks of cancer: Constructing an evolutionary framework for understanding the incidence of cancer. *Philosophical Transactions of the Royal Society B: Biological Sciences, 370*(1673), 20150161.
O'Connor, C. (2019). The natural selection of conservative science. *Studies in History and Philosophy of Science Part A, 76*, 24–29.
Peacock, M. S. (2009). Path dependence in the production of scientific knowledge. *Social Epistemology, 23*(2), 105–124.
Peto, R. (1984). The need for ignorance in cancer research. In R. Duncan & M. Weston-Smith (Eds.), *The Encyclopaedia of medical ignorance* (pp. 129–133). Pergamon Press.
Plutynski, A. (2018). *Explaining cancer: Finding order in disorder*. Oxford University Press.
Plutynski, A. (2019). Cancer modeling: The advantages and limitations of multiple perspectives. In C. D. McCoy & M. Massimi (Eds.), *Understanding perspectivism* (pp. 160–177). Routledge.
Plutynski, A. (2023). On explaining Peto's paradox. *European Journal of Epidemiology, 38*, 1–6.
Prasad, V. K. (2020). *Malignant: How bad policy and bad evidence harm people with cancer*. JHU Press.
Radick, G. (2023). *Disputed inheritance: The battle over Mendel and the future of biology*. University of Chicago Press.
Schaffner, K. F. (2006). Reduction: the Cheshire cat problem and a return to roots. *Synthese, 151*, 377–402.
Schneider, M. D. (2021). Creativity in the social epistemology of science. *Philosophy of Science, 88*(5), 882–893.
Sonnenschein, C., & Soto, A. M. (2018). An integrative approach toward biology, organisms, and cancer. *Systems Biology, 1702*, 15–26.
Sonnenschein, C., & Soto, A. M. (2020). Over a century of cancer research: Inconvenient truths and promising leads. *PLoS Biology, 18*(4), e3000670.
Stanford, P. K. (2019). Unconceived alternatives and conservatism in science: The impact of professionalization, peer-review, and big science. *Synthese, 196*, 3915–3932.
Stephens, Z. D., Lee, S. Y., Faghri, F., Campbell, R. H., Zhai, C., Efron, M. J., et al. (2015). Big data: Astronomical or genomical? *PLoS Biology, 13*(7), e1002195.
Tabery, J. (2023). *Tyranny of the gene*. Knopf.
Varmus, H. E. (1989). Retroviruses and oncogenes I. *Nobel Lectures Physiology or Medicine*, 504–522.
Vogt, H., & Green, S. (2020). Personalised medicine: Problems of translation into the human domain. In *De-Sequencing: Identity Work with Genes* (pp. 19–48).
Wimsatt, W. C. (1994). The ontology of complex systems: Levels of organization, perspectives, and causal thickets. *Canadian Journal of Philosophy Supplementary, 20*, 207–274.
Wimsatt, W. C. (1997). Aggregativity: Reductive heuristics for finding emergence. *Philosophy of Science, 64*(S4), S372–S384.
Wimsatt, W. C. (2007). *Re-engineering philosophy for limited beings: Piecewise approximations to reality*. Harvard University Press.
Wu, J., O'Connor, C., & Smaldino, P. E. (2023). The cultural evolution of science. In J. Kendal, R. Kendal, & J. Tehrani (Eds.), *The Oxford handbook of cultural evolution*. Oxford University Press.

Open Access This chapter is licensed under the terms of the Creative Commons Attribution-NonCommercial-NoDerivatives 4.0 International License (http://creativecommons.org/licenses/by-nc-nd/4.0/), which permits any noncommercial use, sharing, distribution and reproduction in any medium or format, as long as you give appropriate credit to the original author(s) and the source, provide a link to the Creative Commons license and indicate if you modified the licensed material. You do not have permission under this license to share adapted material derived from this chapter or parts of it.

The images or other third party material in this chapter are included in the chapter's Creative Commons license, unless indicated otherwise in a credit line to the material. If material is not included in the chapter's Creative Commons license and your intended use is not permitted by statutory regulation or exceeds the permitted use, you will need to obtain permission directly from the copyright holder.

Chapter 14
Epilogue: Scientific Theory-Change and Rationality – Lakatos and the "Popper-Kuhn Debate"

John Worrall

Abstract Lakatos's career in the West began in the Philosophy of Mathematics. His interest in the Philosophy of Science was kindled by the differences in the accounts of scientific progress given by his mentor, Karl Popper, and by Thomas Kuhn. These differences were highlighted in a debate between Kuhn and Popper at the Bedford College Colloquium in 1965. This paper examines some of the history of that debate and Lakatos's contributions to it. The chief point at issue between Kuhn and Lakatos concerns the 'theory choices' made in 'scientific revolutions' and whether those choices are dictated by principles of scientific rationality.

As is well-known, Imre Lakatos first made his mark (in the anglophone world) as a philosopher of mathematics. His interest in the philosophy of *science* was kindled largely by the apparent clashes between Karl Popper's ideas about theory change in science and those of Thomas Kuhn. In particular, Lakatos's attention was drawn to those clashes by a session in a famous 1965 conference of which he himself was the principal organiser. This was the Bedford College Colloquium, one long session of which consisted of Kuhn's presenting an analysis of the relationship of his own ideas to those of Popper (Kuhn, 1970a), a response from Popper (Popper, 1970) (and from some other commentators) and finally a reply to Popper and other critics by Kuhn (1970b).

The proceedings of this part of the Colloquium were not published until 5 years later (Lakatos & Musgrave, 1970) and by that time Lakatos's thoughts had coalesced in the form of his "Falsification and the Methodology of Scientific Research

---

This paper was written for presentation at another Lakatos Centenary Conference—this one held on November 11, 2022 at the University of Sao Paulo, Brazil. I thank the organisers, particularly Osvaldo Pessoa and Miguel Flach, for the invitation to speak.

J. Worrall (✉)
Department of Philosophy, Logic and Scientific Method, London School of Economics and Political Science, London, UK
e-mail: j.worrall@lse.ac.uk

Programmes" (Lakatos, 1970)—an essay which remains his most significant contribution to the philosophy of science. Lakatos believed that, although Kuhn's account was in many respects closer than Popper's to the historical facts about theory change, Kuhn's account failed to make theory-change a rationally explicable process. And Lakatos claimed that his own position—the Methodology of Scientific Research Programmes (hereafter MSRP)—was in effect a sort of "synthesis" of Popper's and Kuhn's views: it acknowledged the greater historical accuracy of the latter while restoring the idea (shared by Popper) that theory-change is a rational process.

The occasion of the centenary of Imre Lakatos's birth seems an appropriate one on which to revisit the "Popper-Kuhn debate" and Lakatos's response to it.

The actual 1965 debate between Kuhn and Popper seems, in retrospect, rather disappointing. Popper in particular shows no signs of really understanding Kuhn's views. He takes it that Kuhn is applauding the role in science of dogma, of closed-mindedness to new ideas. Popper suggests that Kuhn is correct that dogma plays a role but argues that he is wrong to applaud it: normal science, which for Popper means work done while in the grip of a dogma or set of dogmas is "hack science" and even "a threat to civilisation"(!). (Popper 1970) But although Kuhn did write about the "function of dogma" in science, in fact the correct translation of his image of science into a generally Popperian or testing framework is as a re-presentation of Duhem's insight that what gets tested in science is not a single theory, like Newton's theory of gravitation or Maxwell's theory of electromagnetism, but instead a—usually very large—theoretical system built around that theory. When an "anomaly" arises, that is a refutation, not of the single theory at its centre, but instead of some part of the much wider theoretical system, this means only that at least one of the theories in that theoretical system is false. It is surely no more "dogmatic" for a scientist to address the anomaly by holding on to the central theory and looking to modify one of the necessary secondary or auxiliary assumptions, than it would be to insist (as Popper even sometimes seems to be proposing) that the scientist must *always* give up the central theory. Indeed, Kuhn is surely correct (and importantly correct) that, given that the central theory over time develops around itself certain "puzzle-solving techniques" designed in particular to solve exactly the problems caused by anomalies, the "natural" first move for a scientist is to hold onto that central theory and exploit the available puzzle-solving power rather than make a leap into the dark in search of some new paradigm. (Witness Adams and Leverrier holding onto Newton's theory despite the anomalous data from the planet Uranus—a move which led to the discovery of Neptune and forms an episode which Popper himself elsewhere cites as one of the great successes in the history of science.)

On the other hand, some of Kuhn's own contributions to the debates are, I think, equally disappointing. For one thing, he re-endorses a claim that has always struck me as one of the most mysterious in his famous (1962) book: the claim that events like the discovery of X-rays or, still more surprisingly, the discovery of the planet Uranus count for him as "revolutions". Most readers had and have taken it that Kuhnian revolutions are the big (or at least medium-sized) upheavals, involving at least some conceptual rather than merely empirical change: the Copernican,

Newtonian, Relativistic or Quantum Revolutions; along with some other rather less large-scale but still quite radical changes like the switch from the corpuscular theory of light to the wave theory of light in the early nineteenth century or from phlogiston-based to oxygen-based chemistry in the later 18th. But Herschel discovered Uranus simply through careful, one might say obsessional, observation of the night sky: eventually noticing that what had previously been thought to be one of the fixed stars was in fact moving, moving of course very slowly, against the background of the fixed stars. This was an entirely empirically-based discovery involving nothing more radical in terms of change of belief about the universe than the switch from the view that there are six planets orbiting the Sun to the view that there are seven (Neptune and the anyway subsequently "demoted" Pluto having, of course, not yet been discovered when Herschel identified Uranus.) If the discovery of Uranus counts as a "revolution", a "change of paradigm" then I am afraid that I lose all intuitive grip on the concept.

Moreover, Kuhn confesses in his Bedford Colloquium talks that when he is asked of a certain theory-change in science whether it counts as a revolution or not, he "frequently finds [himself] at a loss for an answer" (1970b) p. 251. But surely the whole of *SSR* is premised on there being a sharp distinction between revolutionary change and normal science. Conceding that the distinction is blurred, seems to make the whole position hard to interpret.

Finally, the case that Kuhn presented in his London remarks for the thesis that the central theories involved in successive paradigms are "incommensurable" (1970b) pp. 266–277) seems to me fundamentally and rather obviously flawed. The basis for the argument, so Kuhn explicitly asserts, is the alleged lack of a neutral language into which at least the empirical consequences of the two theories we are concerned to compare can be translated: "The point-by-point comparison of two successive theories" he writes "demands a language into which at least the empirical consequences of both can be translated without loss or change."(1970b, p. 266) Kuhn denies that this condition is met in cases of revolutionary change. On the contrary:

> "In the transition from one theory to the next words change their meanings or conditions of applicability in subtle ways. Though most of the same signs are used before and after a revolution – e.g. force, mass, element, compound, cell – the ways in which [they] attach to nature has somehow changed. Successive theories are thus ... incommensurable." (*op.cit*. pp. 266–267)

Elsewhere, Kuhn claims that the Ptolemaic and Copernican theories are incommensurable because even such an (allegedly) "observational" term as "planet" changed its meaning in that revolution: the Earth is a planet for Copernicus but not for Ptolemy, while the Sun is a planet for Ptolemy but not for Copernicus! But of course, 'force' 'planet' and especially 'mass' are theoretical terms, not observational ones as Kuhn suggests, and so no wonder that their meanings change somewhat alongside theory-change. Surely, however, Kuhn did not dig deep enough in the search for a theory-neutral comparison between the consequences of the pre- and post-revolutionary theories. *Obviously* 'force' 'mass' and 'planet' carry (some limited) theoretical content. However, it is straightforward to go to a lower, more observational level and compare how the two theories involving those notions fare when tested at that at least theory-neutral level. Instead of planets, we can consider particular spots of light in the night sky and whether or not they change their relative (apparent) positions over time. Or, for another example, given that relativity theory asserts that a body's mass can be increased simply by accelerating that body, while in classical physics the mass of a body is of course constant, the notion of mass did indeed change its meaning in the shift from classical to relativistic physics, but that does not mean that we cannot compare classical and relativistic physics in terms of what they predict about the—observable or at least theory-neutral—shifts in visible fringe patterns in the Michelson-Morley experiment, about the observable—or at least theory-neutral—apparent motions of Mercury, or about the observable or at least theory-neutral tracks in a cloud or bubble chamber.

Despite these disappointments, the central point at issue remains a challenging and fascinating one. This is the issue of the rationality (or otherwise) of theory-change in science. Most of us, I suppose, start out pre-reflectively from the Enlightenment view that modern science has enabled humankind to unlock the secrets of the universe. As the English poet, Alexander Pope, famously put it "Nature and Nature's laws lay hid in night, God said 'Let Newton be' and all was light." Most of us start from the position that, where successful, as it surely has been in physics and elsewhere, science has told us more and more about the structure of the universe. The central challenge to this modernist view is exactly the existence of "scientific revolutions". Newton's theory has been replaced by Einstein's—Newton's theory is contradicted in important parts by Einstein's and so it cannot be true if Einstein's is. (The challenge is reflected in Sir John Squire's almost equally famous riposte to Pope: 'Twas not to last, for Devil shouting 'Ho! Let Einstein be' restored the *status quo*.') Perhaps it can be argued that Newton's theory continues to look, in some clear sense, *approximately* true from the vantage point of Einstein's. But however that issue is resolved, if any substantial part of the modernist picture of science is to be retained, it clearly must be shown that, where accepted theories have changed (in "mature science"), the change has been from one good theory to a still better one—'better' in some objective sense. So that, unlike perhaps, in fashion or in styles of art, changes in scientific theory exhibit rationally accredited *progress* rather than simple change.

Kuhn's celebrated (1962) seemed to many commentators to be challenging that view. Kuhn was interpreted as saying that scientists for the great majority of the time do not (perhaps even cannot) question their paradigm, believing that any experimental or observational anomalies will eventually be dealt with within that approach. However sometimes anomalies accumulate and prove recalcitrant to attempts to "normalise" them. Eventually, a feeling of "crisis" affects the scientific community built around the paradigm; but there are no rules about how many anomalies or how recalcitrant the anomalies must prove to be in order to justify the feeling of crisis. There is, as Kuhn explicitly and repeatedly stated, no criterion higher than the community view and the community either feels a sense of crisis or it does not. If it does, then it will look for another paradigm and switch to it when found: that switch being analogous, he claimed, to a religious conversion rather than anything objectively rule-governed. Furthermore, the conversion will never be made by the whole of the relevant community: in any revolution there are hold-outs, usually elderly scientists who have made significant contributions to the older, pre-revolutionary paradigm and stick to that older paradigm perhaps long after their more mobile colleagues have shifted to the new paradigm; according to Kuhn, those hold-outs cannot be judged to be mistaken or to be failing to make the rational choice. They simply lose the vote, so to speak, and thus eventually either die or define themselves out of the relevant scientific community. Finally, because methodological standards are themselves paradigm-dependent and so also subject to change in revolutions, there is no neutral basis on which we can judge the theories involved in the new paradigm as better than those involved in the old.

Lakatos, echoing other philosophers such as Scheffler (1967) and Shapere (1964), but characteristically expressing it more abrasively, claimed that this account of theory-change in science by Kuhn reduced it to a question of "mob psychology". As noted earlier, Lakatos famously claimed that his MSRP, while conceding that many aspects of the process of science are better described by Kuhn than by Popper (for example, experimental difficulties in science are generally treated more like Kuhnian anomalies than Popperian refutations), saves the rationality of science from Kuhnian relativism by—allegedly—showing that one theory, or rather in Lakatos's terms, one research programme is only ever replaced in science by one that is objectively superior to it in terms of how it stands up to the evidence.

What progress was made in this debate during the 1965 Bedford College discussions? Well, the first thing to be noted is that Kuhn, in responding to his critics, heatedly denied the "mob psychology" charge and in effect insisted that his account is in no need of any injection of rationality from Lakatos or anyone else: it is already an account that gives a central role to rationality. "Does anything in [my] argument" he asks "suggest the appropriateness of phrases like decision by 'mob psychology'?" And he answers: "I think not. "(Kuhn, 1970b, pp. 262–263) Indeed, he continues, "no part of ... [my] ... argument implies that scientists may choose any theory they like so long as they agree in their choice and thereafter enforce it" (*op. cit.* p. 263) Far from the adoption of a new paradigm being, on his account, "mystical" or purely sociological, that account insists that there are "good reasons for theory choice" or better, that good reasons are always involved in choosing the new theory/paradigm (*op. cit.* p. 261). Moreover, says Kuhn, "these are ... reasons of exactly the kind standard in philosophy of science: accuracy, scope, simplicity, fruitfulness and the like".(*ibid.*)

So did Kuhn in effect claim that, when his views are correctly understood, there is no real issue between him and what we might call his objectivist critics as regards theory-change and rationality? No: there still is a difference—one that *seems* at least to be a major one—, and it lies in what Kuhn's account *denies*:

> "What I am denying ... is neither the existence of good reasons nor that these reasons are of the sort usually described. I am, however, insisting that such reasons constitute values to be used in making choices rather than rules of choice." (*ibid.*)

This has the practical consequence that "Scientists who share [belief in the 'good reasons'] may nevertheless make different choices in the same concrete situations." Reason—in the form of the "objective factors" of traditional philosophy of science (empirical accuracy, simplicity, "and the like")—certainly plays a role but it never *dictates* the switch to the new paradigm. Consequently, on Kuhn's view, it is never actually irrational to resist the switch to that new paradigm. There is, claims Kuhn, no "point at which resistance becomes illogical or unscientific". An "elderly" holdout, like Priestley holding out for phlogiston against Lavoisier's oxygen theory, may infuriate his colleagues by what they see as his stubbornness, but cannot legitimately be regarded as mistaken or "irrational".

As just indicated, Kuhn's claims about the failure of holdouts to be irrational were already to be found in *Structure* but in his Bedford College replies to critics he is more explicit about the arguments behind those claims and in particular the relative roles of the good reasons (or "objective or shared factors") and other, "subjective" (or individual) factors in theory choice.

Kuhn cited two ways in which reason in the form of the allegedly standard factors may fail, and invariably or almost invariably *does* fail, to dictate a particular choice of theory or paradigm. The *first* is that two individual 'good reasons' or 'objective factors' may point in different directions: one of them indicating a preference for theory 1 over theory 2, and the other a preference for theory 2 over theory 1. (An alleged example that Kuhn cites more than once is that, at the time it was adopted by Kepler and Galileo, Copernican theory was simpler than Ptolemaic theory; but on the score of detailed empirical accuracy, the Ptolemaic theory was better.) Kuhn writes:

"In many concrete situations, different values, though all constitutive of good reasons, dictate different conclusions, different choices. In such cases of value conflict (e.g. one theory is simpler, the other is more accurate), the relative weight placed on different values by different individuals can [legitimately] play a decisive role in individual choice." (1970b, p. 262)

The *second* way in which 'good reasons' may fail to determine a choice of theory, according to Kuhn, is that individual scientists may—again legitimately on his view—come to different judgments about how an *individual* objective factor applies in a particular case of theory-choice. He writes (*ibid.*):

"More important, though scientists share these values [dictated by 'good reason'] and must continue to do so if science is to survive, they do not all apply them in the same way. Simplicity, scope, fruitfulness and even accuracy can be judged quite differently (which is not to say they may be judged arbitrarily) by different people. Again, they may differ in their conclusions without violating any accepted rule."

It is noteworthy that, while as indicated, Kuhn does cite examples that he sees as exemplifying the first kind of indeterminacy—two objective factors pointing in different directions; so far as I can tell, he cites no examples of the second kind— one objective factor being "interpreted differently" and being *reasonably* interpreted differently.

But, be that as it may, Kuhn's account is definitely, then, at any rate somewhat more nuanced than his early objectivist critics were allowing. The charge of making scientific theory-change a matter of 'mob psychology' does not stand—at any rate not without further elaboration aimed at showing that Kuhn's account still makes scientific theory-change a non-rational affair despite his insistence that objective factors play an ineliminable role in such theory-changes.

Lakatos in fact never replied to this more nuanced Kuhnian account. Of course, there was no opportunity for such a reply within the structure of the Bedford Colloquium debate and hence within the structure of the book *Criticism and the Growth of Knowledge*. That debate—both at the conference and in the book—began and ended with Kuhn; and Lakatos did not "cheat" by presaging Kuhn's reply within his own contribution. It is, nonetheless, perhaps a bit surprising that Lakatos never took the opportunity to reply later—though admittedly, and sadly, he did not have *very* long in which to do so: Kuhn submitted his 'Replies to Critics' only just before *Criticism and the Growth of Knowledge* was published in 1970 and Lakatos died in February 1974.

I will try here to make good on this omission by speculating on how Lakatos might have responded had he ever directly confronted this more elaborate Kuhnian view.

I have no doubt that Lakatos's first reaction would have been that the charge of "mob psychology" or, to put it less tendentiously, of reducing theory-choice in science to a merely sociological affair still stands—indeed that the charge is conceded and confirmed by Kuhn's elaborated account. But to see why, we have to get clear about what exactly it is that Lakatos expects from a methodology that Kuhn's account failed to yield, even in the elaborated form that I have outlined.

A methodology, for Lakatos, needs to produce an objective ordering of theories in the light of the empirical evidence, an ordering that—at least in all normal cases— places the winning side in any scientific "revolution" higher than the deposed theory/paradigm/research programme. It is crucial here to distinguish the objective ordering that Lakatos was looking for and which was to be a denizen of Plato's or Frege's or Popper's objective or logical "World 3" (see for example, Popper, 1972, Chap. 3) from any issues about individual scientists' beliefs or joint beliefs of scientific communities or their decisions about which theories to work on (these are denizens of Popper's psychological "World 2" (*op.cit.*)). Kuhn's account of what he called theory-choice insists, as we saw, that the choices are always dependent on individual or subjective factors as well as shared or objective factors and hence it entails that the theory eventually accepted in any scientific revolution was *not objectively, scientifically superior* to its predecessor theory in the sense that Lakatos believed was required. This being reflected most clearly in Kuhn's insistence that holdouts to the revolutionary theory were not wrong or irrational.

Lakatos cannot, however, here resort to "mob psychology"-style name-calling but must argue that there is something wrong with Kuhn's account of theory-choice and in particular with his account of the objective (or shared) factors underpinning such choices. When correctly identified, objective factors do, Lakatos must argue, determine which is the better of any two theories/research programmes at any particular time and in particular determine theory-preference (for of course the winning theory) at the time of a "revolution".

Kuhn, remember, insisted that his objective factors are ones "standard in philosophy of science", and he listed them several times as "accuracy, scope, simplicity, fruitfulness and the like". He also sometimes added "consistency (both internal and with other accepted theories)" as a further objective factor. Contrary to his claims that these criteria are "standard in the philosophy of science", I in fact know of no philosopher of science of an objectivist kind who would endorse all the items on Kuhn's list as it stands and none who would be happy to leave such a list unstructured as Kuhn does, rather than attaching differing degrees of importance to different factors.

For Lakatos, there is, of course, a dominant criterion which does not even appear on Kuhn's list of "objective factors" (at least it doesn't appear explicitly): namely independent testability and predictive success. This is, for Lakatos, *the* criterion of a progressive research programme: a programme is progressive if, and only if, successive theories produced by it make testable predictions, independent of any empirical results used in the construction of those theories; and at least some of the time those predictions are empirically verified.

Unlike predictive success, which clearly requires the cooperation of Nature, empirical accuracy and scope, which *are* on Kuhn's list, are readily manufactured by scientists: once they know the facts, scientists can readily find a place for them in a system based on any central theory you care to specify. This is a consequence of Duhemian "underdetermination of theory by data" (Duhem, 1906). If one paradigm shows greater empirical accuracy or scope than another, this, then, is standardly a merely historically contingent state of affairs reflecting the lengths of time that the two paradigms have been worked on. Hence, contrary to Kuhn's view, empirical scope/accuracy supplies on its own no telling reason to prefer one paradigm over the other.

For example, the greater empirical accuracy and scope of Ptolemaic theory in the years shortly after the publication of Copernicus's *De Revolutionibus* is cited by Kuhn as an 'objective reason' to choose Ptolemaic theory over Copernican. In fact, however, that greater empirical accuracy and scope is no surprise: it is provable (and anyway obvious) that all the empirical astronomical data then available—the apparent motions of fixed stars, the sun and the known planets—can, with sufficient ingenuity, be fitted within *either* a heliocentric (more accurately, helio*static*) system *or* a geocentric (again more accurately, geostatic) system. Ptolemy started to plot apparent astronomical motions within his geostatic system in the second century AD (and the roots of the geostatic approach go back still further in the Greek, Roman and Babylonian traditions). Copernicus's system, by contrast, was published shortly before his death in 1543. It is therefore no wonder at all that, when Galileo and Kepler began to think about these matters, the Ptolemaic system was ahead in terms of the number and accuracy of the phenomena it had brought within that system.

Kuhn's explicit view, remember, was that the choice between the Ptolemaic and Copernican systems at the time of Kepler and Galileo was not determined by the "objective factors" because while empirical accuracy/scope told in favour of Ptolemy, simplicity (in, as Kuhn puts it, "a special sense") told in favour of Copernicus. But, as we just saw, empirical accuracy/scope arguably carries no weight in underwriting any preference.

As for the special sense of simplicity that Kuhn refers to, this again in fact reflects Lakatos's supreme predictive success criterion. "Simplicity" and "unity"—in the scientifically important sense of these terms—are closely related to predictive success. There are surely no clear-cut intuitions about when one basic theory in science is simpler than a rival. Is, for example, the basic idea of a fixed earth—ahead of any detailed elaboration—simpler or more complicated than the basic claim that it is the sun that is motionless? Or is the idea that light consists of material particles more or less simple than the idea that it consists of waves in a medium? I don't see the slightest reason to think that there's an answer either way. Where we *do* have clear-cut intuitions is in cases where one basic theory has been so hedged around with qualifications and split into so many unrelated subcases that it clearly becomes too complex, not sufficiently simple, to be scientifically acceptable. But, in all such cases, the complexity and disunity have been introduced under the pressure of initially independent or anomalous experimental results. The basic theory has enjoyed no predictive success: it has either turned out to be silent about some phenomenon clearly in its field, or, more often, turned out to yield an incorrect prediction. Special cases and exceptions have therefore had to be introduced to accommodate the facts—at the cost of increased complexity and decreased unity. In short, the theory's becoming complex means, in Lakatosian terms, that the associated research programme has degenerated. A theory's remaining simple and unified means that the associated research programme has progressed.

This is the "special sense of simplicity" that so impressed Kepler and Galileo about Copernicus's theory: phenomena such as planetary stations and retrogressions, or the bounded elongation of Mercury and Venus had to be "worked into" the Ptolemaic theory courtesy of special assumptions—principally of course assumptions about epicycles specified exactly using features of those already known phenomena. By stark contrast, the phenomena of stations and retrogressions and of bounded elongation (and also of the order of the planets in terms of distance from the central, fixed body) fall naturally out of the Copernican approach: they follow from the basic model, without the need for any special *ad hoc* assumptions.

The two other items on Kuhn's unstructured list of objective factors in theory choice are "fruitfulness" and consistency. Under the only precise sense I can make of it, fruitfulness too is intimately connected to simplicity and hence to predictive success (and hence to progressiveness of a research programme). A general theoretical approach (a paradigm or research programme) shows its fruitfulness by supplying ideas for developing specific theories independently of empirical results. Such an approach will be judged on the contrary, barren or lacking in fruitfulness (as Lakatos put it, the research programme's "heuristic" will have "run out of steam"') only when all these ideas have been tried without predictive success; and hence the approach has been reduced to tagging along behind the empirical data, always accommodating that data *post hoc* rather than predicting it in advance.

By the early to mid-1830s, for example, the emission or corpuscular approach to optics had very definitely proved barren—its former "fruitfulness" lay exhausted. The ideas supplied by the general claim that light is a Newtonian particle had all been tried in the attempt to produce specific theories that successfully dealt with optical phenomena. Particles are, of course, in the classical picture, subject to forces; forces could be attractive or repulsive: all the apparent deviations from rectilinear propagation of light—reflection, refraction, interference, and diffraction—*might* be explained by having ordinary "gross" material objects exert forces of various kinds on the light particles. (This was, in essence, the corpuscularist or emissionist programme.) The idea that the "particles" of light are strictly *point* particles always had to be an idealization; so the finite dimensions of the real particles might come in useful: it might for example be proposed that the particles have sides or poles and revolve with respect to these poles as they move along, affecting the way those particles react to the various forces supposed to be exerted on them. Various isolated results could be explained (in *very* rough terms) on the basis of these assumptions—but, when it came to anything like details, the "natural" assumptions about the forces and the polar revolutions unambiguously failed and instead the required theoretical assumptions had always to be "read off" the already given facts. There was never any correct prediction of a phenomenon different from those used in the construction of particular corpuscularist theories. Instead, each new phenomenon required further elaboration of the theoretical assumptions (perhaps another complication in the field of force set up by the diffracting or refracting body or yet another axis of revolution in the particles). As the optical scientist Humphrey Lloyd put it in a famous report on the "Progress and Present State of Physical Optics" produced in 1833:

> "An unfruitful theory may ... be fertilized by the addition of new hypotheses. By such subsidiary principles it may be brought up to the level of experimental science, and appear to meet the accumulating weight of evidence furnished by new phenomena. But a theory thus overloaded does not merit the name. It is a union of unconnected principles.... Its very complexity furnishes a presumption against its truth.... The theory of emission, in its present state, exhibits all these symptoms of unsoundness, ... "(1833, p. 296)

By contrast, there existed within the general wave theoretical approach at the same time (the early to mid-1830s) some hopeful lines of attack on the problems it faced. One such problem emphasised right from the beginning by its opponents, concerned the phenomenon of prismatic dispersion. According to Fresnel's initial theory, the amount of refraction a ray of light undergoes when entering a transparent substance should be dependent only on the refractive index of the substance (or, more properly for the wave theory, the refractive index of the ether as structured within that substance). Hence it entails that if a ray of sunlight enters a transparent body it will be refracted as one ray. But in fact of course a beam of sunlight when, for example, passed through a glass prism spreads out into the familiar spectrum—a phenomenon that had been extensively studied by Isaac Newton, as reported in his *Opticks*. But this initial version of the wave theory of light was based on a very simple theory of the ether—one that involved the assumption that its parts *strictly* obey Hooke's law of the direct proportionality of restoring force to displacement. It was known from studies in mechanics, that not all vibrations in all substances strictly obey Hooke's law and several general ideas were already available concerning how a somewhat more sophisticated theory of the luminiferous ether involving a slightly more complicated expression for the restoring force could be constructed that might yield dispersion. Though none of these had yet borne unambiguous fruit, equally they had not all unambiguously run into sand. This is just a fact about the wave approach: it already possessed potential explanatory resources with respect to dispersion that had not been exhausted in the 1830s.

Moreover, again in contrast to the corpuscular programme, the wave theory of light in the 1830s already had an impressive record of success—in the shape of shifts of theory that *had* proved significantly predictively successful. Wave theorists before Fresnel had all assumed that the ether is an extremely rare and subtle fluid—how else could the planets move so freely through it? It is a theorem of mechanics that fluids transmit only longitudinal (sometimes called pressure) waves. (Longitudinal waves are ones in which the particles of the medium oscillate in the same direction as the overall transmission of the wave through the medium; an example being a sound wave in air.) Fresnel's own initial theory was indeed that light is a longitudinal wave. However, he and his colleague Arago then established experimentally that if, say, the two beams emerging from the two slits in the double-slit experiment are polarized at right angles to one another (by passage through suitably oriented crystal plates), then the interference fringes disappear. It seemed that light beams polarized in mutually orthogonal planes fail to interfere (or, rather, fail to produce interference fringes). Neither Fresnel nor any other wave theorist had, at this stage, any coherent theory of the polarization of light. But, so long as the light waves were assumed longitudinal, the precise account of what happened when light is polarized could make no difference. Assuming that the wave theory is at all correct, the longitudinal assumption alone means that the disturbances in the two coherent and near-parallel beams (the slits are, remember, very close together) must themselves be near parallel and hence must alternately interfere constructively and destructively for different path differences. The Fresnel-Arago experiment which resulted in no interference fringes, therefore, put the wave theory into deep trouble. Fresnel took a still deeper breath and switched to the *transverse* wave theory: to the theory that the ether particles oscillate *at right angles* to the direction of the propagation of light. This yields an easy theoretical account of the process of polarization: the disturbance in an unpolarized beam has components in all planes through the direction of propagation; polarization (linear or plane polarization, that is) consists in restricting the disturbance to one such plane. This explained the apparent "sidedness" of polarized beams, and it also explained the Fresnel-Arago observations. The oscillations in beams that are polarized orthogonally are assumed themselves to be orthogonal. Hence, although the two sets of oscillations certainly interfere or superpose—to produce, in general, elliptically polarised light—they operate at right angles rather than along the same line, and hence can never destructively interfere so as to produce fringes.

Although it straightforwardly dealt with this difficulty over polarized light, the switch to the transverse theory certainly required a deep breath. This was because elastic media can transmit such waves only if they exhibit resistance to sheer, that is, only if they are solids. But how could the planets move completely freely through an elastic solid ether? But whatever the conceptual difficulties, Fresnel's new transverse theory scored stunning empirical successes. Not least when Hamilton showed in 1830 that the transverse theory entails the hitherto entirely unsuspected phenomena of internal and external conical refraction—predictions that were dramatically confirmed by Humphrey Lloyd in 1833.

So "fruitfulness" is again unambiguous: the wave theory was fruitful—it specified avenues of research that had not yet been exhausted and it had a track record of change accompanied by predictive success; the corpuscular theory was not fruitful—it had no unexhausted avenues of development and no track record of predictive success. And again one of Kuhn's "objective factors"—this time "fruitfulness"—crucially involves, when analysed, predictive success.

The remaining item on Kuhn's list of objective factors involved in theory-choice is "consistency (both internal and with other accepted theories)". Well, internal consistency is obviously a logical requirement—no one can "choose" (to use Kuhn's term) an inconsistent theory since that theory, by definition, contradicts itself: so, by accepting it you would also be accepting its negation! (Lakatos has some deep-sounding but in fact rather sloppy remarks about scientists sometimes fruitfully proceeding on "inconsistent foundations" but this always means that those scientists are at least dimly aware of how any inconsistency can be rectified and confident that their positive results will be recoverable within the eventual consistent version of the theory.) So, the interesting question is whether or not consistency between some new theory and ones that are already "established" should be treated as an objective factor in theory-choice, a theoretical virtue that can legitimately count in favour of preferring that theory.

Kuhn of course asserts that it does and again cites the Copernicus/Ptolemy case as one in which it plays a role: while "simplicity in a special sense" counted in favour of Copernicus, not only did empirical scope/accuracy, according to Kuhn, justify a preference for Ptolemy, as discussed earlier, so also did the fact that Ptolemy was consistent with other theories considered well-established at the time—notably Aristotelian physics, while Copernican theory was clearly inconsistent with that physics. But surely this inconsistency was a *virtue* of the Copernican theory, not a vice. The inconsistency supplied an interesting and demanding problem for further research duly addressed by Galileo and later by Newton, indicating the need to develop a different physics to that of Aristotle. Of course, this judgment is premised on the fact that Copernican theory was predictively successful (with, as we saw, planetary stations and retrogressions and the bounded elongations of Mercury and of Venus), while Aristotle's physics had become "well-established", sociologically speaking, despite never enjoying any such predictive success. Scientists in general do, no doubt correctly, downgrade (or more usually ignore) new theories that clash with well-established ones—*but only* when there is no independent evidence for the new theory. To take a relatively trivial but illustrative example, the theory that homeopathic "remedies" are effective (really "more effective than placebo") is multiply inconsistent with accepted theories in physics. This fact is correctly taken as strong evidence against homeopathy, but only because those theories in physics are supported by predictive successes while the hypothesis that homeopathic "remedies" are more effective than placebo has no such support. On the other hand, if I am right that Aristotelian physics had only historical, but no evidential legitimacy, then it follows that it is the preferred candidate for replacement given its inconsistency with the predictively successful Copernican theory, not vice versa: just as Kepler, Galileo and Newton recognised. It is predictive success that flips inconsistency with other accepted theories over from a vice to a virtue.

In sum, then, all the 'objective' criteria that Kuhn cites, either play no real role in theory-preference or reduce to Lakatos' single criterion of progressiveness. On Lakatos's account, in stark contrast to Kuhn's, there is essentially only one criterion of scientific merit and hence there is no possibility of the sort of clash between different objective criteria of scientific merit that Kuhn's Bedford College account sees as requiring individual or subjective factors of theory "choice" to resolve. Moreover, that one criterion is definite: either a research programme makes independently testable predictions some of which are confirmed or it does not. So again the space is not there for subjective factors to play a role—this time in resolving varying applications of single objective factors.

Notice that it is, of course, not true that Lakatos's account *always* underwrites a preference in any dispute between two theories/research programmes. In particular, it might well be the case at some stage in the history of science that neither of two competing programmes is progressive—this was for instance true of Newton's and Hooke's contrasting approaches to optics in the mid- to late- seventeenth century: neither Newton nor Hooke could do any better than accommodate already known phenomena post hoc within his preferred framework. And so the choice between those two frameworks at that time was indeed subjective. But no revolution occurred in optics in the mid- to late-seventeenth century—the scientific community was divided between the two available theories, and when the revolution did occur in the early nineteenth century, the wave programme—now led by Fresnel—was definitely progressive, while the corpuscular programme had definitely degenerated. So, in contrast to Kuhn's account, Lakatos's approach yields a rationalist explanation of the development of science: every change, every scientific revolution has constituted progress—the theory, or rather research programme, displaced in the revolution had degenerated while the new, superseding theory/research programme had proved progressive.

So, this may look like the end of story: even on the amended version that he developed in his London remarks Kuhn's account does make scientific theory change too subjective an affair for the tastes of objectivist philosophers, but Lakatos produced an alternative account that is equally sensitive to the history of science while restoring the objectivity of theory change.

However, my guess is that Kuhn would have regarded this as a pyrrhic victory for Lakatos. The latter was quite clear, especially in his PSA 'Replies to Critics' paper of (1971), that his objective appraisals of the current merits of rival programmes in the light of evidence have no consequences either for scientists' beliefs about which theory, if either, is true or, more significantly for current purposes any consequences for which theory/programme it is rational to work on. He writes (1971, p. 176):

"... my methodology .... only appraises fully articulated theories (or research programmes) but it presumes to give advice to the scientist neither about how to arrive at good theories nor even about which of two rival programmes he should work on."

And he goes on to emphasise (*ibid.*):

"... when it turns out that, on my criteria, one research programme is 'progressing' and its rival is 'degenerating', this tells us only that the two programmes possess certain objective features but does not tell us that scientists must work only in the progressive one. (Indeed, as I constantly stress, degenerating research programmes can always stage a comeback .... But this would, of course, be impossible if no scientist 'worked' on the programme.)"

My guess is that Kuhn found it difficult to see any content at all in Lakatos's "objective appraisals" if they have no implication for what scientists should and should not believe and do in particular situations. (Indeed Kuhn's use of the term 'theory-*choice*' reflects the fact that he is entirely focussed on scientists' decisions—denizens of Popper's psychological 'world 2' rather than of the objective, logical World 3). Lakatos may be right that his "objective criteria" always rank the new programme in any revolution ahead of the old one at the time that the "revolution" occurred, but it is difficult not to have at least some sympathy for Kuhn's explicit view that if Lakatos's methodology has no advice for scientists then it "has told us nothing at all"—a view that was famously echoed by Paul Feyerabend: "Scientific method, as softened up by Lakatos, is but an ornament which makes us forget that a position of 'anything goes' has been adopted" Feyerabend (1970).

Are Kuhn and Feyerabend correct? Well, the issue of the connection, if any, between Lakatosian objective appraisals and the rationality of scientists is certainly not straightforward: philosophers of science have for a long time insisted on a distinction between the 'logic of acceptance' and the 'logic of pursuit'. And for good reason: no one, for example, should ever have supposed that Lakatos should endorse the view that it is rational to work on a research programme if and only if it is progressing. Amongst other defects, that thoroughly naïve rule would pronounce the great innovators in science "irrational". Was the wave optics research programme progressive when Fresnel first chose to work on it? Of course not, it was Fresnel's work on it that made it progressive. Was the relativity programme progressive when Einstein first chose to work on it? Of course not, it was Einstein's work on it that made it progressive. And so on.

On the other hand, if, as Lakatos insisted, the only thing that a scientist in a situation of choice between two programmes needs to do to count as rational is to acknowledge the "current score" between the two and then can choose to believe, and more importantly for present purposes, choose to "work on" the programme with the lower score (the one that is degenerating), then this does indeed seem to provide only a very thin theory of rationality at very best. Most philosophers of science have taken it that in order for a methodology to count as one that makes the actual development of science a rational process, it not only has to yield the consequence that those scientists who accepted the new theory in a revolution were right, but also the consequence that the "hold outs" who continued to adhere to (and try to work on) the old theory were wrong. As we saw, one of the central reasons for counting Kuhn's view as unacceptable for a rationalist about science was the fact that it delivers the verdict that the hold-outs cannot legitimately be characterised as wrong or irrational.

However, depending on exactly what is meant by "continuing to adhere" to the older theory, it seems that Lakatos's view may share this feature. All that Priestley, for example, needs to do to count as "Lakatosian-rational", it seems, is to admit that Lavoisier's oxygen theory was ahead in terms of the support it receives from the phenomena, but then go on to insist that he will continue to work on the phlogiston approach with the intention of turning the evidential tables and eventually making the phlogiston theory the better evidentially supported theory.

Of course, scientists in the history of science did not express themselves in explicitly Lakatosian terms, and I do not know enough about Priestley to judge whether he would have been willing to make this concession (though I suspect, since it is such a minor concession, that he surely would have). However, there is another hold out against a scientific revolution whom I do know well, having studied his work in depth (see, e.g., Worrall (1990)); and, so far as he goes, the situation is clear.

The "hold out" I refer to is David Brewster. Brewster was a significant optical scientist of the early- to mid-nineteenth Century. He was the discoverer of a great many of the properties of polarized light, especially elliptically polarized light; he discovered "Brewster's law," relating the polarizing angle and refractive index of transparent substances; he discovered a whole new class of doubly refracting crystals, the "biaxal crystals"; he discovered that ordinary unirefringent transparent matter can be made birefringent by the application of mechanical pressure; and he discovered the then unknown general phenomenon of selective absorption.

As well as a significant scientist, Brewster was certainly some sort of holdout for the corpuscular or emission theory of light—even though Fresnel's earlier work had made the wave optics programme unambiguously progressive on Lakatos's criteria, while the corpuscular programme had, by Brewster's time, unambiguously degenerated.

In 1831, Brewster presented a "Report on the Present State of Physical Optics" to the British Association for the Advancement of Science, in which he asserted that the undulatory theory is "still burthened with difficulties, and cannot claim our implicit assent," (Brewster, 1833a, p. 318). And in 1883 he reported that: "I have not yet ventured to kneel at the new shrine [that is, the shrine of the wave theory] and I must acknowledge myself subject to the national weakness which urges me to venerate, and even to support, the falling temple in which Newton once worshipped." (1833b, 361) (The corpuscular programme was of course traditionally regarded as having been invented and supported by Newton.)

Brewster believed that, despite all the difficulties that had mounted against it, there was life left in the Newtonian emissionist theory. He echoed and endorsed Herschel's sentiment expressed some 10 years earlier that, were sufficient talent and energy invested in the emission theory, it might yet turn the tables of scientific superiority on its undulatory rival.

But Brewster, while continuing to recommend work on the corpuscular programme, did very definitely accept that the wave optics programme was unambiguously ahead in terms of the objective support it received from the empirical data. He wrote, for example.

> "I have long been an admirer of the *singular* power of this [wave] theory to explain some of the most perplexing phenomena of optics; and the recent discoveries of Professor Airy, Mr Hamilton and Mr Lloyd afford the finest examples of its influence in predicting new phenomena." (1833b, 360; my italics)

The reference to Lloyd and Hamilton here concerns the episode already mentioned: Fresnel switched from the longitudinal to the transverse theory; Hamilton showed that the transverse theory entails the entirely unexpected phenomena of conical refraction; and finally Lloyd experimentally verified these predictions.

So, Brewster the hold-out would definitely have counted as rational on Lakatos's view: he accepted the "objective [current] score" was in favour of his wave opponent, but, reflecting Lakatos's concession that "degenerating research programmes can always stage a comeback", Brewster continued to encourage work on his favoured corpuscular approach.

So, what is the conclusion of this long and rather convoluted story? I have imagined Lakatos and Kuhn continuing their debate starting from Kuhn's replies to critics at the Bedford College Colloquium. I have argued that, although it might seem that Kuhn's insistence that objective factors always play a role in what he calls "theory choice" was a conciliatory move in the debate, Lakatos could in fact successfully argue that Kuhn misidentified the objective factors: there is in fact at root only one objective factor—progress and degeneration. And that objective factor always pronounces the winning theory in any case of scientific change objectively superior to the displaced theory.

However, Kuhn clearly regarded Lakatos's objective theory preference as in effect just so much hot air. If we concentrate on what Kuhn held really matters so far as rationality is concerned, namely scientists' beliefs and their consequent decisions about which paradigm/programme to try to develop, then Lakatos, through his concession that it is always possible for a degenerating programme to make a comeback and turn the evidential tables on its rival, automatically further conceded that subjective or individualist factors always play a role in any decision about which theory a scientist "chooses".

Fresnel and many others regarded the wave optics programme as progressive and therefore superior to its degenerating corpuscularist rival, and chose to continue to work on it in the attempt to make it even more progressive. Brewster and a *very few* others, accepted that the wave optics programme was predictively progressive and the corpuscular programme degenerating and so accepted that the wave programme was, in Lakatosian terms, objectively superior as things currently stood, but nonetheless chose to work on the corpuscular approach in the hope, perhaps even expectation, that it would eventually become even more progressive than its rival. Both Fresnel and Brewster were perfectly rational according both to Kuhn *AND—perhaps more surprisingly—to Lakatos*. It seems in the end, and rather disappointingly, that both Lakatos and Kuhn were right.

Having meticulously stuck to the task of interpreting the debate between Kuhn and Lakatos, let me end by indicating my own view—albeit very briefly. While disagreeing with Kuhn that Lakatos's objective appraisals telling us "nothing" if not connected at all to advice to scientists, I think that Lakatos could have done better: I do not agree with his claim that his appraisals, when properly articulated, do not have any consequences concerning scientists' decisions about which programmes to work on. More particularly, I do not agree with Lakatos's famous remark that 'degenerating research programmes can always stage a comeback'. Sometimes it is correct to allow that they might stage a comeback; generally it clearly is not.

To see why, let's return to my "hold out" Sir David Brewster and his view that the monopoly enjoyed in his time by the wave optics programme was a mistake and that if sufficiently many, sufficiently talented scientists worked on the corpuscular programme, it could turn the evidential tables, "stage a comeback" in Lakatos's phrase. The fact is that it is entirely unclear what "working on" the corpuscular programme in the 1830s would have involved. The crucial factor, and what after all was meant to be special in analysing science in terms of research programmes rather than just theories, is the *heuristic*. The heuristic of the corpuscular programme was essentially to exploit the already massively developed mechanical theory of particle motion. This heuristic therefore supplies an array of factors whose variability might be exploited to explain optical effects: masses and velocities of the different particles of light, suppositions about the forces acting on those particles in different situations, de-idealisations from point particles to particles with finite dimensions, perhaps with something like magnetic poles. All these ideas had been tried and had not even moved the programme toward anything like adequate theories of basic phenomena such as reflection, partial reflection and refraction, let alone diffraction and interference. No predictive success had been scored and none was remotely in sight. The heuristic was objectively exhausted: there were no ideas left to try. A scientist who followed Brewster's advice to choose to work on the corpuscular optics programme in the early nineteenth century would, therefore, be entirely at a loss as to what actually to do. On this extended Lakatosian analysis, Brewster (and I would strongly conjecture other 'elderly holdouts') were, contrary to Kuhn's view wrong and irrational.

So, a much less "thin" account of the rationality of theory-change in science than the one officially endorsed by Lakatos in his 1971 "Replies to Critics" can, I think, be developed; but can be developed using his ideas—the criteria of progress and degeneration are of course involved, but so also should be the crucial, but underdeveloped idea of the heuristic appraisal of programmes (which he alluded to many times in his work but seemed to be forgetting about in his 1971 "Replies to Critics" paper). An appraisal of the remaining heuristic power of a programme at any stage should be part of the objective appraisal of its merits at that stage. A programme may be degenerating at some particular time, but still have unexhausted heuristic resources. This would I think be the correct appraisal of corpuscular optics in 1666 when Newton was working on it. If so, then the programme might definitely stage a comeback -and further work on it was therefore reasonable. But if a programme is both degenerating *and* its heuristic is exhausted, then there is no sensible work to be done on it and so to choose to work on it would definitely be irrational: there is then no possibility of the programme's staging a comeback.

So the main improvement that I think is necessary in MSRP is a fuller account of heuristic progress and degeneration in science somewhat analogous to what Lakatos provides in the case of progress in mathematics in his *Proofs and Refutations.*

However, the fact that, a hundred years after his birth, and nearly fifty years after his death, we are still debating how to improve on Lakatos's ideas is a reflection of just how significant those ideas are.

References

Brewster, D. (1833a). Report on the recent Progress of optics. *British Association for the Advancement of science. Report of the first and second meetings 1831 and 1832.*
Brewster, D. (1833b). Observations on the absorption of specific rays in reference to the Undulatory theory of light. *Philosophical Magazine 3rd Series, 2,* 360–363.
Duhem, P. (1906). *The aim and structure of physical theory. (English edition).* Princeton University Press, 1954.
Feyerabend, P. K. (1970). Consolations for the specialist. In Lakatos & Musgrave (Eds.).
Kuhn, T.S. (1962). The structure of scientific revolutions.
Kuhn, T. S. (1970a). Logic of discovery or psychology of research? In Lakatos & Musgrave (Eds.).
Kuhn, T. S. (1970b). Reflections on my critics. In Lakatos & Musgrave (Eds.).
Lakatos, I. (1970). Falsification and the methodology of scientific research programmes. In Lakatos & Musgrave (Eds.).
Lakatos, I. (1971). Replies to critics. In R. C. Buck & R. S. Cohen (Eds.), *PSA 170. Boston studies in the philosophy of science* (Vol. 8). Reidel.
Lakatos, I., & Musgrave, A. E. (Eds.). (1970). *Criticism and the growth of knowledge.* Cambridge University Press.
Popper, K. R. (1970). Normal science and its dangers. In Lakatos & Musgrave (Eds.).
Popper, K. R. (1972). *Epistemology without a knowing subject. Chapter 3 of objective knowledge.* Clarendon Press.
Scheffler, I. (1967). *Science and subjectivity.* Bobbs-Merrill.
Shapere, D. (1964). The structure of scientific revolutions. *Philosophical Review, 63,* 383–384.
Worrall, J. (1990). Scientific revolutions and scientific rationality: The case of the 'elderly hold-out. In C. Wade Savage (Ed.), *Scientific theories.* University of Minnesota Press.

Open Access This chapter is licensed under the terms of the Creative Commons Attribution-NonCommercial-NoDerivatives 4.0 International License (http://creativecommons.org/licenses/by-nc-nd/4.0/), which permits any noncommercial use, sharing, distribution and reproduction in any medium or format, as long as you give appropriate credit to the original author(s) and the source, provide a link to the Creative Commons license and indicate if you modified the licensed material. You do not have permission under this license to share adapted material derived from this chapter or parts of it.

The images or other third party material in this chapter are included in the chapter's Creative Commons license, unless indicated otherwise in a credit line to the material. If material is not included in the chapter's Creative Commons license and your intended use is not permitted by statutory regulation or exceeds the permitted use, you will need to obtain permission directly from the copyright holder.

Index

A
Ad hoc, 101, 109, 110, 139, 166, 207, 208, 211–219, 221, 222, 232, 241, 243, 255
Adversariality, 3, 28, 31, 37–41
Aristarchus, 120–124
Aristotle, 50, 54–58, 60, 65, 93, 156, 258
Axioms, 3, 28, 31, 48–56, 60, 62–66, 91, 96, 97, 130, 151, 152

B
Basic *vs.* applied science, 232
"Big" science, 233
Black box v. mechanistic approaches to science, 239
Bohr, N., 76, 101, 102, 172, 187
Brahe, T., 112–116, 185
Bueno, O., 76, 80, 91–102

C
Cardio-vascular disease (CVD), 5, 205–229
Carnap, R., 57, 128, 129, 185
Chang, H., 192, 194, 232, 233
Cholesterol, 5, 205–229
Collins, F.S., 234, 235
Collins, P.J., 137
Completeness, 64, 65, 100, 192
Complex numbers, 14–17
Comprehension principle, 95–97
Concept-stretching, 86, 177–179
Conditionals, indicative, 133, 136, 138, 141
Confirmation, 51, 110, 131, 132, 134–136, 209–212, 216, 219, 223
Conservativity principle, 4, 137–139, 141

Contradictions, 3, 69–79, 87, 94–99, 120
Cooperation, 3, 28, 31, 37–41, 43, 254
Copernican revolution, 107, 108, 124, 150, 186
Copernicus, 3, 106–108, 112–121, 150, 156, 249, 254, 255, 258
Counterexample, 3, 29–32, 34–37, 39, 43, 59, 72, 75, 77, 78, 82–86, 92–95, 98–101, 110, 167, 177, 197, 198
Counterexample: local, global, 30, 34–37, 72, 75, 85, 110, 167, 177, 197, 198

D
da Costa, N.C.A., 76, 95–98, 100
Deduction, 3, 28, 29, 40, 42, 44, 53, 57, 58, 70, 169, 179
Deductive system, 48, 51–53, 55, 57, 111, 151
Definitions, 17, 29, 47, 49, 56, 58, 59, 63, 81, 82, 84, 86, 110, 137, 154, 185, 189, 239, 258
Degenerating research programmes, 5, 166, 205–229, 232, 240, 241, 260, 262, 263
Demarcation, 109, 147–149, 155, 156, 161, 185, 188, 201
Dependency, 65
Descartes, R., 12, 19, 21, 24, 50, 54, 56, 60, 63, 65
Dialectic, 3, 27–31, 39–44, 69, 70, 73–75
Dialetheic, 70, 73–75, 80, 86
Discourse on Method, 50
Discovery, 2, 8, 36, 37, 55, 72, 86, 87, 91–102, 112, 140, 150, 153, 159, 168, 169, 176, 177, 185, 194, 195, 199, 200, 208, 234–236, 248, 249
Dorling, J., 129

Duhem Problem, 129
Dummett, M., 101
Dutch book arguments, 130
Dutilh Novaes, C., 3, 27–44, 93

E
Elements, 3, 28, 59, 60, 74, 80, 82, 83, 85–87, 109, 110, 138, 166, 174, 187, 193, 199, 210
Empirically progressive, 232
Entailment, 58, 77
Epistemic activities, 233
Epistemic good, 52, 53, 61, 62
Epistemic luck, 3, 105–124
Etchemendy, J., 93
Euclid, 11, 18, 47, 50, 59, 60
Euclideanism, 48, 51, 57–59
Evidence, structural, 133, 138

F
f-divergence, 137, 138, 141
Finitude, 53–54, 63, 64
Flows, 3, 48–53, 61, 62, 66, 151
Formalism, 1, 16, 28, 49, 57
Formality, 41, 57–58, 93
Forster, T., 96
Foundationalism, 48, 53, 56, 57, 59
Frege, G., 9, 12, 13, 15, 65, 93, 94, 151, 253
French, S., 99, 100

G
Galilei, G., 106–108, 112, 113, 117–120
Games, 3, 13, 18, 19, 23–24, 30–33, 37–40, 69–87, 200, 241
Game theory, 3, 24, 80–82, 84, 86, 87
Generality, 8, 34, 63, 93, 174, 192, 196
Genomics, 232, 234–238, 240
Geometry, 9–12, 19, 20, 22, 24, 47, 49, 50, 52, 54, 57, 58, 64, 76, 83, 97, 98, 151
Gillies, D., 28, 42
Glymour, C., 131, 139

H
Hegel, G.W.F., 41, 70, 74, 75
Hegelian idealism, 29
Heliocentrism, 107, 109, 118–124, 150
Heuristics, 4, 28, 30, 41, 69, 74, 81, 85, 86, 91–102, 128, 165–180, 191, 199, 231–233, 237, 239–242, 255, 263, 264
Hilbert, D., 9, 64, 65, 78, 97, 98

Hintikka, J., 30, 36–39, 70, 72, 80, 85, 87
History/historiography, 4, 8, 28, 49, 73, 92, 106, 128, 146, 166, 183, 206, 232, 248
History of mathematics, 10, 24, 28, 49, 74
Howson, C., 2, 129, 139, 193

I
Inconsistency, 3, 69–77, 79–82, 86, 95, 97, 98, 100, 101, 150, 159, 258, 259
Independence, 64, 93
Independent testability, 208, 254
Indubitable, 50, 52, 151
Infallible, 48, 51, 55, 58, 60, 185
Internal and external history, 5, 128, 140, 148–151, 153, 185–187, 189–191, 196, 206, 223–229
Intuitionistic logic, 40, 101

J
Jech, T., 96
Jeffrey conditionalisation, 131, 137, 138, 141
Justification, 2, 3, 36, 52, 53, 61, 106, 108, 130, 154, 156, 214

K
Kant, I., 65, 146, 161, 184
Kelley, J., 96
Kepler, J., 3, 4, 106–108, 113, 117–120, 172, 185, 252, 254, 255, 259
Kneale, M., 93
Kneale, W., 93
Krause, D., 100
Kuhn, T., 107, 121, 199, 242, 247
Kuhn, T.S., 146, 148–150, 152, 153, 157, 183, 185–187, 189, 190, 193, 198, 199, 247–255, 258–263
Kullback-Leibler divergence
  Lakatos, 137, 138

L
Lakatos, I., 1, 7, 27, 48, 69, 91, 108, 127, 165, 183, 206, 231, 247
Larvor, B., 28, 31, 37, 41, 42, 47, 59, 70, 73–75, 187
Lemma-incorporation, 29, 35, 71, 81, 85, 167
Liberal naturalism, 146, 147, 160
Logic, 2, 8, 30, 53, 71, 92, 107, 128, 159, 169, 185, 210, 232, 260
Lorenzen, P., 30, 36–40
Lukács, G., 41

Index

M
Machine, P., 52, 61
Mathematical knowledge, 3, 8, 9, 15, 27, 39, 41–43, 49, 59, 73, 87, 177
Mathematical languages, 9, 23
Mathematical progress, 15–16
Mathematical proof, 3, 28, 30–33, 41, 44, 78
Mathematics, 1, 7, 27, 47, 78, 91, 167, 247
Mayo, D.G., 129
Meaning
Methodology, 2, 8, 60, 74, 92, 108, 129, 146, 165, 185, 206, 253
Methodology of scientific research programmes, 4, 108, 127, 141, 165–168, 185, 189, 197, 210, 223
Monster-barring, 29, 35, 59, 71, 76, 78, 83, 84, 167
Multi-scalar modelling, 168, 170–175

N
National Human Genome Research Institute (NHGRI), 234, 236
National Institutes of Health (NIH), 234, 236
Naturalism, 4, 145–162
Non-classical logic, 71, 82, 85, 87, 93, 98
Normative/norms, 4, 8, 9, 84, 138, 145–162, 188–191, 200, 233

P
Paraconsistency, 3, 70, 71, 73–87, 92, 94, 95, 97, 98, 100, 101
Paraconsistent logic, 73, 75, 79, 86, 94, 95, 97, 98, 100, 101
Pascal, B., 19, 23, 49, 50, 54, 57, 60
Peano Arithmetic, 54
Peto, R., 238, 239
Philosophy of mathematical practice, 27–44
Physics avoidance, 4, 165–180
Pólya, G., 41
Polyhedron, 29, 35, 42, 75, 78, 82, 84, 98, 177
Popper, K.R., 2, 4, 6, 8, 36, 41, 66, 70, 91, 132, 148, 151, 157, 165, 166, 188, 190, 191, 197–201, 207, 211, 247, 248, 251, 253, 260
Port-Royal Logic, 49
Positive and negative heuristics, 100, 128, 166
Positive heuristic, 4, 74, 92, 100, 166–168, 172–176, 178, 179, 191, 199, 231, 242
Positron, 99, 101, 102
Possibility, 15, 32, 35, 40, 54, 56, 92, 94, 95, 97, 98, 100–102, 121, 139, 155, 161, 188, 192, 220, 239, 259, 264
Posterior Analytics, 50, 54–56
Predictivism, 109–111
Priest, G., 74, 76, 77, 81, 94
Primitive terms, 3, 48, 49, 55–57, 61
Problem of old evidence, 108, 131, 139, 141
Proclus, 64
Progressive research programs, 129, 140, 254
Proofs, 1, 7, 27, 48, 69, 92, 118, 135, 167, 214, 264
Proofs and Refutations, 1–3, 7, 24, 27–32, 39–42, 44, 57–59, 69–87, 99, 100, 102, 167, 176–177, 179, 187, 264
Ptolemy, 108, 112, 113, 117, 119, 121, 123, 124, 249, 254, 255, 258

Q
Quasi-empiricism, 1, 2, 52–55, 58, 78, 146, 159
Quaternions, 15–18
Quine, W.V., 8, 96, 152

R
Rationality, 4–6, 15, 16, 41, 71, 72, 76, 86, 87, 127, 129, 146–162, 166, 185, 189, 190, 197–201, 247–264
Rational reconstructions, 4, 5, 9, 28, 33, 44, 49, 55, 59, 62, 73, 74, 128, 140, 145–162, 166, 183–187, 191, 197, 200, 223, 232
Real analysis, 13
Reductive heuristics, 233, 237, 239, 240, 242
Regress, 48, 56, 57, 81
Research program, 2, 74, 92, 108, 127, 150, 165, 185, 206, 231, 248
Russell, B., 8, 9, 51, 57, 65, 78, 95–97, 100, 151

S
Scaffolding, 43, 233, 236, 237, 241
Scepticism, 55, 56, 147, 175
Schweber, S., 99
Scientific revolutions, 152, 250, 253, 259, 261
Self-evidence, 53, 55, 60, 61
Set theory, 8, 93, 95–98, 100, 101
Statins, 219, 222, 226–228
Systems of practice, 233

T
Theoretically progressive, 231, 233
Theory choice, 4, 188, 199, 242, 251–253, 258, 260, 262
Three worlds, 4, 157–160, 162
Trade-offs, 5, 231–243

Trivial, 30, 48, 54, 55, 66, 76, 79, 85, 152, 207, 258
Truth, 1, 3, 48–56, 59–66, 78–80, 85, 86, 91, 93–95, 116, 118, 151, 152, 154, 157, 188, 226, 227, 256
Truth-value, 48, 50, 51, 53, 55, 94

U
Urbach, P., 2, 129
Use-novelty, 109–111, 118, 200

V
van Heijenoort, J., 93

W
Wandering significance, 177
Wilson, M., 4, 167, 168, 170–175, 177–180
Wimsatt, W.C., 233, 240
Worrall, J., 1–7, 102, 110, 112, 114, 121, 129, 140, 167–169, 176, 180, 200, 201, 205–229, 247–264

The manufacturer's authorised representative in the EU is Springer Nature Customer Service Centre GmbH, Europaplatz 3, 69115 Heidelberg, Germany. If you have any concerns regarding our products, please contact ProductSafety@springernature.com

Printed and bound by CPI Group (UK) Ltd, Croydon, CR0 4YY

26/03/2026

02078916-0005